Corrosion Mechanisms

CHEMICAL INDUSTRIES

A Series of Reference Books and Textbooks

Consulting Editor
HEINZ HEINEMANN
Heinz Heinemann, Inc.,
Berkeley, California

Corrosion Mechanisms

edited by

Florian Mansfeld

Department of Materials Science
University of Southern California
Los Angeles, California

CRC Press
Taylor & Francis Group
Boca Raton London New York

CRC Press is an imprint of the
Taylor & Francis Group, an **informa** business

First published 1987 by Marcel Dekker, Inc.

Published 2019 by CRC Press
Taylor & Francis Group
6000 Broken Sound Parkway NW, Suite 300
Boca Raton, FL 33487-2742

First issued in paperback 2019

ISBN-13: 978-0-367-45155-4 (pbk)
ISBN-13: 978-0-8247-7627-5 (hbk)

Visit the Taylor & Francis Web site at
http://www.taylorandfrancis.com

and the CRC Press Web site at
http://www.crcpress.com

Library of Congress Cataloging-in-Publication Data

Corrosion mechanisms.

    (Chemical industries ; 28)
    Includes index.
    1. Corrosion and anti-corrosives. I. Mansfeld,
Florian. II. Series.
TA462.C65573 1986   620.1'1223   86-23940
ISBN 0-8247-7627-5

PREFACE

The purpose of this book is to discuss the mechanisms that have been proposed for the main corrosion phenomena. The authors have attempted to provide a thorough discussion of the pros and cons of the various mechanisms with support by experimental and theoretical results rather than try to outline the "true" mechanism. Citing of the appropriate references was considered important to allow the reader to study additional details.

Chapter 1 discusses the anodic dissolution of iron group metals and summarizes the various mechanisms of this well-researched subject. In addition, it proposes a new mechanism for iron, which also has applications for nickel and cobalt. Chapter 2 provides a comprehensive summary on alloy dissolution, which is of great practical importance.

The major approaches to corrosion protection, including the use of corrosion inhibitors and the application of coatings, are discussed in detail for various types of corrosion phenomena in Chapters 3 and 4. Electrochemical mechanisms for corrosion inhibition in acid, neutral, and alkaline media are discussed in Chapter 3 and examples are given of applications of inhibitors in areas of practical importance such as localized corrosion. Chapter 4 focuses on protective coatings and gives an overview of metallic, polymeric, conversion, and cementitious coatings, illustrating their practical applications.

Atmospheric corrosion is gaining more attention now due to the problems with acid rain; Chapter 5 presents mechanistic information regarding the atmospheric corrosion phenomena. Localized corrosion is one of the most important failure mechanisms, and Chapter 6 thoroughly discusses this area, emphasizing the initiation and growth of pits.

Chapter 7 discusses the effects of hydrogen on metals and the degradation of the mechanical properties of metal by hydrogen, attempting to provide some insight into the process of hydrogen embrittlement. Another type of corrosion of major importance is corrosion fatigue. Chapter 8 explores the possible mechanisms associated with this subject.

Finally, Chapter 9 focuses on high-temperature corrosion and the mechanisms of diffusion-controlled scale growth, which is of central importance to virtually all aspects of high-temperature oxidation and corrosion.

Florian Mansfeld

CONTRIBUTORS

Hans Böhni   Professor, Institute for Materials, Chemistry, and Corrosion, Federal Institute of Technology, Zurich, Switzerland

David J. Duquette, Ph.D.   Professor, Materials Engineering Department, Rensselaer Polytechnic Institute, Troy, New York

Konrad E. Heusler, Ph.D.   Professor, Metallurgy and Materials Science, Department of Corrosion and Corrosion Protection, Institute for Metallurgy and Metal Physics, Technical University of Clausthal, Clausthal-Zellerfeld, Federal Republic of Germany

Hermann Kaiser, Ph.D.   Professor, Department of Corrosion and Surface Technology, Institute for Materials Science, Friedrich-Alexander-University Erlangen-Nürnberg, Erlangen, Federal Republic of Germany

Vladimir Kucera, Ph.D.   Director of Research, Swedish Corrosion Institute, Stockholm, Sweden

Henry Leidheiser, Jr., Ph.D.   Professor, Department of Chemistry, Sinclair Laboratory No. 7, Lehigh University, Bethlehem, Pennsylvania

Wolfgang J. Lorenz   Professor, Faculty of Chemistry, Institute of Physical Chemistry and Electrochemistry, University of Karlsruhe, Karlsruhe, Federal Republic of Germany

M. R. Louthan, Jr.   Professor, Department of Materials Engineering, Virginia Polytechnic Institute and State University, Blacksburg, Virginia

Florian Mansfeld*, Ph.D.   Department Manager, Surface and Electrochemical Processes, Rockwell International Science Center, Thousand
Oaks, California

Einar Mattsson, Ph.D.   Professor, Managing Director, Swedish
Corrosion Institute, Stockholm, Sweden

Giordano Trabanelli   Professor, Department of Chemistry, Corrosion
Study Center "Aldo Daccò," University of Ferrara, Ferrara, Italy

Gregory J. Yurek, M.D., Ph.D.   Associate Professor, Department
of Materials Science and Engineering, Massachusetts Institute of Technology, Cambridge, Massachusetts

Current affiliation
*Professor, Materials Science Department, University of Southern Cal-
ifornia, Los Angeles, California

CONTENTS

Corrosion Mechanisms

1

ANODIC DISSOLUTION OF
IRON GROUP METALS

WOLFGANG J. LORENZ

*Institute of Physical Chemistry and Electrochemistry, University of
Karlsruhe, Karlsruhe, Federal Republic of Germany*

KONRAD E. HEUSLER

*Institute for Metallurgy and Metal Physics, Technical University of
Clausthal, Clausthal-Zellerfeld, Federal Republic of Germany*

1.1 INTRODUCTION

Knowledge of the dissolution kinetics and mechanisms of the iron group
metals in the active state is important for corrosion research and practical applications. Identification of corrosion damage, determination of
corrosion rates by electrochemical means, and carefully directed corrosion protection require an understanding of the corrosion processes
taking place under a variety of conditions.

Most corrosion processes are of an electrochemical nature. The interpretation of corrosion phenomena is based on knowledge of the electrochemical kinetics resulting from recent progress in measuring techniques. The increasing amount and variety of experimental data lead
to a more detailed knowledge of the kinetics and mechanisms of corrosion processes. Although simple systems are fairly well understood at
present, not all aspects of the complicated kinetics of transition metal
electrodes have been elucidated, especially for corroding systems under practical conditions. In the case of the iron group metals the
main difficulties arise because of the great number of parameters influencing the kinetics of anodic metal dissolution:

1. The rates depend on the surface structure of the substrate, i.e.,
    the density of steps and kinks on the surface. The surface struc-
    ture is determined by the orientation of crystal faces exposed to
    the electrolyte, by dislocations and grain boundaries in the metal,
    by segregation of impurities from the metal, by chemisorption of
    various substances from the electrolyte, etc. Electrochemical pro-
    cesses such as selective dissolution of a component from an alloy
    or absorption of hydrogen into the bulk metal may change the

dislocation density. Details of the influence of structure on
the dissolution kinetics of metals and alloys are widely unex-
plored.

2. Further complications arise from the possibility that chemi-
    sorbed substances from the electrolyte change the structure
    of the interphase metal/electrolyte, catalyze or inhibit metal
    dissolution, and may change the reaction path. Changes of
    the interphase may also arise by the formation of uniform or
    nonuniform two-dimensional (2-D) films and of porous or non-
    porous three-dimensional (3-D) films of intermediates and re-
    action products on the substrate surface.

Because of these complications, the dissolution kinetics of iron
group metals continue to be the subject of many papers. Most studies concerning the fundamentals of the anodic dissolution of iron group
metals were carried out with relatively simple systems in order to eliminate or keep constant some of the parameters just mentioned. Simple
systems ideally consist of single-crystalline or polycrystalline, highly
pure metals in acid aqueous solutions free of dissolved oxygen, surface-active substances, and complexing agents. The dissolution behavior in alkaline or neutral media, even in deaerated or aerated aqueous solutions free of surface-active and complexing substances, is difficult to study because of the easy formation and slow dissolution of
protective layers. Also, the influence of surface-active substances
on the dissolution kinetics and mechanisms of iron group metals is
better known in acid than in neutral or alkaline environments.

The aim of this chapter is to discuss critically the present state of
knowledge on the active dissolution of the iron group metals. The
history of research will be described through some highlights only.

1.2 IRON

1.2.1 Historical Background

In 1939 the first kinetic study of the Fe/Fe^{2+} electrode was carried
out by Roiter et al. [1], using large-signal pulse polarization. Under galvanostatic conditions, they observed a nonmonotonous potential-time response of the system with a maximum value after charging
of the electrode capacitance. Now this phenomenon is often called
superpolarization. The results were interpreted in terms of a rate-
determining transfer of Fe(II) in one step influenced by crystalliza-
tion phenomena:

$$Fe \rightleftarrows Fe_{sol}^{2+} + 2e^- \qquad\qquad (1)$$

The stimulation of reaction (1) with increasing pH was first observed
by Kabanov et al. [2--5]. The kinetic data obtained in alkaline media

were interpreted in terms of a "consecutive mechanism" of iron dissolution with hydroxyl ions participating in the formation of intermediates and products:

$$Fe + OH^- \rightleftarrows Fe(OH)_{ads} + e^- \qquad\qquad (2a)$$

$$Fe(OH)_{ads} + OH^- \rightleftarrows FeO_{ads} + H_2O + e^- \qquad\qquad (2b)$$

$$FeO_{ads} + OH^- \rightleftarrows HFeO_2^- \qquad\qquad (2c)$$

$$HFeO_2^- + H_2O \rightleftarrows Fe(OH)_2 + OH^- \qquad\qquad (2d)$$

The kinetics in acid solutions were studied by Heusler and Bonhoeffer [6–10] and Hoar and Hurlen [11–13] starting in the middle of the
1950s. Anodic and cathodic polarization curves were measured under
steady-state and transient conditions at different pH values and Fe^{2+}
concentrations in the bulk electrolyte. From Tafel slopes and from
electrochemical reaction orders a *catalytic mechanism* involving transfer of Fe(II) in one step was postulated.

In the beginning of the 1960s, Bockris and co-workers [14–16] investigated the kinetics of the Fe/Fe^{2+} electrode in acid solutions using
different iron samples. The kinetic data obtained differed from those
of Heusler. Therefore, they adopted the noncatalytic *consecutive iron
dissolution mechanism* analogous to Eq. (2) for their interpretation.

Later, many other authors reinvestigated the kinetics of the Fe/Fe^{2+}
system under various experimental conditions, but mainly two different sets of steady-state kinetic data were obtained, fitting either the
Heusler mechanism [17–24] or the Bockris mechanism [25,26]. A comprehensive review of the experimental data has been published in 1982
[27].

Lorenz and co-workers [28–35] showed that both sets of kinetic
data can be observed, depending on the substructure of the electrode
material. The influence of crystal imperfections on the reaction kinetics was established by comparing studies of iron samples by X-ray
analysis and electron microscopy with electrochemical steady-state
and transient measurements. These findings were confirmed by other
authors [36,37]. The possibility was discussed that the reaction path
may depend on the surface structure.

The kinetics of iron dissolution change at potentials in the "active"
range before reaching the passive state of the metal. This observation was made by Epelboin et al. [38–48], Lorenz et al. [49–54], Bech-
Nielsen et al. [37,55–66], and Allgaier and Heusler [67,68]. At least
two current density maxima can be observed in aqueous acid solutions
at $2 \leq pH \leq 6$, as schematically shown in Fig. 1. It will be demonstrated in this chapter that the current density of the first maximum
increases with decreasing pH. Therefore, in many previous investigations, which were carried out in strongly acid solutions, the first

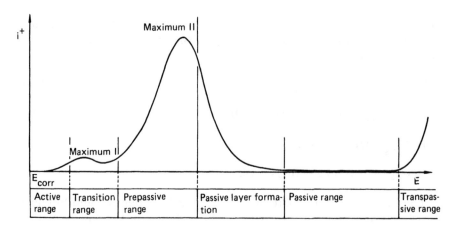

FIGURE 1 Schematic polarization curve of anodic iron dissolution in acid solutions between the active and the passive states of the metal [51–54,95,130,131].

maximum was not detected and the iron dissolution was considered to be active in the entire potential range up to the onset of passivation. However, due to the different kinetics and corresponding changes of the mechanism, this overall "active iron dissolution range" was sub-divided by Lorenz et al. [49–54] into the active range (I_1 reaction*), transition range, prepassive range (I_2 reaction*), and passive layer formation range as a result of the formation of 2-D and 3-D films on the electrode surface consisting of $Fe(OH)_{n,ads}$ with $n = 1$, 2, 3 [38–66,69]. These surface layers can be either stimulating ($n = 1$) or retarding ($n = 2, 3$) as interface (2-D) or interphase (3-D) inhibitors (see Section 1.2.5). Furthermore, because of the continuous polarization curve between the active and passive states, it was postulated that the active-passive transition of iron in weakly acid solutions should be considered as a continuous process not exhibiting a discontinuity of the current at a certain critical potential for passivation [70,71].

Fundamental research on the dissolution kinetics and morphology of low-index planes of iron single crystals was carried out by Heusler and Allgaier [67,68,72–74]. It was demonstrated that a strong correlation exists between the electrochemical kinetics and the atomistic

---

*According to Bech-Nielsen and co-workers [55,56].

structure of the electrode surface. Steady-state dissolution corresponds to a steady-state surface morphology. The atomic structure of the surface was described quantitatively in terms of monoatomic steps and kinks. The surface concentration of kinks strongly depends on the electrode potential, while the step density is determined by the pH of the solution. The dissolution process was assumed to proceed via kinks only, and not via adatoms. The atomistic structure of the surface was shown to agree quantitatively with the prediction of the catalytic mechanism and cannot be understood by the consecutive mechanism. On the other hand, a rate-determining surface diffusion step of adatoms in the iron dissolution mechanisms was proposed by Bignold and Fleischmann [75], but the paper was criticized by others [76].

From ac impedance measurements in systems consisting of polycrystalline substrates and aqueous solutions free of oxygen and surface-active substances in the range $0 \leq pH \leq 6$, Lorenz et al. [77−79] and Keddam et al. [47,48] concluded that the surface concentration of kink sites should play an important role in the iron dissolution kinetics. The observed inductive loops in the low-frequency range of Nyquist diagrams cannot be explained by a consecutive charge transfer mechanism analogous to Eq. (2). It should be mentioned, however, that for a long time ac impedance results had been interpreted erroneously in terms of a consecutive mechanism [38−46].

1.2.2 Iron Dissolution in Acid Solutions

Experimental Conditions

Fundamental studies on the iron dissolution kinetics were carried out in acid aqueous solutions free of oxygen and surface-active or complexing substances. Pure iron in acid sulfate or perchlorate solutions deaerated by purified nitrogen, hydrogen, or rare gases was mainly investigated. The reproducibility of the experimental data strongly depends on the purity of the iron and the electrolyte, the solid-state properties of the iron sample, and the pretreatment of the electrode surface.

Highly pure polycrystalline iron recrystallized under high vacuum often served as the electrode. It was polished mechanically and chemically or electrochemically, thoroughly cleaned, and separately pre-etched prior to each measurement [30,33]. In other studies cathodically deposited iron was used as the electrode material [10,31,33]. Investigations with iron single-crystal faces require special pretreatment techniques controlled by morphological observations [67,68,72−74].

Alkali perchlorates or sulfates were found to be almost inert electrolytes; i.e., neither the cations nor the anions of these electrolytes influence the dissolution process significantly, and the electrochemical kinetic data are mainly independent of the activity of the supporting electrolyte [6—10,33,52,80,81]. Contradictory results [82] could be explained by experimental errors [80].

Iron is considered active at electrode potentials negative to the equilibrium potential of the iron-magnetite electrode [83]. In this region, the iron electrode is not covered by a 3-D passivating oxide film. However, porous layers may be present, allowing direct contact between the iron and the electrolyte solution. The dissolution kinetics of active iron differs significantly in the different potential regions mentioned above, namely the active, transition, prepassive, and passive layer formation ranges, which will be discussed separately.

The Active Range

The active iron dissolution range is considered to extend from negative electrode potentials up to the first maximum of the anodic polarization curve (see Fig. 1). Usually, anodic Tafel lines are observed in this potential region, exhibiting different slopes depending on the iron substrate and the polarization technique used.

Under steady-state and quasi-steady-state* conditions different Tafel slopes of $b_{ss}^+ = (\partial E_{ss}/\partial \log i_{ss}^+)_a = +30 \pm 2$ mV or $+40 \pm 2$ mV were observed at T = 298 K (Fig. 2). On the other hand, "transient" or "initial" polarization curves of the iron dissolution yielded Tafel slopes of $b_t^+ = +60 \pm 5$ mV at T = 298 K independent of the iron substrate used and the steady-state starting potential (Fig. 2). The transient polarization curves were obtained from transient measurements using the superpolarization values of galvanostatic large-signal pulse polarization or the corresponding minima in the current-time response of potentiostatic pulse experiments as shown in Fig. 3. The same steady state must be reestablished before application of any new pulse. The characteristic time of the relaxation process between nonsteady state and steady state was found to be of the order of tenths of a second, independent of the starting potential and the pH value but depending on the temperature [6,9,10,15,27,41,75—79,84]. The transition time can increase to many minutes if the steady-state morphology of the substrate has not already been established [27,67].

The increase of the dissolution rate with increasing pH of the electrolyte can be expressed by a positive electrochemical reaction order with respect to hydroxyl ions, according to

$$y_{(OH^-)}^+ = \left(\frac{\partial \log i_{ss}^+}{\partial pH} \right)_{a_i \neq a_{H^+}, E}$$

---

*Quasi-steady-state conditions were established by cyclic or sweep voltammetry with slow sweep rates $|dE/dt| \leq 10$ mV sec^{-1}.

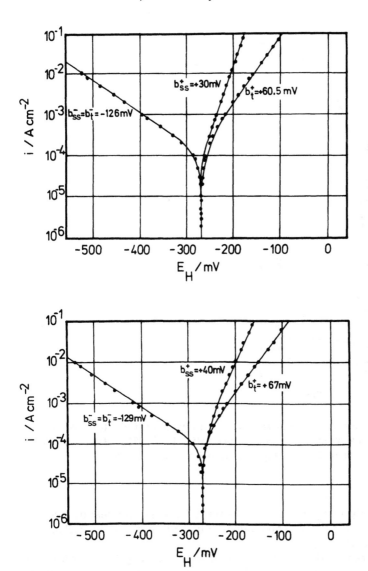

FIGURE 2 Steady-state and transient current density-potential curves [30--35]. System: Fe/0.5 M H_2SO_4, pH = 0.35, T = 298 K. a) Cold-worked iron; b) recrystallized iron.

Under steady-state conditions, values of $1 \le y(OH-)^+ \le 2$ were usually observed. The higher values corresponded to the smaller steady-state Tafel slopes mentioned above. The determination of electrochemical reaction orders with respect to hydroxyl ions from transient measurements

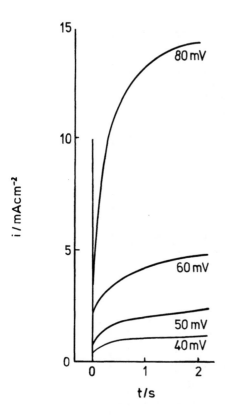

FIGURE 3 Anodic transients following anodic potential steps with respect to the steady-state corrosion potential [6,8,27]. System: Fe/ x M Na_2SO_4 + y M H_2SO_4, x + y = 0.5, pH = 1.9, T = 298 K.

yields physically insignificant parameters, because the results are determined, e.g., by the pH-dependent initial state of the iron electrode [10,27,30,31,33].

In unbuffered weakly acid deaerated solutions at pH $\geq$ 5, the steady-state and transient anodic Tafel slopes were found to increase to b_{ss}^+ = +60 ± 5 mV and b_t^+ = +120 ± 5 mV at T = 298 K, respectively [85]. Simultaneously, the electrochemical reaction orders with respect to the hydroxyl ions decreased.

The concentration of ferrous ions in the electrolyte does not influence the dissolution kinetics of iron. The electrochemical reaction order with respect to iron(II) ions is zero [27].

Experimental kinetic data from dc measurements at T = 298 K are summarized in Table 1. The two different sets of kinetic dc data in

TABLE 1 Kinetic dc Data for Iron Dissolution in the Active Range at T = 298 K[a]

Substrate	pH of the electrolyte	Steady state			Transient state	References
		b_{ss}^+ (mV)	$y(OH^-)^+$	$y(Fe^{2+})^+$	b_t^+ (mV)	
Polycrystalline iron (recrystallized)	$0 \leq pH \leq 5$	40 ± 2	1.0 ± 0.1	0	60 ± 7	[14–16, 25, 26, 28–54, 66, 75–79] and references given in [27] and [85]
Zone-melted iron						
Electrodeposited iron	$5 \leq pH \leq 7$	65 ± 5	0.7 ± 0.1		115 ± 10	
Polycrystalline iron (cold-worked)						[6–10, 17–24, 28–37, 66–68, 72–74] and references given in [27]
Electrodeposited iron						
Iron single crystals	$0 \leq pH \leq 5$	30 ± 2	2.0 ± 0.3	0	60 ± 7	[27]

[a]Systems: Fe/x M acid + y M salt with acid = $HClO_4$ or H_2SO_4 and salt = $NaClO_4$, Na_2SO_4, or K_2SO_4.

acid solutions were found to exist with iron of the same chemical composition, depending on the density of crystal imperfections and lattice distortions [28–37]. The smaller steady-state Tafel slopes were obtained with cold-worked polycrystalline iron and the higher ones with the same iron after recrystallization. On the other hand, the opposite change in the kinetic data could also be demonstrated by heavy cold-working of recrystallized iron. These electrochemical results were compared with X-ray measurements to determine the density of subgrain boundaries and the degree of lattice distortion. It was supposed that the iron dissolution reaction is strongly influenced by the surface density of different dislocations emerging to the surface [28–34]. This interpretation could also be confirmed by electron microscopic studies [31–35].

Morphological and electrochemical dc investigations of iron single crystal faces showed a strong correlation between the dissolution morphology and the dissolution kinetics [27,67,68,72–74]. The single crystal faces used were macroscopically oriented vicinal to (112), having a misorientation of only a few degrees. A steady-state surface morphology was developed during the anodic dissolution, exhibiting flat three-sided pyramids enclosed by almost ideal (112) faces (Fig. 4). These faces intersect in <113> directions. Pairs of monoatomic steps in the corresponding <113> directions were formed on

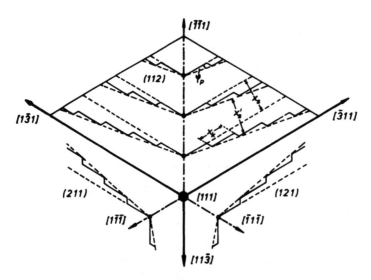

FIGURE 4   Schematic model of steps and kinks on crystallographic (211), (112), and (121) planes projected into a (111) plane [27,67, 68,72–74].

each (112) face. The distance of such monoatomic steps in acid per-chlorate solution at pH 0.96 was found to be about 0.3 μm indepen-dent of the electrode potential. The angle between the pairs of mono-atomic steps corresponds nearly to the angle between two <311> di-rections on a (112) face. This angle increased with increasing anodic potential; i.e., the mean distance between equivalent kink sites lo-cated at the monoatomic steps decreased. The surface concentration of kink sites, Γ_k, was found to follow the relation

$$(\partial E / \partial \ln \Gamma_k)_{a_i} = RT/F$$

in agreement with predictions from electrochemical data [27]. Similar behavior was observed with iron single crystal faces of (110) orienta-tion [86]. On these surfaces, the step distance x_s decreased with in-creasing pH [86] according to

$$\left(\frac{\partial \log x_s}{\partial pH} \right)_{a_i \neq a_{H^+}, E} = -0.6 \quad \text{at } T = 298 \text{ K}$$

Ac impedance measurements in the active range of the dissolution of polycrystalline iron showed a depressed capacitive semicircle at relatively high frequencies and different inductive loops at low fre-quencies in the Nyquist representation [48, 77–79]. As shown in Fig. 5, the number of inductive loops increased from one at electrode potentials close to the corrosion potential to three at the most positive electrode potentials in the active range. The resonance frequencies of the inductive loops are relatively low, being in the range of hertz to millihertz.

The interpretation of the electrochemical, morphological, and other results in terms of an iron dissolution mechanism in the active range is still controversial. Many authors agree that structural surface ef-fects and the surface concentration of kink sites, from which the iron dissolution reaction may start, play a very important role for a mech-anistic description. Therefore, the inhomogeneity of the active iron surface is taken into account in many papers [10, 28–37, 47, 48, 60–68, 72–79, 86]. On the other hand, it is an open problem whether all ex-perimental results obtained in the active range can be explained in terms of the same mechanism of iron dissolution. In the following, an attempt is made to summarize the recent concepts of the mechanisms, taking into consideration new results with other metal-ion electrodes.

Heusler's concept is represented by a rate-determining charge transfer step of ferrous ions from kink sites on the electrode surface

$$FeFe_k \rightleftarrows Fe_k + Fe_{sol}^{2+} + 2e^- \tag{3}$$

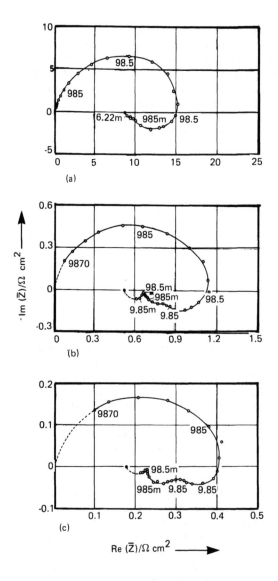

FIGURE 5 Impedance diagrams in the active range of iron dissolution [77−79]. System: Fe(RDE)/0.5 M H_2SO_4, pH = 0.3, ν_{rot} = 70 sec^{-1}; frequency f (Hz); •, polarization resistance Rp (f→0) calculated from Tafel slope; T = 298 K. a) E_H = -225 mV, i_{ss}^+ = 2 mA cm^{-2}; b) E_H = -203 mV, i_{ss}^+ = 8 mA cm^{-2}; c) E_H = -160 mV, i_{ss}^+ = 100 mA cm^{-2}.

One elementary act shifts the kink site to a neighboring position. Thus, step (3) does not change the surface concentration of kinks. The rate of reaction (3) is given by

$$i^+ = k^+ a_k \exp\left[\frac{2\alpha^+ FE}{RT}\right] \tag{4}$$

where a_k denotes the surface activity of kink sites. Both k^+ and a_k depend on the crystallographic orientation of the iron surface and may also depend on the crystal imperfection density and the degree of lattice distortion. For $a_k = \text{const}$, Eq. (4) represents the transient Tafel line with a slope of $b_t^+ = +60$ mV for $\alpha^+ = 0.5$ at $T = 298$ K. For given crystallographic and solid-state conditions, a_k is proportional to the surface concentration Γ_k of kinks. Using the experimental dependence of Γ_k on the electrode potential

$$\Gamma_k = \Gamma_k^0 \exp\left[\frac{F(E - E^0)}{RT}\right] \tag{5}$$

where E^0 and Γ_k^0 are the equilibrium potential of the Fe/Fe^{2+} electrode and the equilibrium concentration of kinks, respectively, one obtains for the steady state

$$i_{ss}^+ = k_{ss}^+ \exp\left[\frac{(1 + 2\alpha^+)FE}{RT}\right] \tag{6}$$

with $k_{ss}^+ \propto k^+ \Gamma_k^0 \exp(-FE^0/RT)$. This equation describes the steady-state Tafel line with a slope of $b_{ss}^+ = +30$ mV for $\alpha^+ - 0.5$ at $T = 298$ K.

The catalytic mechanism of iron dissolution is supported by measurements of the kinetics of cathodic iron deposition yielding steady-state and transient polarization curves with Tafel slopes of $b_{ss}^- = -30$ mV and $b_t^- = -60$ mV, respectively [10,11,27,34,87]. This means that the anodic and cathodic polarization curves of the iron electrode are symmetric with respect to E^0. Consequently, the cathodic iron deposition reaction should also follow Eq. (3) under the assumption that the surface concentration of kink sites increases with increasing anodic as well as cathodic overvoltage:

$$\Gamma_k = \Gamma_k^0 \exp\left[\pm\frac{F(E - E^0)}{RT}\right] \tag{7}$$

where $\pm$ stands for $E \gtrless E^0$.

The observed increase of step density, $1/x_s$, with pH partially explains the pH dependence of the dissolution rate. The remaining part could be due either to a decrease of the mean distance of kinks in steps at constant electrode potential with pH or, more probably [86], to FeOH$^+$ being the species participating in the charge transfer reaction.

The formation of steps and kinks requires iron ion transfer from specific sites of the iron surface, e.g., from the intersection of steps and from the apex of intersecting planes. These steps proceed in parallel to iron ion transfer at kinks. Along these lines, a detailed mechanism was proposed [68,73,74,86].

Similar iron dissolution mechanisms were proposed by Lorenz et al. [77—79] and Keddam et al. [48] to explain experimental dc and ac results obtained with recrystallized polycrystalline iron exhibiting the dc data thought to be characteristic of the consecutive mechanism (see Table 1). In accordance with previous arguments [27], both groups of authors agree that an interpretation in terms of the conventional consecutive mechanism according to Bockris et al. [14—16]

$$Fe + H_2O \rightleftarrows Fe(OH)_{ads} + H^+ + e^- \tag{8a}$$

$$Fe(OH)_{ads} \rightleftarrows FeOH^+ + e^- \tag{8b}$$

$$FeOH^+ + H^+ \rightleftarrows Fe_{sol}^{2+} + H_2O \tag{8c}$$

is not compatible with experimental ac results. The calculated transfer function of this model, assuming that step (8a) is always close to equilibrium and step (8b) is rate-determining, predicts a purely capacitive behavior of the ac impedance, in contradiction to the experimental findings of inductive loops at low frequencies (Fig. 5). Therefore, the concept of a one-step charge transfer of ferrous ions starting at active surface sites was adopted by both authors—however, in different ways.

Both interpretations start with the normal assumption of Langmuir adsorption behavior of the intermediate Fe(OH)$_{ads}$. Schweickert et al. [77—79] postulated a potential dependence of the active surface centers, Fe*, which are assumed to be kink sites:

$$\Gamma_k = \Gamma_k{}^0 \exp\left[\frac{\beta F(E - E^0)}{RT}\right] \tag{9}$$

where β denotes an effective charge number, which may differ from unity. Moreover, a surface relaxation process of these active centers has been proposed, following a kinetic first-order law given in the Laplace domain as

$$\Delta\psi(s) = \Delta E(s)\frac{\partial\psi/\partial E}{1 + s\tau} \tag{10}$$

where $\psi = \Gamma_k / \Gamma_k^0$ and τ denotes the relaxation time. Under these conditions, the electrochemical dc and ac results were explained by Schweickert et al. [77--79] in terms of a modified catalytic iron dissolution mechanism:

$$H_2O \rightleftharpoons (OH)_{ads} + H^+ + e^- \tag{11a}$$

$$FeFe_k + (OH)_{ads} \rightleftharpoons FeFe_k(OH)_{ads} \tag{11b}$$

$$FeFe_k(OH)_{ads} + (OH)^-_{ads} \rightleftharpoons Fe_k(OH)_{ads} + FeOH^+ + 2e^- \tag{11c}$$

$$FeOH^+ + H^+ \rightleftharpoons Fe_{sol}^{2+} + H_2O \tag{11d}$$

The overall reaction corresponds to the reaction sequence (11b)--(11d). The potential-dependent surface concentration of the intermediate $Fe(OH)_{ads}$ is formed in the fast reaction (11a) running in parallel. In order to explain the dc data, step (11b) was considered as rate-determining for the dissolution of active iron with a relatively low density of active centers at the surface.

Keddam et al. [47,48] combined the classical catalytic charge transfer mechanism and a modified consecutive one in a parallel reaction scheme:

$$\tag{12}$$

In this scheme, the formation of $Fe(I)_{ads}$ from $Fe(0)$ corresponds to chemisorption of OH according to Eq. (8a). Dissolution of Fe_{sol}^{2+} with rate constant k_2 is equivalent to reactions (8b) and (8c). In contrast to the findings of all other authors, both steps representing a consecutive mechanism are assumed to be totally irreversible.

The dissolution of Fe_{sol}^{2+} via active sites, $Fe^*(I)_{ads}$, can be identified with the catalytic mechanism.

At the active sites $Fe^*(II)_{ads}$, two OH were believed to be adsorbed. Thus, the iron in these sites is formally divalent. The second catalytic path leading to Fe_{sol}^{2+} was introduced in order to explain the number of inductive loops and the behavior in the transition and prepassive ranges.

Using scheme (12), the ac behavior was simulated in good agreement with experimental results [48]. However, the kinetic parameters used were mostly unrealistic, e.g., asymmetric electrochemical charge transfer coefficients being nearly one or zero. Moreover, a calculated steady-state Tafel slope of $b_{ss}^+ \geq$ +60 mV at T = 298 K is not in agreement with experimental results (Table 1 and [77−79]).

The changes in the kinetic dc data for the iron dissolution reaction at higher pH values (Table 1) have been explained by a Frumkin adsorption behavior of the intermediate Fe(OH)$_{ads}$ in terms of a consecutive mechanism [85]. This interpretation must be changed in view of the results already discussed and will be reconsidered later (Section 1.2.4).

Bech-Nielsen et al. [37,55−66] proposed another modified consecutive mechanism. Apart from the specific cosorption of other anions, it can be rewritten as

$$Fe + H_2O \rightleftharpoons Fe(OH)_{ads} + H^+ + e^- \tag{13a}$$

$$Fe(OH)_{ads} + H_2O \rightleftharpoons Fe(OH)_{2,ads} + H^+ + e^- \tag{13b}$$

$$Fe(OH)_{2,ads} \rightleftharpoons Fe(OH)_{2,des} \tag{13c}$$

$$Fe(OH)_{2,des} + 2 H^+ \rightleftharpoons Fe_{sol}^{2+} + 2 H_2O \tag{13d}$$

Steps (13b) and (13c) were assumed to be rate-determining for substrates of high and low crystal imperfection densities, respectively. A similar iron dissolution mechanism including a desorption step for FeOH$_{ads}$ was proposed by Drazic and Hao [88]. These interpretations, however, are inconsistent with experimental dc and ac results.

At present, it seems that Heusler's concept according to Eqs. (3)−(7) represents the most probable explanation for the active iron dissolution mechanism. However, an open problem is how to explain the variable potential and pH dependences of the dissolution rate. Different mechanisms yielding monoatomic steps with various kink spacings are conceivable. Etch pits at screw or edge dislocations, two-dimensional nucleation, and intersections of nonequivalent crystal faces may behave in different ways as sources of steps and kinks and could be responsible for the influence of the iron structure on the dissolution kinetics. These processes modifying the surface structure are taken into account formally by the factor β different from unity in Eq. (9). This concept also explains the various experimental data for iron deposition [14−16,32−34,87].

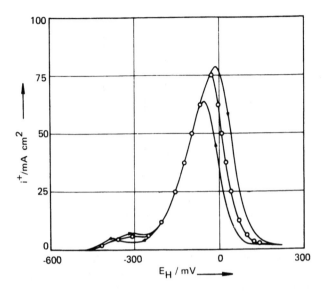

FIGURE 6 Anodic current density-potential curves [81,95]. System: Fe(RDE)/x M NaClO$_4$ + y M HClO$_4$, x = 1.0, pH = 3.9, ν_{rot} = 69 sec^{-1}, T = 298 K. —, Cyclic voltammetric measurement, $|dE/dt|$ = 3 mV sec^{-1}; ○, steady-state measurement.

Macroscopic surface inhomogeneities caused by morphological or sorption phenomena were also found in the Ag/Ag$^+$ and Cu/Cu^{2+} systems [89].

The Transition Range

The transition range of active iron dissolution is characterized by a current maximum I and a subsequent minimum in the anodic polarization curve (Figs. 1 and 6) measured under quasi-steady-state conditions, i.e., using potential sweeps with relatively low sweep rates of 0.1 mV sec^{-1}, < $|dE/dt|$ < 10 mV sec^{-1} [48,49,51−66,77−79,81]. Obviously, the current maximum I degenerates to a flat current plateau under steady-state conditions [67,77]. The transition range was also reflected in typical transients measured by large-signal pulse polarization under potentiostatic or galvanostatic conditions [50], as shown in Figs. 7 and 8.

Both the potential and the current density of maximum I, E_{maxI} and i_{maxI^+}, respectively, were found to depend on pH but not on the anion activity in the electrolyte, as shown in Figs. 9 and 10. The kinetic data obtained at different temperatures are listed in Table 2.

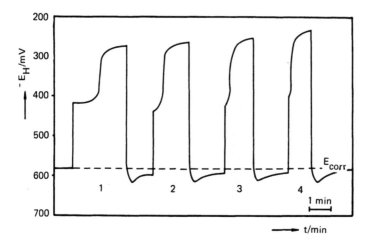

FIGURE 7 Galvanostatic transients starting from E_{corr} for $i^+ > i_{max}I$
[50]. System: Fe/x M Na_2SO_4 + y M H_2SO_4, x = 0.5, pH = 5.05, magnetic stirring, T = 298 K. i^+ (mA cm^{-2}): 1, 10; 2, 15; 3, 20; and 4,
30.

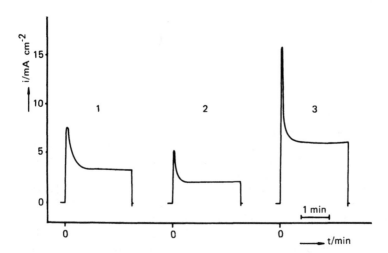

FIGURE 8 Potentiostatic transients starting from E_{corr} for $E > E_{max}I$
[50]. System: Fe/x M Na_2SO_4 + y M H_2SO_4, x = 0.5, pH = 5.45,
magnetic stirring, T = 298 K. E (mV): 1, -410; 2, -390; and 3,
-300.

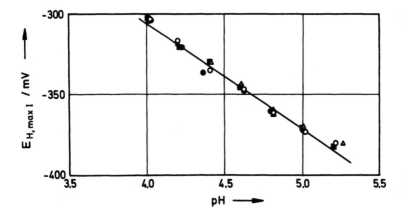

FIGURE 9 Potential of maximum I vs. pH [52,81]. System: Fe(RDE)/ x M Na_2SO_4 + y M H_2SO_4, $4.0 \leq pH \leq 5.3$, ν_{rot} = 69 sec^{-1}, $|dE/dt|$ = 5 mV sec^{-1}, T = 298 K. x: ■, 0.4; ○, 0.5; ●, 0.7; and △, 0.9.

The absolute values of the partial derivatives increase with rising temperature, but the following expression remained nearly constant [53, 81]:

$$\frac{F}{(\ln 10)RT} \left(\frac{\partial E_{maxI}}{\partial pH} \right)_{a_i \neq a_{H^+}} = 1.09 \pm 0.04$$

Furthermore, the pH dependence of E_{maxI} was found not to be influenced by the ionic strength (Fig. 11), whereas that of $i_{maxI}+$ decreased with increasing ionic strength of the electrolyte (Fig. 12).

Ac impedance measurements in the transition range of iron dissolution exhibit a capacitive behavior with a negative or infinite value of the polarization resistance

$$R_p = \lim_{\omega \to 0} Re\{\overline{Z}_f\}$$

as shown in Fig. 13 for point E_s, where $Re\{\overline{Z}_f\}$ denotes the real part of the measured faradic impedance [38−48,77−79].

It was noted by Allgaier and Heusler [73] that the spacing of kinks at monoatomic steps approaches its minimum value comparable to interatomic spacing just where the transition range begins. Consequently,

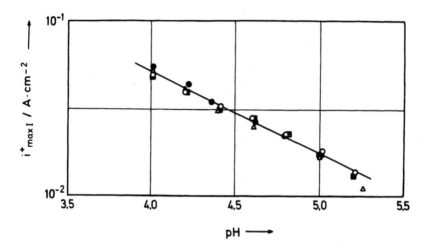

FIGURE 10   Current density of maximum I vs. pH [52,81]. System:
Fe(RDE)/x M Na$_2$SO$_4$ + y M H$_2$SO$_4$, 4.0 $\leq$ pH $\leq$ 5.3, ν_{rot} = 69 sec^{-1}, $|$dE/
dt$|$ = 5 mV sec^{-1}, T = 298 K. x: ■, 0.4; ○, 0.5; ●, 0.7; and △, 0.9.

the iron dissolution reaction should be changed at more positive potentials (see Section on the prepassive range).

The kinetic results in the transition range can be explained by assuming that a formally divalent iron oxide species is adsorbed at the
electrode surface. Its surface concentration will increase with the

TABLE 2   Kinetic dc Data for Iron Dissolution in the Transition
Range at Different Temperatures [46–51, 74–76, 78][a]

T/K	$\left(\dfrac{\partial E_{maxI}}{\partial pH}\right)_{a_i \neq a_{H^+}}$ (mV)	$\left(\dfrac{a \log i_{maxI}^{+}}{\partial pH}\right)_{a_i \neq a_{H^+}}$
283	-59 ± 3	-0.38 ± 0.02
298	-66 ± 3	-0.47 ± 0.02
313	-69 ± 4	-0.63 ± 0.03
333	-71 ± 4	-0.72 ± 0.03

[a]Systems: Fe/x M acid + y M salt, deaerated, with acid = HClO$_4$
or H$_2$SO$_4$ and salt = NaClO$_4$, Na$_2$SO$_4$, or K$_2$SO$_4$, 0 $\leq$ pH $\leq$ 6.

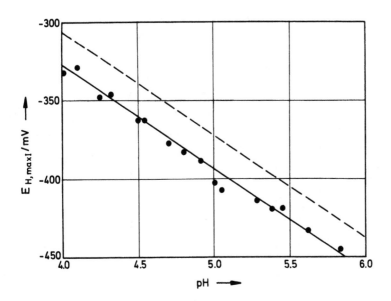

FIGURE 11 Potential of the maximum I vs. pH [53,81]. System: Fe(RDE)/x M Na_2SO_4 + y M H_2SO_4, $4.0 \leq pH \leq 5.7$, $\nu_{rot} = 69$ sec^{-1}, $|dE/dt| = 5$ mV sec^{-1}, T = 298 K. •, x = 0.1; ----, $0.4 \leq x \leq 0.9$.

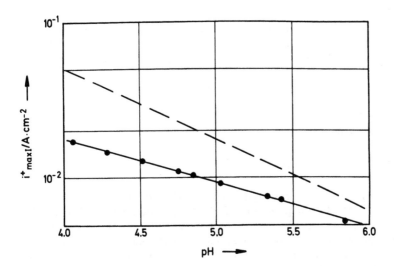

FIGURE 12 Current density of maximum I vs. pH [53,81]. System: Fe(RDE)/x M Na_2SO_4 + y M H_2SO_4, $4.0 \leq pH \leq 5.7$, $\nu_{rot} = 69$ sec^{-1}, $|dE/dt| = 5$ mV sec^{-1}, T = 298 K. •, x = 0.1; ---, $0.4 \leq x \leq 0.9$.

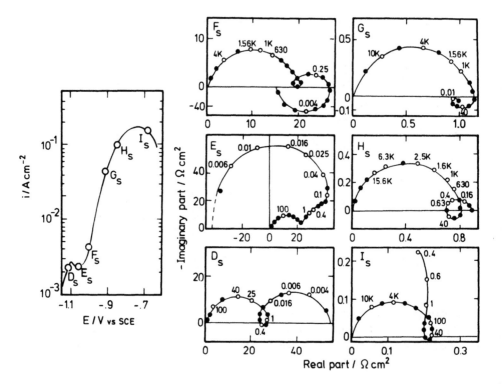

FIGURE 13 Impedance diagrams at different polarizations of the corresponding current density-potential curve [48]. System: Fe(RDE)/
x M Na$_2$SO$_4$ + y M H$_2$SO$_4$, x + y = 1, pH = 5, νrot = 27 sec^{-1}, T = 298 K.

electrode potential and the pH of the electrolyte [47−66,77−79,81].
This reaction is described by

$$Fe(OH)_{ads} \rightleftharpoons Fe(O)_{ads} + H^+ + e^- \qquad (14)$$

A distinction between the hydrated and dehydrated adsorbed intermediates of Fe(II) is not possible by electrochemical means only. If
the degree of coverage of this divalent iron oxide species increases,
the degree of coverage of the precursor Fe(OH)$_{ads}$ has to decrease.
Blocking of the active sites by (O)$_{ads}$ inhibits the active dissolution
of iron in the transition range.

The potential of current maximum I and its pH dependence (Fig. 9)
correspond approximately to the equilibrium potential for the formation
of magnetite [83]. Therefore, the formation of a trivalent iron oxide

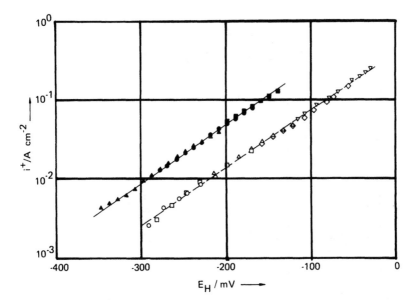

FIGURE 14 Current density-potential curves in the prepassive range [52,81]. System: $Fe(RDE)/x$ M Na_2SO_4 + y M H_2SO_4, x = 0.5, ν_{rot} = 69 sec^{-1}, $|dE/dt|$ = 3 mV sec^{-1}, T = 298 K, pH: ■, 4.4; ●, 5.0; and ▲, 5.6. System: $Fe(RDE)/x$ M $NaClO_4$ + y M $HClO_4$, x = 1.0, ν_{rot} = 69 sec^{-1}, $|dE/dt|$ = 5 mV sec^{-1}, T = 298 K; pH: ▽, 2.5; ◇, 2.9; △, 3.7; □, 4.4; and ○. 4.9.

species at more positive potentials in the prepassive range must be taken into account.

Previously, the kinetics of the dissolution process in the transition range were mainly explained on the basis of the noncatalyzed consecutive charge transfer mechanism occurring in the active dissolution range according to Eqs. (8a)–(8c) [38–46,49–66,81]. With respect to the new results and their interpretations, this concept should be replaced by the catalytic mechanism described in the preceding section.

The Prepassive Range

In the prepassive range of active iron dissolution, a Tafel behavior of the polarization curves between the minimum of the transition range and a second current maximum II (Figs. 1, 6, and 14) is observed [37,48,49,51–66,77–79,81]. All kinetic dc data obtained at different temperatures are summarized in Table 3.

TABLE 3 Kinetic dc Data for Iron Dissolution in the Prepassive Range at Different Temperatures[a]

T/K	b^+ (mV)	$\left(\dfrac{\partial E_{maxII}}{\partial pH}\right)_{a_i \neq a_{H^+}}$ (mV)	$\left(\dfrac{\partial \log i_{maxII}^+}{\partial pH}\right)_{a_j \neq a_{H^+}}$	References
283	+130 ± 10	-60 ± 5	-0.44 ± 0.02	[53,81]
298	+130 ± 10	-66 ± 3	-0.47 ± 0.02	[53,81]
	+60 ± 8	—	—	[77,78]
	+120 ± 20	—	—	[55—66]
	+120	—	—	[38—48]
313	+137 ± 15	—	-0.55 ± 0.02	[53,81]
333	+110 ± 15	—	-0.61 ± 0.03	[53,81]

[a]Systems: Fe/x M acid + y M salt with acid = $HClO_4$ or H_2SO_4 and salt = $NaClO_4$, Na_2SO_4, or K_2SO_4, $0 \leq pH \leq 6$.

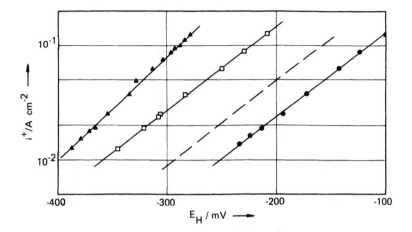

FIGURE 15 Tafel lines in the prepassive range as a function of temperature [53,81]. System: Fe(RDE)/x M Na_2SO_4 + y M H_2SO_4, x = 0.5, 4.0 ≤ pH ≤ 5.5 at T = 298 K, ν_{rot} = 69 sec^{-1}, |dE/dt| = 5 mV sec^{-1}. T (K): ●, 283; ----, 298; □, 313; and ▲, 333.

The slope of this Tafel line depends on the polarization routine. Under quasi-steady-state conditions, using potential sweep rates of 0.1 mV sec^{-1} ≤ |dE/dt| ≤ 10 mV sec^{-1}, Tafel slopes of b^+ = +120 ± 20 mV at T = 298 K were usually measured, whereas under steady-state conditions the Tafel slope was b_{ss}^+ = +60 ± 8 mV [67,77–79]. The slope increased [38–46] while the current approached the second maximum.

The Tafel slopes were found to be almost independent of temperature but increased with decreasing ionic strength as shown in Figs. 15 and 16.

Many experimental studies showed that the dissolution rate of iron in the prepassive range is independent of both pH (Fig. 14) and anion activity in the electrolyte (Figs. 17 and 18) at temperatures 283 K ≤ T ≤ 333 K.

Both the potential and the current density of maximum II, E_{maxII} and i_{maxII}^+, respectively, were found to depend on the pH of the bulk electrolyte, but no influence of the anion activity was observed, as shown in Figs. 19 and 20. The absolute values of the partial derivatives,

$$\left(\frac{\partial E_{maxII}}{\partial pH}\right)_{a_i \neq a_{H^+}} \quad \text{and} \quad \left(\frac{\partial \log i_{maxII}^+}{\partial pH}\right)_{a_i \neq a_{H^+}}$$

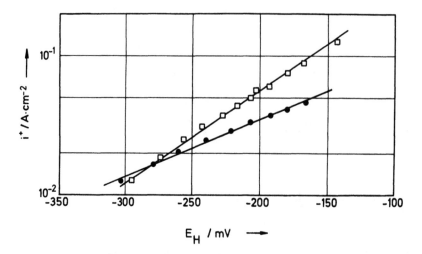

FIGURE 16 Tafel lines in the prepassive range as a function of ionic strength [53,81]. System: Fe(RDE)/x M Na_2SO_4 + y M H_2SO_4, 0.01 ≤ x ≤ 0.05, 4.0 ≤ pH ≤ 5.5, ν_{rot} = 69 sec^{-1}, $|dE/dt|$ = 5 mV sec^{-1}, T = 298 K. x: •, 0.01; □, 0.05.

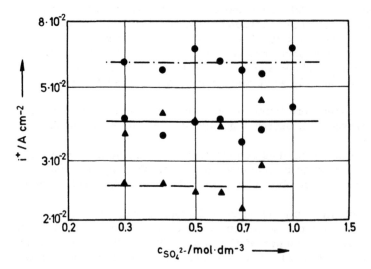

FIGURE 17 Current densities in the prepassive range depending on the sulfate ion concentration [52,81]. System: Fe(RDE)/x M Na_2SO_4 + y M H_2SO_4, 0.3 ≤ x ≤ 1.0, ν_{rot} = 69 sec^{-1}, $|dE/dt|$ = 5 mV sec^{-1}, T = 298 K. pH: •, 4.8; ▲, 5.0. E_H (mV): -·-, -188; ——, -218, and ----, -248.

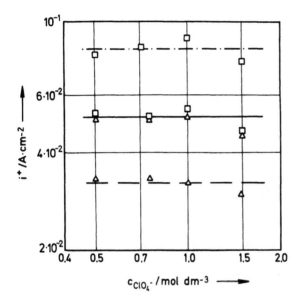

FIGURE 18 Current densities in the prepassive range depending on the perchlorate ion concentration [52,81]. System: Fe(RDE)/x M NaClO$_4$ + y M HClO$_4$, $0.5 \leq x \leq 1.5$, ν_{rot} = 69 sec^{-1}, $|dE/dt|$ = 5 mV sec^{-1}, T = 298 K. pH: □, 2.4; △, 2.8. E$_H$ (mV): —·—, -88, ——, -118; ----, -148.

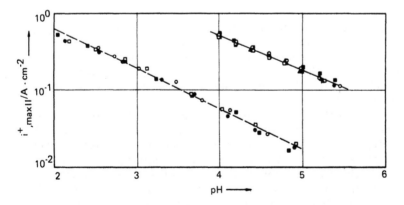

FIGURE 19 Current densities of maximum II as a function of pH and anion concentration [52,81]. System: Fe/x M Na$_2$SO$_4$ + y M H$_2$SO$_4$, $0.3 < x < 1.0$, $4.0 < $ pH $ < 5.5$. x: □, 0.3; ■, 0.4; ○, 0.6; ●, 0.7; △, 0.8; ▲, 1.0. System: Fe(RDE)/x M NaClO$_4$ + y M HClO$_4$, $0.5 \leq x \leq 1.5$, $2.0 \leq$ pH ≤ 5.0, ν_{rot} = 69 sec^{-1}, $|dE/dt|$ = 5 mV sec^{-1}, T = 298 K. x: ○, 0.5; ●, 0.75; □, 1.0; ■, 1.5.

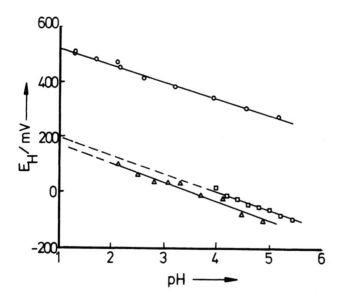

FIGURE 20 pH dependence of the potential of maximum II [52,81] and of the Flade potential (○) [52,54,70,71,81,95]. (○,□) System: Fe(RDE)/x M Na$_2$SO$_4$ + y M H$_2$SO$_4$, 0.3 ≤ x ≤ 1.0. (△) System: Fe(RDE)/x M NaClO$_4$ + y M HClO$_4$, 0.5 ≤ x ≤ 1.5. (○) Steady-state conditions, ν_{rot} = 69 sec^{-1}. (□,△) Cyclic voltammetry, ν_{rot} = 69 sec^{-1}, |dE/dt| = 5 mV sec^{-1}, T = 298 K.

increased with increasing temperature. The pH dependence of i_{maxII}^+ grew with ionic strength.

Ac impedance measurements in the prepassive range exhibited various capacitive and inductive loops in Nyquist diagrams, depending on the potential and the pH of the system. Examples are given in Fig. 13, points F_S to I_S.

Usually, the iron oxide species Fe(O)$_{ads}$ was believed to participate in the dissolution reaction of iron in the prepassive state [38–54,60–71,81]. Recently, it was proposed that this dissolution reaction takes place only at active surface sites Fe$_k$(O)$_{ads}$ and follows either a catalytic [48] or a consecutive [77–79] charge transfer mechanism. The rate constant for dissolution of Fe$_{sol}^{2+}$ from kinks with (O)$_{ads}$ is assumed to be much lower than that from kinks with (OH)$_{ads}$. As explained above, the spacing of kinks cannot be changed at the positive potentials of the prepassive range. The rate-determining step can be described in terms of the catalytic mechanism as follows:

$$FeFe_k(O)_{ads} \rightarrow Fe_k(O)_{ads} + Fe^{2+} + 2e^- \qquad (15)$$

Under steady-state conditions, a Tafel slope of $b_{ss}^+ = +60$ mV at T = 298 K is expected. Besides this dissolution reaction (15), the formation of a trivalent iron oxidic species at the surface must be assumed according to the more negative equilibrium potential of the magnetite formation (see the preceding section).

On the other hand, a rate-determining chemical dissolution step (I_2 reaction) occurring at random surface sites was postulated by Bech-Nielsen et al. [60—66]. Surface complexes containing several anions were assumed. Assignment of a physical meaning to these complexes seems to be impossible. Moreover, a special shape of the adsorption isotherm had to be assumed in order to fit the dc polarization curves.

The appearance of current maximum II was ascribed to an inhibiting effect of oxidic Fe(III) at the iron surface on the dissolution reaction of Fe(II) [51]. It was assumed that the degree of coverage with oxidic Fe(III) increases rapidly in the vicinity of E_{maxII}, causing current maximum II [51—54,81]. The oxidic Fe(III) can be considered as a constituent of an oxide film, which is 2-D in the vicinity of E_{maxII}. As the potential becomes more positive, this film thickens to a 3-D film in the passive layer formation range. As long as the film is still 2-D, dissolution of Fe(II) will proceed according to reaction (15) at active sites.

The Passive Layer Formation Range

The iron dissolution reaction in the passive layer formation range is characterized by a continuous polarization curve with a negative Tafel slope of $b_{ss}^+ = -100 \pm 20$ mV at T = 298 K, extending from current maximum II to the onset of passivation as shown in Figs. 1 and 6 [17, 51,52,70,81,90—95]. In this range, a negative electrochemical reaction order with respect to hydroxyl ions $y_{(OH^-)}^+ < 0$ was observed [70,71,96].

Discontinuous [97] and z-shaped [98—100] polarization curves measured in very acidic media under different potentiostatic conditions may be due to measuring artifacts caused by the ohmic drop [70,71, 101].

The potential of current maximum II was found to be about 0.35 V more negative than the Flade potential and about 0.33 V more positive than the equilibrium potential of the formation of magnetite, as shown in Fig. 20.

The passive layer formation range can be described as a transition from a 2-D oxidic film, still existing at the potential of current maximum II, to a 3-D nonporous oxidic film in the passive metal state with a steady-state stoichiometry determined by the electrode potential through the equilibrium of oxygen ions at the oxide/electrolyte interphase and the equilibrium of electrons within the film [27]. The reactions taking place in the passive layer formation range obviously involve iron dissolution, increased formation of oxidic Fe(III) at the surface, and 3-D film growth. A kinetic interpretation of the complicated electrode behavior requires more experimental results than were obtained previously. Kinetic descriptions based on either a 2-D or a 3-D film approach are only of a qualitative nature [70,71].

1.2.3 Iron Dissolution in Alkaline Solutions

Knowledge of the electrochemical behavior of iron in alkaline solutions is very important for technical applications, e.g.,

1. Technology of alkaline accumulators and batteries with iron electrodes
2. Electrochemical manufacturing of oxide films
3. Corrosion of iron and steels and their protection

Rather few investigations have been carried out in this field compared to those in acid solutions. The studies started at the beginning of this century [102—104]. Decisive progress was made by Kabanov et al. [2—5]. Later work was usually performed with polycrystalline iron in mixtures of KOH with K_2SO_4 or NaOH with Na_2SO_4 [105—120]. The nature of oxide films was studied by X-ray measurements, electron diffraction, and optical methods [121—126].

The solubilities of ferrate(II) and ferrate(III) in alkaline solutions are rather low. Therefore, the kinetics of the iron electrode were found to be complicated by precipitation of hydrated oxide layers. The electrochemical properties of such layers change with ageing and oxidation or reduction. Thus, several authors [109—114,117,119] considered steady-state dc measurements not useful. They attempted to circumvent the experimental difficulties by employing fast methods such as large-signal pulse polarization or single sweep and cyclic voltammetry at relatively high sweep rates. As an example, Fig. 21 shows a typical voltammogram of an iron electrode preactivated at negative electrode potentials in the region of hydrogen evolution. Three anodic current maxima (I, II, and III) and two cathodic ones (II' and III') were observed. In the passive range a further pair of maxima was found by some authors [114,117,119].

Since Fig. 21 is qualitatively similar to Fig. 1 for acid solutions, the different regions were named in the same way. However, no mechanistic information can be extracted from measurements like those in Fig. 21. Even the qualitative assignment of the maxima to defined processes is not entirely certain. Anodic maximum I was found to appear after preceding hydrogen evolution and therefore was tentatively attributed to oxidation of adsorbed hydrogen [114,117]. Other authors assumed that maximum I corresponds to a first stage of iron(II) oxide formation [113,119], perhaps dissolution of ferrate(II).

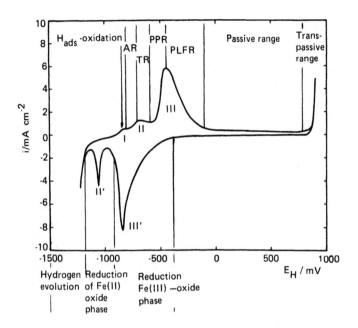

FIGURE 21 Cyclic voltammetric polarization curve of iron dissolution in alkaine solutions between the active and the passive state of the metal [117, 134]. System: $Fe/0.1$ M $NaOH$ + 0.4 M Na_2SO_4, ν_{rot} = 0 Hz, $|dE/dt|$ = 40 mV sec^{-1}, T = 298 K. AR, Active range; TR, transition range; PPR, prepassive range; and PLFR, passive layer formation range.

All the authors [106—117,119] agree on a strong correlation between maxima II and II' and, on the other hand, III and III'. It was concluded from thermodynamics [83] that maxima II and II' are due to formation and reduction of a porous iron(II) oxide film and maxima III and III' to oxidation of this film to an iron(III) film and its reduction, respectively [106—120].

Steady-state polarization curves of iron dissolution and deposition were measured by Kabanov et al. [2—5], avoiding supersaturation of ferrate(II) and oxidation of iron(II). Anodic and cathodic Tafel slopes b_{ss}^+ = +30 to +40 mV and b_{ss}^- = -40 mV and an electrochemical reaction order $y_{(OH^-)}^+$ = 2 with respect to hydroxyl ions were observed. More recently [118,120], a higher Tafel slope of b_t^+ = +60 ± 10 mV and a lower reaction order $y_{(OH^-)}^+$ = 1.1 were found under transient conditions. The authors [2—5,118,120] explained their data on the basis of a consecutive charge transfer mechanism. A Frumkin isotherm for the adsorbed intermediate was assumed by Drazic and Hao

[120]. However, there is no reason to assume a reaction mechanism different from that in acid solutions. Evidence for the catalytic mechanism came from the anodic and cathodic Tafel slopes both being small. A possible modification of any mechanism is the direct participation of hydroxyl ions in the reaction steps. Further reaction steps not encountered in acid and neutral solutions result in the formation of ferrate(II), e.g., $HFeO_2^-$. So far, there is no experimental evidence for such modifications of the reaction mechanism.

1.2.4 Iron Dissolution in Neutral Solutions

The electrochemical behavior of iron and steel in neutral aqueous solutions in the absence or in the presence of dissolved oxygen is of great importance in the majority of practical corrosion processes. However, relatively few investigations have been made, and these deal with the iron dissolution kinetics under such conditions [54,81,85,127–133].

In deaerated weakly acid and neutral unbuffered aqueous solutions, steady-state and transient measurements have been carried out in the system

Fe(polycrystalline)/x M H_2SO_4 + y M K_2SO_4

at constant or variable ionic strength and T = 298 K [85]. The iron dissolution kinetics in the active range at $5 \leq pH \leq 7$ are characterized by an increase of the steady-state anodic Tafel slope to $b_{ss}^+ = +65 \pm 5$ mV and a decrease of the electrochemical reaction order with respect to hydroxyl ions to $y_{(OH^-)}^+ \approx 0$ and $y_{(OH^-)}^+ = 0.7 \pm 0.1$ for unstirred and stirred solutions, respectively, as seen in Table 1. Pulse measurements yielded transient Tafel lines with slopes of $b_t^+ = +115 \pm 10$ mV. These results can be explained by taking into account an increased surface pH and a certain buffer capacity of Fe(II) hydrolysis products.

An increase of the surface pH is generally caused by diffusion control of hydrogen evolution or oxygen reduction processes. In this case, the difference between surface and bulk pH increases with decreasing electrode potential. It can easily be shown that

$$(\partial pH/\partial E)_{a_i \neq a_{H^+}} = -\alpha^- nF/(\ln 10)RT$$

if the cathodic process is limiting diffusion-controlled. Here, α^- represents the charge transfer coefficient of the cathodic process.

Therefore, the rate of anodic dissolution increases above the rate expected for constant pH at the surface as the potential becomes more negative. With $\alpha^- = 0.5$, the steady slope $b_{ss}^+ = +65$ mV would correspond to a slope of about $b_{ss}^+ = +35$ mV for constant pH at the surface. The experimental transient slope $b_t^+ = 115$ mV may be explained in a similar way, if the pH at the surface changes faster than the superpolarization values are attained.

In aerated weakly acid and neutral unbuffered aqueous solutions, steady-state and quasi-steady-state measurements were carried out in the systems

Fe(polycrystalline)/x M salt + y M acid + z M O_2

with x = 0.5, $0 \leq z \leq 8.4 \times 10^{-4}$, salt = Na_2SO_4 or $NaClO_4$, and acid = H_2SO_4 or $HClO_4$, respectively, in the range $2 \leq pH \leq 9$ at T = 298 K [54,81,131] and

Fe(polycrystalline)/aerated California tap water + 0.1 M $NaClO_4$

at T = 298 K [132]. A strong influence of the cathodic oxygen reduction process on the iron dissolution kinetics was found. In order to establish defined hydrodynamic conditions for the diffusion-controlled oxygen reduction and the time-dependent formation of 3-D porous oxidic layers on the iron surface, the experiments were performed on rotating disk [54,81,131] or cylinder [132] electrodes. The results were found to depend strongly on the pretreatment and prepolarization of the electrode, its exposure time, the polarization routine, the hydrodynamic conditions, and the content of dissolved oxygen in the bulk electrolyte.

Diffusion-controlled oxygen reduction was found to predominate over the hydrogen evolution reaction in aerated solutions at a bulk pH ≥ 4.2 at T = 298 K, as seen in Fig. 22. The diffusion coefficient of oxygen was determined from the dependence of the cathodic limiting diffusion current density on the rotational speed of an iron disk electrode at pH 4.4, giving a value of

$D(O_2) = 1.6 \times 10^{-5}$ cm^2 sec^{-1}

which could be converted into that valid for pure water, $D(O_2)_{H_2O} = 1.9 \times 10^{-5}$ cm^2 sec^{-1} by using the Stokes–Einstein relation [54,81].

The anodic polarization curves exhibited clearly all the different ranges shown in Fig. 1 and discussed above at bulk pH < 4.2. Steady-state anodic Tafel slopes were determined from either the measured anodic polarization curves or the dependence of the corrosion potential on the angular velocity of the disk electrode [54,81].

The b_{ss}^+ values were found to increase strongly within a relatively small pH range, from $b_{ss}^+ = +40 \pm 2$ mV at pH 3.8 to $b_{ss}^+ = +119 \pm 10$ mV at pH 4.2 (Table 4). Simultaneously, the electrochemical reaction order with respect to OH$^-$ ions decreased from $y_{(OH^-)}^+ = +1$ to nearly zero. This change of the kinetic data in aerated solutions at relatively

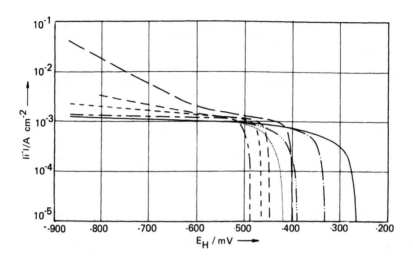

FIGURE 22 Steady-state cathodic current density-potential curves for various pH values in aerated electrolytes [54,81]. System: Fe(RDE)/ 0.5 M Na₂SO₄ + y M H₂SO₄ + 1.76 × 10⁻⁴ M O₂, 3.0 ≤ pH ≤ 5.0, ν_{rot} = 69 sec⁻¹, T = 298 K. pH: — —, 3.0; - - -, 3.8; ----, 4.0; -·-, 4.2; ····, 4.4; -·····-, 4.8; ——, 5.0.

low bulk pH values can be explained by the same arguments used for deaerated neutral media.

At pH ≥ 4.2, no current maximum I could be observed in aerated solutions [54,81]. At the same time, a drastic shift of the corrosion potential in the positive direction was observed, as shown in Fig. 23, whereas the kinetic data b_{ss}^+ = +120 ± 10 mV and $y_{(OH^-)}^+$ = 0 remained constant [54,81,132]. This behavior was explained by a change of the iron dissolution from the active range at pH ≤ 4.2 (i_{maxI}^+ > i_{corr}) to the prepassive range at pH > 4.2 (i_{maxI}^+ < i_{corr}) as demonstrated schematically in Fig. 24 [54,81]. As expected, this critical pH value at the jump of the corrosion potential was found to be shifted

TABLE 4 Anodic Tafel Slopes of Rotated Iron Disk Electrodes in Aerated Aqueous Solutions [81]

pH	3.8	3.9	4.11	4.14	4.2
b_{ss}^+ (mV)	40 ± 2	44 ± 3	80 ± 5	95 ± 8	119 ± 10

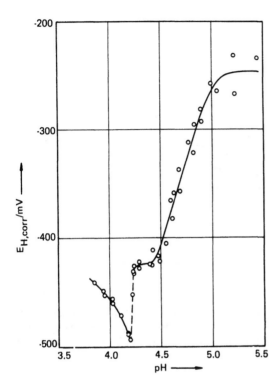

FIGURE 23 Iron corrosion potentials vs. pH [54,81]. System: Fe-
(RDE)/0.5 M Na_2SO_4 + y M H_2SO_4 + 1.76 × 10^{-4} M O_2, 3.8 ≤ pH ≤ 5.5,
ν_{rot} = 69 sec^{-1}, T = 298 K.

to higher values with decreasing bulk concentration of dissolved oxygen, as demonstrated in Fig. 25.

The influence of the exposure time of the disk electrode at the corrosion potential, t, on cyclic voltammetric studies in aerated neutral
solutions is represented in Fig. 26 [131]. The polarization curves indicated irreversible processes characterized by different current maxima II, III, and II', III' in the anodic and cathodic sweeps, respectively. These maxima were found to grow with increasing sweep rate
(Fig. 26). Anodic maximum II appeared only after a relatively long
exposure time at the corrosion potential and at relatively high sweep
rates.

A comparison of the results presented in Fig. 26 with those obtained in alkaline solutions (Fig. 21) seems to be appropriate. The
potentials of the anodic current maxima II and III in Fig. 26 are more
positive than those measured in alkaline media [117,119]. This shift

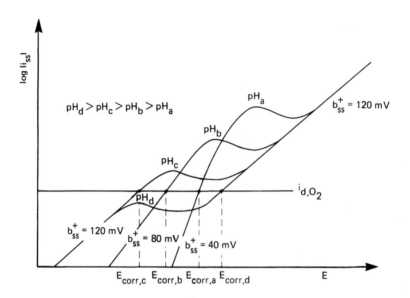

FIGURE 24 Schematic representation of steady-state polarization curves of iron dissolution in the active, transition, and prepassive ranges depending on the pH under limiting diffusion-controlled oxygen reduction conditions [54].

is expected from the different bulk pH values of the systems studied. However, the potentials of cathodic maxima II' and III' in Fig. 26 are nearly compatible with those in Fig. 21 because of the increase of the surface pH in aerated neutral systems. The observed current in Fig. 26 in the range $E_{maxIII'} \leq E \leq E_{corr}$ can be attributed to the dominating oxygen reduction process. The rise of the cathodic current at $E < E_{maxII}$ corresponds to the hydrogen evolution reaction with the decomposition of water. According to the investigations in alkaline solutions free of oxygen [117,119], current maxima II and II' can be attributed to the formation and reduction of 3-D iron(II) hydroxide layers, respectively. Current maxima III and III' represent the formation and reduction of 3-D oxide layers consisting of $Fe(OH)_3$ and the spinel phase, $\gamma\text{-}Fe_2O_3/Fe_3O_4$.

The shift of the corrosion potential in the negative direction with increasing exposure time of the electrode held at E_{corr}, noticeable in Fig. 26, is demonstrated in more detail in Fig. 27. Simultaneously, the cathodic oxygen reduction process became no longer limiting diffusion-controlled, and the anodic Tafel slope decreased. The corrosion current density decreased with increased exposure time of the

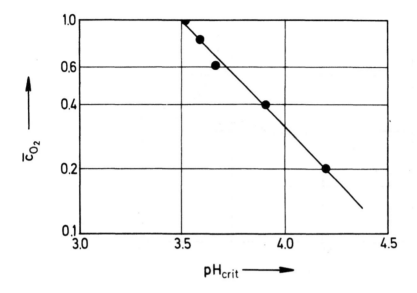

FIGURE 25 pH_{crit} in dependence of the bulk oxygen concentration c_{O_2} [54, 81]. System: $Fe(RDE)/0.5$ M $Na_2SO_4 + y$ M $H_2SO_4 + z$ M O_2, $pH = pH_{crit}$, $\nu_{rot} = 69$ sec^{-1}, $T = 298$ K; c_{O_2}, bulk oxygen concentration related to the saturation value of 8.38×10^{-4} mol dm^{-3} at $p = 1.013$ bar.

electrode, and inhomogeneous corrosion of the surfaces was observed optically [131]. This behavior was explained by the formation of differential aeration cells at the corroding surface, where the oxygen reduction and iron dissolution obviously take place at different surface areas. The surface inhomogeneity was assumed to arise through the time-dependent formation of 3-D porous oxide layers acting as an interphase inhibitor. The drastic shift of the corrosion potential and the decrease of the anodic Tafel slope were attributed to a change of the dissolution reaction from the prepassive range to the active one with increasing exposure time due to a local decrease of the surface pH at the anodic areas.

Summarizing the results in neutral media, it is emphasized that the iron dissolution mechanism obviously remains unchanged; however, the kinetic data are strongly influenced by the time-dependent formation of 3-D porous oxidic layers, nonuniform corrosion phenomena, the hydrodynamic conditions, and the content of oxygen in the electrolyte.

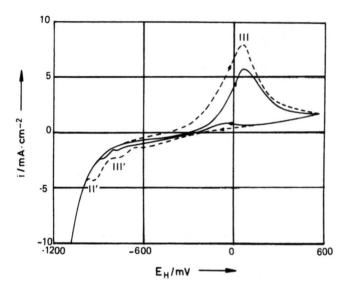

(a)

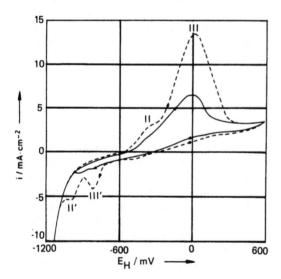

(b)

FIGURE 26 Cyclic voltammetric polarization curves [131,134]. System: Fe(RDE)/0.5 M Na_2SO_4 + 1.76 × 10^{-4} M O_2, pH = 7.0, ν_{rot} = 69 sec^{-1}, T = 298 K. $|dE/dt|$: ———, 5 mV sec^{-1}; — —, 40 mV sec^{-1}. a) t at E = E_{corr} ≤ 10 min; b) t at E = E_{corr} ≈ 5 hr; t = waiting time.

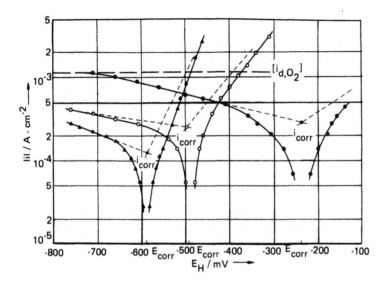

FIGURE 27 Current density-potential curves in the vicinity of the corrosion potential depending on the exposure time, t, at $E = E_{corr}$ [131,134,139]. System: $Fe(RDE)/0.5\ M\ Na_2SO_4 + 1.76 \times 10^{-4}\ M\ O_2$, pH = 7.0, ν_{rot} = 69 sec^{-1}, T = 298 K. $-\ -$, i_{d,O_2}; t (hr): $\bullet$, 0; $\circ$, 2; $\blacktriangle$, 12.

1.2.5 Iron Dissolution in the Presence of Surface-Active Substances

In the preceding sections the iron dissolution kinetics were considered not to be influenced by any electrolyte constituent except water or hydroxyl ions. Now the effect of surface-active substances will be discussed. Such substances may participate in the dissolution kinetics, leading to acceleration or retardation of the reaction rate, i.e., stimulation or inhibition effects, respectively. Under these conditions, the kinetic data and the mechanism of the iron dissolution reaction may change. Surface-active substances are either ions or neutral molecules dissolved in the electrolyte.

From a practical point of view, most investigations were devoted to the inhibition of iron corrosion. The effect of surface-active substances on the iron dissolution kinetics will be discussed on the basis of two different retardation mechanisms, called interface and interphase inhibition [134,135].

Interface and Interphase Inhibition

Interface inhibition presumes a strong interaction between the corroding substrate and the inhibitor [130,136–139]. In this case,
there is a potential-dependent adsorption of the inhibitor. The formation of a 2-D adsorbate can affect the basic corrosion reactions in
different ways, which may be discussed in terms of the inhibition efficiency ε being a function of the electrode potential E, the inhibitor
concentration c_I, and the exposure time t:

$$\varepsilon(E,c_I,t) = \frac{i(E,t) - i_I(E,c_I,t)}{i(E,t)} \tag{16}$$

where i and i_I represent the current densities of the electrode reaction in the absence and in the presence of inhibitor, respectively.
   Theoretically, three cases can be distinguished:

1. Geometric blocking effect of the electrode surface by an inert ad-
   sorbate described by the relative coverage θ. In this case $i_I = (1 - \theta)i$, and therefore

   $$\varepsilon = \theta \tag{17}$$

   where $\theta = \Gamma/\Gamma_s$; Γ denotes the potential-dependent surface concentration of inhibitor and Γ_s its saturation value.
2. Blocking of active surface sites by an inert adsorbate. In this case
   Eq. (17) holds again, if Γ_s is replaced by a surface concentration
   $\Gamma_{a,s}$ necessary for complete blocking of all active surface sites.
   Strong inhibition with $\varepsilon \to 1$ is possible at low mean coverages.
3. The adsorbate is not inert but reactive. Two different cases may
   be distinguished. First, the inhibitor acts as a positive or negative electrocatalyst on the corrosion reaction, and second, the adsorbate itself undergoes an electrochemical redox process. In the
   latter case, primary and/or secondary inhibition can occur, depending on the retardation effects caused by the original adsorbate and/or its reaction product, respectively. In the case of reactive coverage, ε will be a more complex function of θ than Eq.
   (17) and can also be negative:

   Inhibition: $0 \le \varepsilon(\theta) \le 1$

   Stimulation: $\varepsilon(\theta) < 0$ $\tag{18}$

   These three types of interface inhibition are mostly observed in
corrosion systems with the bare metal surface in contact with the corrosive medium, a condition met for active metal dissolution in acid solutions.

Interphase inhibition presumes a 3-D layer between the corroding substrate and the electrolyte [54,131,135,140]. Such 3-D layers generally consist of weakly soluble compounds such as oxidic corrosion products or inhibitors forming porous layers. The inhibition efficiency depends strongly on the properties of the 3-D layer, especially its porosity and stability. This type of inhibition is often seen in neutral media in the absence or in the presence of oxygen. In this case of interphase inhibition, the inhibition efficiency can be correlated with the rate of the transport-controlled cathodic corrosion reaction taking place within the pores of the 3-D layer; $i \approx i^-$ and $i_I \approx i_I^-$ for $E < E_{corr}$.

Since the cathodic current densities depend strongly on the corrosion time, the protective property of the 3-D layer also becomes a function of time.

Inhibition and Stimulation of Iron Dissolution

Most of the studies published in the literature during the past two decades dealt with interface inhibition in acid solutions. The inhibition efficiency was calculated from dc and ac polarization experiments. For example, steady-state current density-potential curves and ac impedance measurements of iron corrosion in acid solution are shown in Figs. 28 and 29, respectively [139]. The inhibitor triphenyl-benzylphosphonium cation (TPBP$^+$) in acid chloride solutions clearly acts as an interface inhibitor, shifting the Tafel lines, strongly enlarging the polarization resistance R_p, and giving negative electrochemical reaction orders of the metal dissolution process with respect to the bulk activity of the inhibitor I:

$$\left(\frac{\partial \log i_{ss}^+}{\partial \log a_I} \right)_{a_i \neq a_I, E} < 0$$

The inhibition efficiency decreases with increasing electrode potential. This agrees well with the assumption that a strong interaction between the adsorbed cation and the metal surface is necessary for effective inhibition. With increasing electrode potential the surface charge becomes more positive and an electrostatic repulsion will weaken the adsorption bond. Moreover, kinetic effects must be taken into account. The higher the metal dissolution rate, the more adsorbate can be desorbed. At the so-called desorption potential, the degree of coverage drops to zero because the desorption rate becomes higher than the adsorption rate.

A geometric blocking effect of an adsorbed inert interface inhibitor does not affect the slope of the anodic Tafel line at potentials lower than the desorption potential, as seen in Fig. 28. In this case,

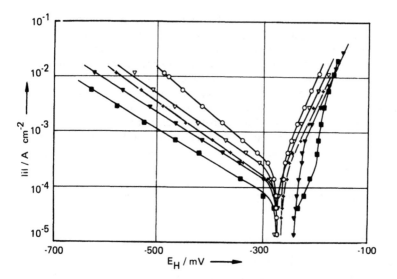

FIGURE 28 Steady-state galvanostatic current density-potential curves
[139]. System: Fe(RDE)/0.5 M H₂SO₄ + x mM TPBP⁺, deaerated; ν_{rot} =
30 sec⁻¹, T = 298 K. x: ○, 0; ▽, 0.1; +, 1; ▼, 5; ■, 10.

the iron dissolution reaction occurs unchanged at uncovered active
sites of the electrode surface. The current density-potential relation
of the iron dissolution reaction remains the same as in inhibitor-free
solutions but must be multiplied by the factor (1 - θ) representing the
active surface area.

A distinction between geometric blocking and blocking of active
sites is generally not possible by simple steady-state or quasi-steady-
state dc measurements. One of the essential pieces of information,
the coverage isotherm of the inhibitor as a function of the electrode
potential, was rarely measured. Estimates resulting from a comparison of the faradic corrosion fluxes in the absence and in the presence of the inhibitor according to Eqs. (16) and (17) were usually
based on the assumption of geometric blocking. Isotherms may be
obtained, e.g., from measurements of the double layer capacity, but
direct analytical determination of the adsorbed amount is preferable.

The distinction between indifferent and reactive coverage of interface inhibitors often requires sophisticated measuring techniques.
Only in some simple cases is a reactive coverage obvious. A clear
indication is a positive electrochemical reaction order with respect to
the activity of the adsorbed species. In this case, the reactive coverage has a stimulating effect on the iron dissolution kinetics as observed for OH⁻ and SH⁻ ions.

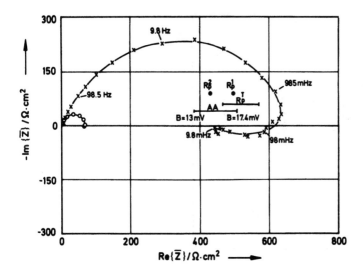

FIGURE 29 Impedance diagrams [139]. Systems: Fe(RDE)/0.5 M H_2SO_4 + x mM $TPBP^+$, aerated; ν_{rot} = 69 sec^{-1}, T = 298 K. x: (○), 0; (x), 10. Calculated polarization resistances Rp: Rp^T, from cyclic voltammetry; Rp^1, from steady-state galvanostatic current density-potential curves; Rp^2, from steady-state potentiostatic current density-potential curves; AA, from solution analysis by atomic adsorption; B = $b^+b^-/ln(10)(b^+ + b^-)$ with b = Tafel slope.

On the other hand, a reactive coverage can also be identified in cases of an apparent negative electrochemical reaction order connected with a significant change of the kinetic data. An example is iron dissolution in the presence of halide ions in the electrolyte. Specific adsorption of halide ions leads either to inhibition or to stimulation, depending on the pH and the halide ion concentration [10,23,27, 33,141−147]. For example, in solutions of $0.3 \leq$ pH ≤ 1 and 10^{-1} M $\leq$ $c_{Cl^-} \leq 2$ M, chloride acts as an inhibitor, yielding a steady-state Tafel slope of b_{ss}^+ = +60 mV at T = 298 K, $y_{(OH^-)}^+$ = +1 ± 0.1, and $y_{(Cl^-)}^+$ = -0.7 ± 0.1. These kinetic data were explained by assuming a rate-determining dissolution reaction starting at covered active areas of the surface [10,23,27,33,141−144]:

$$FeFe_k(X^-)_{ads} + (OH)_{ads} \rightarrow Fe_k(OH)_{ads} + FeX^+ + 2e^- \qquad (19)$$

where iron from the kink site, Fe_k, is dissolved. The kink thereby is reproduced at the neighboring site and the adsorbed halide ion X^- is simultaneously desorbed. The steady-state current density-potential relation of this reaction is given by

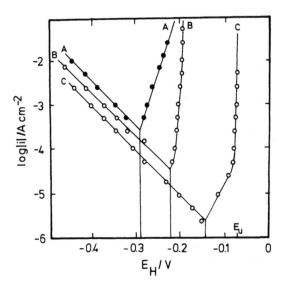

FIGURE 30 Steady-state polarization curves [141]. System: Fe/0.5 M H_2SO_4 + x M KI. x: A, 0; B, 5.5 × 10^{-4}; C, 5.5 ×10^{-2}, T = 298 K.

$$i^+ = k^+ \theta_{X^-} \theta_{OH} \exp\left(\frac{2\alpha^+FE}{RT}\right) \tag{20}$$

Assuming Langmuir behavior of both adsorbates and $\theta_{OH} \ll \theta_{X^-}$ ≈ 1, it follows $\theta_{OH} \approx 1 - \theta_{X^-} \approx$ const/a_{X^-}, representing a measured negative electrochemical reaction order and a Tafel slope of $b_{ss}^+ =$ +60 mV for $\alpha^+ = 0.5$ at T = 298 K. Taking into account the potential dependence of the desorption of X^- according to Eq. (19), Heusler derived an expression for the desorption potential* at which θ_{X^-} drops to zero, as shown in Figs. 28 and 30 for $TPBP^+$ and I^-, respectively [23,27,141].

Stimulation of the iron dissolution reaction was observed in solutions at high concentrations of both chloride ions and protons [144, 147]. Steady-state measurements yielded anodic Tafel slopes of $b_{ss}^+ =$ +100 mV at 298 K and electrochemical reaction orders of $y_{(OH^-)}^+ = 1.1$ and $y_{(Cl^-)}^+ = +0.6$. These kinetic data were interpreted in different

---

*Because in the steady state the potential becomes almost independent of current density, this desorption potential was also called potential of unpolarizability.

ways. Either a consecutive charge transfer mechanism, Frumkin adsorption condition, and a rate-determining topochemical step forming an adsorbed ion pair [138] were assumed according to

$$Fe + Cl^- \rightleftharpoons Fe(Cl)_{ads} + e^- \tag{21}$$

$$Fe(Cl)_{ads} + H^+ \rightleftharpoons Fe(ClH^+) \tag{22}$$

$$Fe(ClH^+)_{ads} \longrightarrow FeCl^+ + H^+ + e^- \tag{23}$$

$$FeCl^+ \rightleftharpoons Fe^{2+} + Cl^- \tag{24}$$

or a rate-determining ferrous ion transfer in one step and Langmuir adsorption behavior [147] were proposed:

$$Fe + Cl^- \rightleftharpoons Fe(Cl^-)_{ads} \tag{25}$$

$$Fe(Cl^-)_{ads} + H^+ \rightleftharpoons Fe(Cl^-H^+) \tag{26}$$

$$Fe(Cl^-H^+) + H^+ \longrightarrow FeCl^+ + 2H^+ + 2e^- \tag{27}$$

$$FeCl^+ \rightleftharpoons Fe^{2+} + Cl^- \tag{28}$$

A distinction between primary and/or secondary inhibition often requires additional nonelectrochemical analysis and transient measurements [148]. Previously it was found that onium ions as corrosion inhibitors are mostly electrochemically reduced [139,148]. For example, TPBP$^+$ was found to be reduced to triphenylbenzylphosphine (TPP) having a very similar inhibition efficiency, since the structure of the adsorbate is essentially maintained. Pulse polarization experiments in the absence and in the presence of TPBP$^+$ are shown in Fig. 31. The interface inhibitor was added to the acid solution during anodic polarization of the iron, where a reduction process can be neglected. Addition of TPBP$^+$ increases the overvoltages of both the anodic and cathodic reactions due to primary inhibition. Similar results have been obtained using the reduction product TPP. Therefore, TPBP$^+$ acts as a primary and TPP as a secondary interface inhibitor.

The second type of inhibition, interphase inhibition, is less well studied. Interphase inhibition is difficult to measure by electrochemical means. The electrode behavior depends on time, on the properties of the surface layer, and on the hydrodynamic conditions. Therefore, rotating disk experiments are necessary in order to get exact information about the electrode kinetics in the absence and in the presence of inhibitors. An example is given in Fig. 27 for iron corrosion in aerated neutral Na_2SO_4 solution. The formation of a porous 3-D layer is strongly time-dependent, the cathodic oxygen reduction

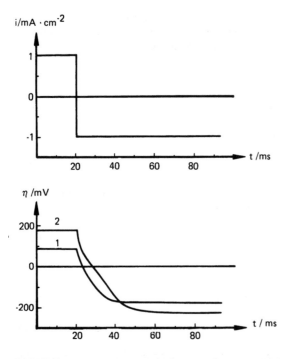

FIGURE 31 Galvanostatic large-signal pulse polarization [148]. System: $Fe/1$ M $HCl + x$ M $TPBP^+$. x: 1, 0; 2, 10^{-3}. T = 298 K.

is no longer characterized by a pure limiting diffusion current density, and simultaneously the anodic metal dissolution process changes from the prepassive range to the active one. The influence of the hydrodynamic conditions on the cathodic process is shown in Fig. 32. With varying rotation speed of the disk, a hysteresis can be observed which can be attributed to changing mechanical stability and porosity of the oxidic layer.

It is well known that many interface inhibitors lose their efficiency in the presence of surface layers. This is obviously due to a weaker interaction between the inhibitor species and the surface in the presence of 3-D coverage resulting in a lower inhibition efficiency. Effective corrosion inhibition in neutral medium in the presence of oxygen is only possible by retardation of the transport-controlled oxygen reduction taking place within the pores of the oxidic layer. Therefore, an effective inhibitor must drastically decrease the density of pores of the layer and increase its mechanical stability. Moreover, effective interphase inhibitors must be incorporated into the oxidic

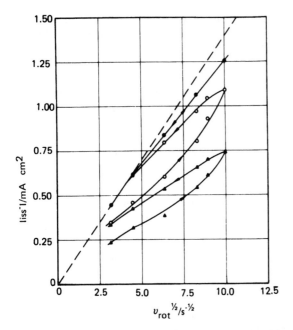

FIGURE 32 Steady-state cathodic current densities, i_{ss}^-, depending on the rotation frequency, ν_{rot}, and the exposure time, t, at $E = E_{corr}$ [131,134]. System: $Fe(RDE)/0.5\ M\ Na_2SO_4 + 1.76 \times 10^{-4}\ M\ O_2$, pH = 7.0, $10 \le \nu_{rot} \le 100\ sec^{-1}$, $E_H = -700\ mV$; T = 298 K. — —, i_{d,O_2}. t (hr): •, 0; ○, 2; ▲, 12.

layer and should react chemically in order to form a better protecting surface network [149,150]. Recent investigations have shown that sodium dihydrogen phosphate, different bisulfonic acids, and different commercial inhibitor packages represent effective interphase inhibitors, leading to more homogeneous and compact protecting layers [131]. As an example, the inhibition efficiencies of hexan (1,6)-bis-phosphonic acid (HBP) and sodium dihydrogenphosphate (SDHP) are listed in Table 5 as functions of the corrosion time t. Here ε_1 and ε_2 correspond to the efficiencies related to $t \to 0$ and $t \to \infty$ at $c_I = 0$, respectively.

Retardation of the iron dissolution reaction by inhibitors is a complicated surface process including the formation of 2-D and 3-D surface layers. The protective properties of such a surface layer depend on many parameters: interaction between the inhibitor and the substrate, incorporation of the inhibitor within 3-D layers, chemical

TABLE 5 Inhibition Efficiencies of Hexan (1,6)-Bisphosphonic Acid (HBP) and Sodium Dihydrogen Phosphate (SDHP) Depending on Exposure Time t at E = E_{corr} [131,134--136][a]

	t	ε_1	ε_2
Inhibitor-free electrolyte, pH 8.0	10 min	0.74	
	2 hr	0.77	
	12 hr	0.87	
HPB (10^{-2} M)	10 min	0.46	
	2 hr	0.70	
	5 hr	0.97	0.78
SDHP (10^{-2} M)	10 min	0.60	
	2 hr	0.97	0.78

[a] ε_1 and ε_2 are related to the limiting diffusion current density i_{d,O_2} = 1.15×10^{-3} A cm^{-2} at t→0 and the corrosion current density i_{corr} at t→∞ in inhibitor-free solutions, respectively.

reactions, electrode potential, inhibitor concentration, temperature, hydrodynamic conditions, corrosion time, properties of the corroding metal surface, etc. Only the combination of various electrochemical steady-state and non-steady-state measurement techniques with non-electrochemical surface and solution analysis will allow an exact determination of the iron dissolution reaction and its type of inhibition.

1.3 COBALT

Steady-state polarization curves of cobalt in acid sulfate and perchlorate solutions follow Tafel lines. With 0.5 M sulfuric acid Kravtsov and Pikov [151,152] observed Tafel slopes $+5 \leq b_{ss}^+ \leq 30$ mV at T = 298 K.

In a later work [153] current densities i^+ normalized to constant electrode capacitance were used to calculate Tafel slopes of b_{ss}^+ = $+40 \pm 3$ mV for cobalt in 1 M $CoSO_4$ with 0.3 M $B(OH)_3$ at pH 3.5 and 4.2. The authors believed that the capacitance is a measure of true surface area. However, as discussed below, there is a potential-dependent chemisorption capacitance that increases with the anodic potential. Thus the normalization is not justified and the actual slope,

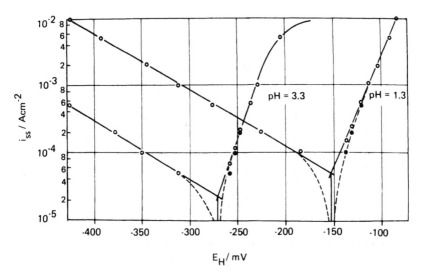

FIGURE 33 Steady-state polarization curves [10,154]. System: Co/ x M NaClO₄ + y M HClO₄, x + y = 0.5, pH = 1.3 and 3.3, T = 293 K.

which is not quoted in the paper, must have been smaller than b_{SS}^+ = +40 mV. Transient polarization curves corresponding to maximum polarization after disturbance of a given steady state yielded Tafel slopes of b_t^+ = +70 ± 10 mV [151,152] and b_t^+ = +64 mV [153]. Kravtsov and Pikov believed that their data are in agreement with a mechanism of consecutive charge transfer involving chemisorbed CoOH as an intermediate. However, this interpretation cannot be correct, since, among other things, the slopes of transient polarization curves are not compatible with the consecutive mechanism. Further arguments will be discussed below.

Heusler [10,154,155] investigated cobalt in perchlorate solutions. Steady-state polarization curves at T = 293 K are shown in Fig. 33. The Tafel slope was b_{SS}^+ = +29 ± 5 mV. From the pH dependence of the anodic current density at constant electrode potential, a reaction order $y_{(OH^-)}^+$ = 2.0 ± 0.3 with respect to hydroxyl ions was calculated, in agreement with Iofa and Wei [156] and Schwabe and Voigt [19]. At high current densities the Tafel slope increased with electrode potential. The deviation from a straight Tafel line occurred at lower current densities when the pH value was increased. Similar effects were observed with iron (see Section 1.2.2) and nickel.

Figure 34 shows transients of the electrode potential for different current pulses starting from the steady-state corrosion potential.

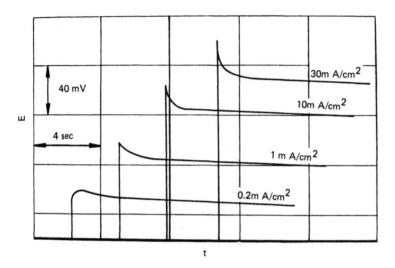

(a)

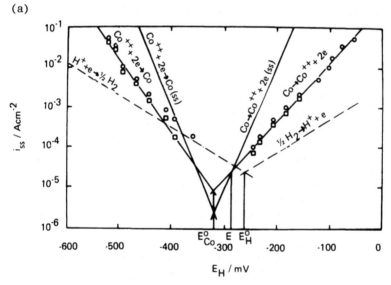

(b)

FIGURE 34 a) Galvanostatic transients [10,154]. System: Co/x M
NaClO₄ + y M HClO₄, x + y = 0.5, pH = 1.3, T = 293 K. Steady-state
starting potential E_H = -153 mV. b) Anodic and cathodic transient polarization curves [10,154]. System: Co/0.4 M Co(ClO₄)₂, pH = 4.4,
T = 293 K. Results of the measurements (○) were corrected (□) for
the influence of hydrogen deposition when necessary. E_{Co}^0, equilibrium potential of cobalt electrode; E_c, corrosion potential; E_H^0,
equilibrium potential of hydrogen electrode for p_{H_2} = 1 bar.

The maximum potential differences yielded transient Tafel lines with slopes of $b_t^+ = +64 \pm 10$ mV, in agreement with Kravtsov's data. One example is shown in Fig. 34. Parallel transient polarization curves are obtained by starting the pulses from any other steady state. Each member of such a family of transient Tafel lines intersects the steady-state polarization curve at the steady state selected to start the transient measurements.

Important arguments relevant to the reaction mechanism were based on the kinetics of cobalt deposition. It can be studied without much interference from the parallel reaction of hydrogen deposition in weakly acid solutions in the range $4 < $ pH $ < 7.5$. Solutions containing $Co(ClO_4)_2$ have a sufficient buffer capacity due to the homogeneous equilibrium

$$Co_{aq}^{2+} + H_2O \rightleftharpoons CoOH^+ + H^+ \tag{29}$$

with an equilibrium constant [157] around $K = 10^{-9}$ M. Cathodic steady-state polarization curves had Tafel slopes of $b_{ss}^- = -26 \pm 4$ mV at T = 293 K. The pH dependence of the current density at constant electrode potential yielded a reaction order close to $y_{(OH^-)}^- = 2$ as for the anodic rates. The reaction order with respect to cobalt ions was $y_{(Co^{2+})}^- = 2$. Transient polarization curves had Tafel slopes of $b_t^- = -48 \pm 7$ mV. It was observed that the same transient polarization curve in both the cathodic and anodic ranges could be measured starting either from a steady state on the cathodic side or from a corresponding one on the anodic side. These transient Tafel lines for given steady states intersected at the equilibrium potential of the cobalt electrode. The transient exchange current densities in i_t^0 thus obtained increased roughly in proportion to the square root of the steady-state current density. In a particular experiment $(\partial \ln i_t^0 / \partial \ln i_{ss})_{ai} = 0.45$ was found.

After a step of the electrode potential, the time dependence of the current from the minimum to the steady state followed a simple law. The difference between the transient current i at any time t and the steady-state current i_{ss} decayed exponentially according to

$$1 - (i/i_{ss}) = \exp(-K) \tag{30}$$

The time constant τ did not depend on electrode potential and composition of the solution and was the same for dissolution and deposition of cobalt. From the temperature dependence of the time constants in Fig. 35 an activation enthalpy of $\Delta H^* = 45$ kJ mol^{-1} was obtained. At room temperature the time constant is about 80 times smaller for cobalt than for iron.

The capacitance of cobalt electrodes depends on electrode potential and pH of the solution. Capacitances were obtained from galvanostatic transients following the electrode potential up to its maximum within

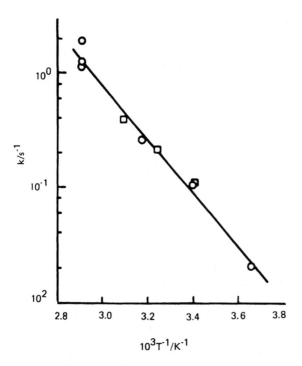

FIGURE 35 Time constant k in Eq. (30) vs. reciprocal temperature T for approach to the steady-state current density after a potential step [10]. □, Anodic dissolution of cobalt; system: Co/x M NaClO$_4$ + y M HClO$_4$, x + y = 0.5, pH = 1.5, T = 293 K. ○, Cathodic deposition of cobalt; system: Co/0.1 M Co(ClO$_4$)$_2$ + 0.3 M Ba(ClO$_4$)$_2$, pH = 7.5, T = 293 K.

milliseconds. There were two methods for evaluating the capacitance: The capacitance was calculated either from the initial slope $(dE/dt)_{t \to 0}$ or from an analysis of the whole transient. Both methods yielded the same potential-dependent capacitances, if the analysis of the whole transient was based on the assumption of capacitive and faradaic currents flowing in parallel. The faradaic currents i_F depended on electrode potential as in the respective transient polarization curve. The potential-dependent capacitance follows from

$$C(E) = (i - i_F)/(dE/dt) \qquad (31)$$

if one measures the slope dE/dt in any electrode potential range covered by the transient. The qualitative difference of potential transients taken in the regions of hydrogen evolution and cobalt dissolution

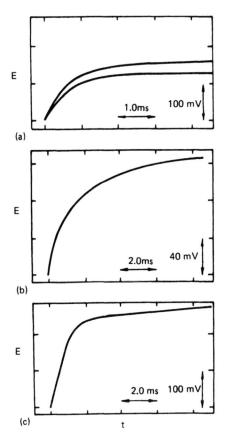

(a)

(b)

(c) t

FIGURE 36 Dependence of electrode potential E on time t after a current density step $\Delta i = i - i_j$ [10]. a) During anodic dissolution of cobalt; system: $Co/0.1$ M $HClO_4$, pH = 1.1, T = 293 K; $\Delta i = +10^{-2}$ A cm^{-2}; $i_j = 0$ (upper curve), $+10^{-4}$ A cm^{-2} (lower curve). b) During cathodic deposition of cobalt; system: $Co/0.1$ M $Co(ClO_4)_2$ + 0.3 M $Ba(ClO_4)_2$, pH = 7.55, T = 293 K; $\Delta i = -5 \times 10^{-3}$ A cm^{-2}, $i_j = -2 \times 10^{-5}$ A cm^{-2}. c) During hydrogen deposition; system: $Co/0.1$ M $HClO_4$, pH = 1.1, T = 293 K; $\Delta i = -10^{-2}$ A cm^{-2}, $i_j = 0$.

or deposition is immediately apparent from Fig. 36. In the region of hydrogen evolution, the capacitance was close to C = 20 µF cm^{-2} independent of the electrode potential. In concentrated electrolyte solutions such a value is expected for the double-layer capacitance C_{dl}.

In the regions of cobalt dissolution and deposition the whole electrode capacitance C was the sum of the double-layer capacitance and an adsorption capacitance C_{ads}

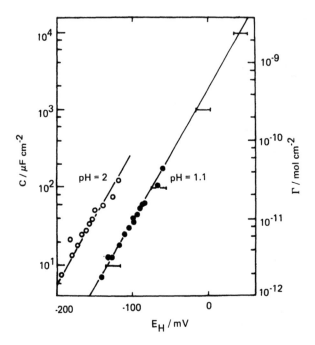

FIGURE 37 Adsorption capacitance corresponding to the reaction H_2O
$\rightleftharpoons$ $(OH)_{ads} + H^+ + e^-$ and surface concentrations $\Gamma_{(OH)ads}$ as functions of the electrode potential E_H during anodic dissolution of cobalt
[10,154]. System: Co/x M $NaClO_4$ + y M $HClO_4$, x + y = 0.1, pH =
1.1 and 2, T = 293 K.

$$C = C_{dl} + C_{ads} \tag{32}$$

As shown in Fig. 37, the adsorption capacitance rose exponentially
with the pH value of the solution and the electrode potential in the region of cobalt dissolution according to

$$C_{ads} = ga_{(OH^-)}\exp(F\Delta E/RT) \tag{33}$$

where g is a constant depending on the selection of a reference electrode potential for which $\Delta E = 0$.

At negative overvoltages, the potential dependence of the adsorption capacitance had the opposite sign, as is apparent from a comparison of Figs. 38 and 37. The adsorption capacitances in the region of
cobalt deposition were approximately proportional to the activities of

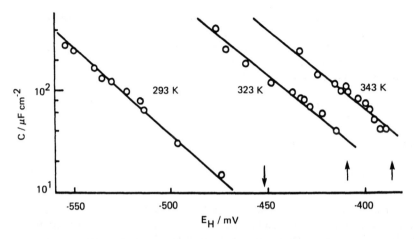

FIGURE 38 Adsorption capacitance C corresponding to the reaction $Co(H_2O)_m^{2+} + e^- \rightleftarrows [(OH)Co(H_2O)_{m-1}^{2+}]_{ads} + H^+$ as a function of the electrode potential E_H at different temperatures during cathodic deposition of cobalt [10,154]. System: $Co/0.1$ M $Co(ClO_4)_2 + 0.2$ M $Ba(ClO_4)_2$, pH (T = 293 K) = 7.65, T = 293, 323, and 343 K. Arrows signify steady-state electrode potentials at $i_i = -5 \times 10^{-5}$ A cm^{-2} (T = 293 and 323 K) and -10^{-4} A cm^{-2} (T = 343 K).

hydroxyl and cobalt ions. The adsorption capacitances increased strongly with temperature, corresponding to an apparent enthalpy of about $\Delta H = 78$ kJ mol^{-1} at $E_H = -0.45$ V.

The anodic adsorption capacitances in Fig. 37 did not depend on the current density up to the order of A cm^{-2}. It follows that the adsorption capacitances were always in equilibrium. The exchange current densities were estimated to be much larger than several A cm^{-2}. However, on the cathodic side, kinetic data for the chemisorption reaction could be obtained from an analysis of galvanostatic transients at high current densities. Under these conditions, the initial slope yielded the double-layer capacitance. From the delayed charging of the adsorption capacitance close to equilibrium the transfer resistance was derived. The corresponding exchange current densities of the adsorption reaction at T = 333 K were in the range of 0.1 to 1 A cm^{-2} for nearly neutral solutions containing up to 1 M cobalt perchlorate. The transfer coefficient obtained from the potential dependence of the exchange current density was $\alpha = 0.5$ for a charge number z = 1 of the electrosorption reaction.

The kinetics of the cobalt electrode are quite analogous to the kinetics of the iron electrode and provide further arguments favoring the catalytic mechanism. It was definitively proved that the adsorption capacitance is charged in a reaction proceeding in parallel and not consecutive to the charge transfer of cobalt ions. From the dependence of the adsorption capacitance on electrode potential and pH, one finds for positive overpotentials the adsorption reaction

$$H_2O \rightleftharpoons OH_{ads} + H^+ + e^- \tag{34}$$

and for negative overpotentials

$$Co^{2+} + H_2O + e^- \rightleftharpoons (CoOH)_{ads} + H^+ \tag{35}$$

Since the chemisorption reaction (34) is very fast, it is not possible to obtain information on its mechanism. However, it is quite probable that at least in acid solutions the adsorbed OH is formed by releasing a proton into the electrolyte from water already being adsorbed. In alkaline solutions there may also be direct participation of hydroxyl ions, but experimental evidence is lacking. The OH adsorbed on the cobalt surface is formally identical to a surface compound CoOH with cobalt bound in the metal phase. Since there is no information on the local distribution of charge, OH and CoOH cannot be distinguished. On the other hand, the surface compound $(CoOH)_{ads}$ formed in reaction (35), in which cobalt ions in the electrolyte react with the metal surface, is clearly different from OH_{ads} formed in reaction (34). Reaction (35) does not involve transfer of cobalt ions across the double layer, which is much slower, but this reaction is an electron transfer in which one electron passes from the metal to Co_{aq}^{2+} or $CoOH^+$, thereby establishing the chemisorption bond. The cobalt ion in $(CoOH)_{ads}$ must be attached to the cobalt surface via an OH bridge.

Chemisorption of OH or CoOH does not change the rate of cobalt ion transfer immediately. While charging the electrode capacitance, this rate is a function of electrode potential only, independent of how fast the electrode potential is changed. Thus, ion transfer proceeds in one step. The rates become time-dependent only some time later, when the electrode capacitance is already charged completely. From the slopes of the anodic transient Tafel lines one finds (ln 10)(RT/ the slopes of the anodic transient Tafel lines one finds (LN 10)(RT/F) $(\partial \log i_t^+/\partial E)_{ai} = \alpha^+n = 0.91 \pm 0.15$ and from the cathodic transient Tafel lines $\alpha^-n = 1.21 \pm 0.18$. Since the sum of the anodic and modynamics, the charge number n of cobalt ion transfer is n = 2.12 ± 0.20, which within the limits of experimental accuracy is n = 2 as expected for the overall reaction

$$Co \rightleftharpoons Co_{sol}^{2+} + 2e^- \tag{36}$$

In the present case, Eq. (36) also describes the rate-determining step of ion transfer. The anodic transfer coefficient $\alpha^+ = 0.45 \pm 0.07$ and the cathodic transfer coefficient $\alpha^- = 0.60 \pm 0.09$ both are close to one-half.

Slow changes of the rate of cobalt ion transfer, while the steady state is approached from the transient state, must correspond to changes of some reaction partner in the surface. This reaction partner influences both the anodic and cathodic rates in the same way, since a given transient polarization curve corresponds to two steady states, one at positive and the other at negative overpotentials. Kinks in monoatomic steps on a crystalline surface have this property. One must conclude that the transfer reaction described by Eq. (36) proceeds at kinks only. At any other sites on the surface, the rate of cobalt ion transfer is orders of magnitude lower. From the dependence of transient exchange current densities on the steady-state potential it follows that the surface concentration Γ_k of the catalyzing kinks grows exponentially with overpotential according to Eq. (7) with the positive and negative signs for positive and negative overvoltages, respectively. Relation (7) was directly observed at iron during anodic dissolution as shown above. In the steady state, Γ_k increases with the surface concentrations of adsorbed OH and CoOH. The increase of Γ_k with the surface concentration of OH was less than linear in the case of iron, but for cobalt a linear relationship must be inferred. Then at constant electrode potential the surface concentration of kinks is a linear function of the activity of hydroxyl ions, and the remaining pH dependence of the dissolution rate at constant Γ_k indicates that $CoOH^+$ is the complex participating in the charge transfer reaction. This explains the experimental value $y_{(OH^-)} = 2$.

The reaction proceeding during the approach to the steady state by changing the surface concentration of kinks is formally described by

$$(OH)_{ads} \rightleftarrows K(OH)_{ads} \tag{37}$$

where $K(OH)_{ads}$ denotes a kink with OH adsorbed to it and $(OH)_{ads}$ is hydroxyl adsorbed at any other sites. Introducing the dependence of Γ_k described by the rate constants k^f and k^b for the forward and back reaction (37) into the rate equation

$$i^+ = k^+ a_{(OH^-)} \Gamma_k \exp(2\alpha^+ FE/RT) \tag{38}$$

following from the mechanism of cobalt ion transfer, one finds for the time dependence of the anodic current density

$$1 - (i^+/i_{ss}^+) = [1 - \exp(-F\,\Delta E/RT)]\,\exp[-(k^f + k^b)t] \tag{39}$$

after a change of the electrode potential by ΔE from the steady-state starting potential E_{ss}. The current densities i^+ and i_{ss}^+ refer to the electrode potential $E = E_{ss} + \Delta E$. As in the case of iron, structural changes in the cobalt surface summarized by reaction (37) will involve several steps. It was demonstrated [158] that the mean distance of kinks in surface steps on cobalt must be large compared to the distance of neighboring atoms. However, direct measurements of kink distances are not available. A detailed mechanism for structural changes on the cobalt surface is lacking. From a comparison of the time constants k in Eq. (30) for cobalt and iron one finds that not only is the activation enthalpy for reaction (37) 20 kJ mol^{-1} larger than for iron, but also the activation entropy by 30 J mol^{-1} K^{-1}. The charge Q stored in the adsorption capacitance is easily obtained by integration of Eq. (33), which is valid for small relative coverages of OH. One finds

$$Q = (RT/F) \, C_{ads}^0 \, \exp[F(E - E^0)RT] \tag{40}$$

where C_{ads}^0 is the adsorption capacitance at an arbitrary reference potential E^0. The surface concentration $\Gamma_{OH} = Q/F$ thus obtained is also given in Fig. 37. For the highest capacitance observed, $\Gamma_{OH,s}$ is about 2.5×10^{-9} mol cm^{-2}, which is close to the number of sites available on the surface.

Theoretically, the capacitance is expected to grow with increasing relative coverage up to $\theta = 0.5$, but to decrease at higher coverages. Proof of this expectation was obtained from measurements of the minimum charge required for passivation [159]. At high current densities, the charge necessary to passivate cobalt grew linearly with the time t_p of passivation. The minimum charge for $t_p \to 0$ corresponded to about one monolayer of OH, if the initial potential was rather negative. For more positive potentials or correspondingly high rates of active dissolution the charges Q became smaller than the charge Q_p observed for the negative potentials. If OH adsorbed in the initial steady state is used in the monolayer sufficient to initiate passivity, one expects the relation

$$(Q_p/Q)-1 = (t_p/t)-1 = \exp[-F(E_{ss}-E)/RT] \tag{41}$$

with E_{ss} being the steady-state potential before application of the passivating current pulse. Experiments shown in Fig. 39 agree with this expectation. Also, the potential E' at which $Q/Q_p = 0.5$ turns out to be the same estimated for half-coverage from capacitance measurements. The potential dependence of the capacitance corresponding to Eq. (41) is

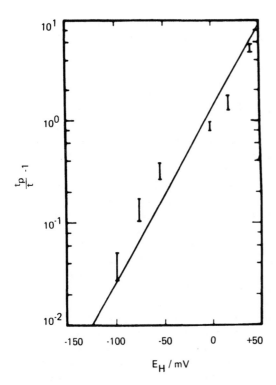

FIGURE 39 Influence of prepolarization at an electrode potential E_H on the time t of passivation relative to the time t_p of passivation for negative values of the prepolarization potential [159]. System: Co/1 M phosphate solution, pH = 2.0, T = 293 K. Solid line represents the theoretical slope given by Eq. (41).

$$C_{ads} = \frac{Q_p F/RT}{\exp[-F(E-E')/RT] + 2 + \exp[F(E-E')/RT]} \quad (42)$$

which essentially becomes equal to Eq. (33) for negative potential differences E-E'. Within the limits of accuracy, the potential E' = 0.111 ± 0.015 V vs. the reversible hydrogen electrode in the same solution is equal to the reversible potential of formation of $Co(OH)_2$ or CoO. Similarly, the potential of half-coverage with OH on iron is also close to the potential of formation of the magnetite equilibrium oxide [27]. Thus, the oxide phase will grow at E > E', if enough time is available, and the decrease of the adsorption capacitance described by Eq. (42) at positive potentials normally cannot be observed.

Despite the experimental proof that the adsorption capacitance is charged in a reaction proceeding in parallel to the charge transfer of cobalt ions, Epelboin et al. [160] assumed $CoOH_{ads}$ to be the intermediate in a consecutive mechanism. In sulfuric acid at concentrations between 0.5 and 2 M they observed Tafel slopes b^+ = +60 mV at rotating disk electrodes upon potentiodynamic polarization at about 0.5 mV sec^{-1}. From the pH dependence of the rates an apparent reaction order $y_{(OH^-)}^+$ = 1 was deduced. It is clear from the above discussion of the transients that an apparent reaction order thus obtained is an artifact of the potentiodynamic method and cannot be used for mechanistic interpretations. At current densities between about 25 and 100 mA cm^{-2} the Tafel slope was found to increase while the pH dependence disappeared. The increase of the slope was also observed at pH $\geq$ 4.5 and can be explained by a mechanism similar to that for iron discussed above. The hysteresis observed in the potentiodynamic polarization curves is due to the fact that a steady state was never attained, but Epelboin et al. [160] attribute it to an inhibiting effect of, e.g., chemisorbed hydrogen. No experiment was described to support that speculation.

The active cobalt electrode apparently was never studied in alkaline solutions from the mechanistic point of view. In acid solutions the effects of halide ions and of other surface-active substances on the dissolution kinetics of cobalt are similar to those for iron, but they are weaker [161]. Chloride ions at concentrations exceeding 1 mM and bromide ions at concentrations exceeding 10 μM increased the slope of the steady-state Tafel line to about b_{ss}^+ = +60 mV, as shown in Figs. 40 and 41. In a certain range of concentrations the slope remained constant, but the Tafel lines were shifted to higher overvoltages with growing concentration. Reaction orders with respect to chloride or bromide ions were $-0.7 \leq y_{(X^-)}^+ \leq -0.4$ and with respect to hydroxyl ions close to $y_{(OH^-)}^+$ = +1. The pH dependence in chloride solution is demonstrated in Fig. 41.

At high concentrations of halide ions, the polarization curves became steeper with rising current density and approached a steady-state potential almost independent of current density [161–164]. With chloride ions this effect is barely seen at 1 M, with bromide it is apparent at 10 mM, but with iodide it is already fully developed at 1 μM. This desorption potential is a linear function of pH and of the logarithm of halide ion concentration. It shifts to more positive values with decreasing pH and with increasing halide ion concentration, as shown in Fig. 42 for iodide. Addition of tetrabutylammonium sulfate did not increase the inhibiting effect of iodide, in contrast to the large synergistic effect found in the case of iron [163].

The shape of the polarization curves in the presence of chemisorbed substances can be described by the same mechanism already discussed for iron. Differences arise from the fact that the retardation

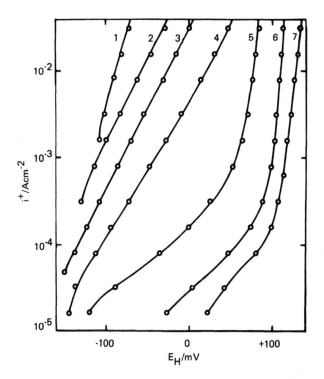

FIGURE 40 Steady-state polarization curves for dissolution of cobalt depending on the concentration of bromide [10,161]. System: Co/x M $NaClO_4$ + y M $HClO_4$ + z M NaBr, x + y = 0.3, pH = 1.1, T = 293 K. z: 1, 0; 2, 10^{-5}; 3, 10^{-4}; 4, 10^{-3}; 5, 10^{-2}; 6, 10^{-1}; 7, 1.

of the reaction is smaller in the case of cobalt or the rate of the reaction path via adsorbed halide ions is faster than the rate of the path catalyzed by hydroxyl ions only. It should be mentioned that hydrogen sulfide also catalyzes the dissolution of both cobalt and iron [156, 163,165]. It is not known whether this effect is due to $FeSH^+$ and $CoSH^+$ complexes participating in the charge transfer reactions or to an increase of the surface concentration of kinks by chemisorbed SH analogous to OH.

1.4 NICKEL

There is much disagreement among the various workers regarding the kinetics of nickel dissolution. The main reasons for this disagreement are the relatively slow approach toward the steady state and the

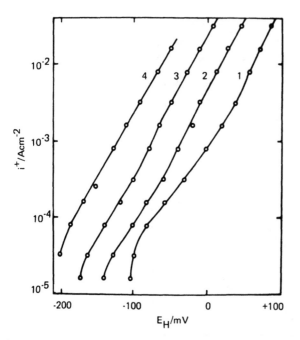

FIGURE 41 Steady-state polarization curves for dissolution of cobalt depending on the pH value [10,161]. System: Co/x M NaCl + y M HCl, x + y = 0.3, T = 293 K. pH: 1, 0.65; 2, 1.25; 3, 1.90; 4, 2.55.

stronger tendency for formation of oxide layers in the vicinity of the corrosion potential compared to iron and cobalt.

Figure 43 is a semiquantitative plot of transient currents after stepping the electrode potential to a more positive value. The charging of the electrode capacitance takes several milliseconds until the minimum current is attained, which changes very little during about three orders of magnitude of the time scale. The characteristic time to approach the steady state is of the order of minutes. Because of this long time scale, transient polarization curves as defined for iron and cobalt often were addressed as steady-state polarization curves for nickel. Due to the sluggish approach to steady state, potentiodynamic measurements with nickel electrodes usually yielded Tafel slopes close to the slopes of transient polarization curves obtained by pulse methods, if the rate of voltage change was chosen properly [166]. Reaction orders from transient measurements are fortuitous, as was already explained in the sections on iron and cobalt.

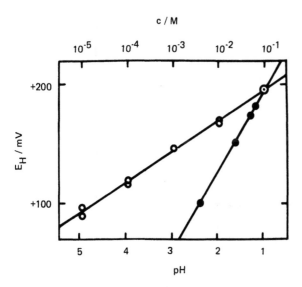

FIGURE 42 Steady-state desorption potentials for the dissolution of cobalt depending on the iodide concentration and the pH [10,161]. (○) System: Co/x M NaClO$_4$ + y M HClO$_4$ + z M NaI, x + y = 0.1, $10^{-5} \leq z \leq 10^{-1}$, T = 293 K. (●) System: Co/0.1 M Co(ClO$_4$)$_2$ + 0.2 M Ba(ClO$_4$)$_2$ + 0.1 M NaI, $1 \leq pH \leq 2.4$, T = 293 K.

At T = 338 K the characteristic time for the transition from the transient to the steady-state polarization curves is about 10 sec in acid perchlorate or sulfate solution, and thus one order of magnitude larger than that for cobalt at the same temperature. This ratio grows as the temperature is lowered, indicating a much higher activation enthalpy for nickel than for cobalt. At T = 298 K, the characteristic time for nickel is roughly two orders of magnitude longer than that at T = 338 K [167,168]. Thus, measurements of steady-state polarization curves at elevated temperatures are usually more reliable.

However, steady-state polarization curves at T = 298 K were measured by Kravtsov et al. [168,169]. They obtained a steady-state Tafel slope of b_{ss}^+ = +30 mV in 0.5 M sulfuric acid. When they normalized the current densities at any potential to a constant electrode capacitance, a slope of b_{ss}^+ = +39 mV was calculated. This procedure was believed to account for changes in the surface roughness, but is incorrect due to the adsorption capacitance, as already pointed out for the case of cobalt.

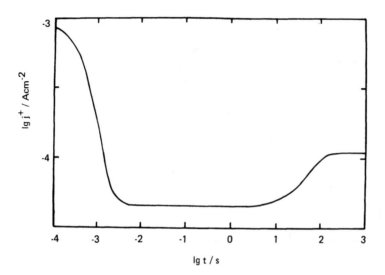

FIGURE 43 Potentiostatic transient on a double-logarithmic scale for active nickel in acid perchlorate or sulfate solution estimated from literature data [10,171,175] with $\Delta E = +20$ mV, showing characteristic times $\tau_1 \approx 1$ msec for charging the adsorption capacitance and $\tau_2 \approx 26$ sec for establishing the steady state after the minimum of i^+.

Sato and Okamoto [170] found a steady-state Tafel slope of about $b_{ss}^+ = +33$ mV in 0.5 M sulfuric acid at $T = 313$ K for current densities below several milliamperes per square centimeter. At higher current densities the slope increased, which is in agreement with results obtained by various authors [10,171–175]. The Tafel lines started to bend at lower current densities if the pH of the solution was increased, as shown in Fig. 44 for nickel in 0.5 M perchlorate solutions at $T = 338$ K [10,171]. In this work steady-state Tafel slopes of $b_{ss}^+ = +37 \pm 4$ mV were observed for nickel dissolution and similar slopes of $b_{ss}^- = -36 \pm 4$ mV for nickel deposition. In later work, Kolotyrkin et al. [172] measured steady-state Tafel slopes of $b_{ss}^+ = +33$ mV, Agladze et al. [176] measured $b_{ss}^+ = +39 \pm 3$ mV in sulfate solutions, and Dvorkina et al. [173] measured $b_{ss}^+ = +39$ mV in phosphate solutions, all at $T = 298$ K.

Transient polarization curves for a given steady state exhibited Tafel slopes of $60 \le b_t^+ \le 120$ mV. At $T = 298$ K, Kravtsov et al. [167,169] found transient Tafel slopes of $b_t^+ = +87 \pm 5$ mV in sulfuric acid. Their results are shown in Fig. 45. Piatti et al. [174] measured $b_t^+ = +55 \pm 5$ mV in acid perchlorate solution. At $T = 313$ K,

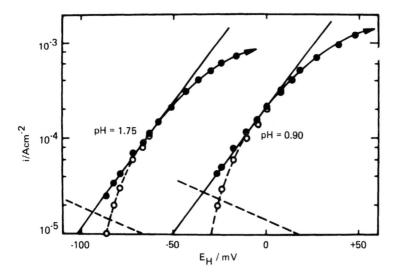

FIGURE 44 Steady-state polarization curves for the dissolution of nickel [10,171]. System: Ni/x M NaClO$_4$ + y M HClO$_4$, x + y = 0.5, pH = 0.90 and 1.75, T = 338 K. Solid points were obtained after correction for hydrogen deposition.

the slope was b_t^+ = +105 ± 13 mV in acid sulfate solution according to Sato and Okamoto [170]. At T = 318 K, Kronenberg et al. [177] found b_t^+ = +118 ± 9 mV in acid sulfate and perchlorate solutions. At T = 338 K, Heusler and Gaiser [171] observed b_t^+ = +67 ± 8 mV in acid perchlorate solutions. There may be several reasons for the large scatter of the data. It was speculated that the larger slopes are found if a thin oxide layer covers the nickel surface [166]. In fact, the passivating oxide film on nickel dissolves very slowly [178]. The dissolution rate decreases by about half a decade per pH, and for values of pH > 3 a film about 1 nm thick is not removed within several hours. On the other hand, smaller slopes of the transient polarization curves were usually observed when special precautions were taken to reduce any film at very negative potentials or to prevent film formation by never exposing the nickel to oxidizing conditions. However, under such conditions transient Tafel slopes around 2RT/F were also observed if measurements were performed relatively quickly. Ionization of hydrogen dissolved in the solution or in the bulk metal was suggested as a possible explanation for such high anodic Tafel slopes [177], but slow chemisorption of OH is an alternative explanation.

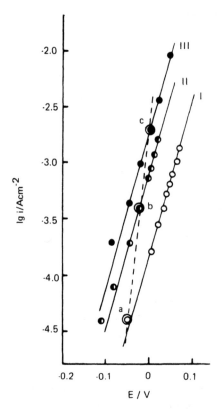

FIGURE 45 Steady-state polarization curve (dashed line) and transient polarization curves corresponding to the steady states a, b, and c of nickel dissolution [168,169]. System: Ni/0.5 M H_2SO_4, T = 298 K.

The relatively slow kinetics of OH adsorption according to

$$Ni(H_2O)_{ads} \rightleftarrows Ni(OH)_{ads} + H^+ + e^- \tag{43}$$

was investigated by Agladze et al. [176] in acid sulfate solutions. After a potential step, the current decayed exponentially from a larger current established after charging the double layer toward the minimum current as indicated in Fig. 43. The time constants $k = k^f + k^b a_{(H^+)}$ or the relaxation times $\tau = 1/k$ shown in Fig. 46 grew with the electrode potential at constant pH according to $(\partial E / \partial \log k)_{pH} = +1.20 \pm 5$ mV and decreased with the pH value at constant electrode potential according to $(\partial \log k/\partial pH)_E = -0.9 \pm 0.1$,

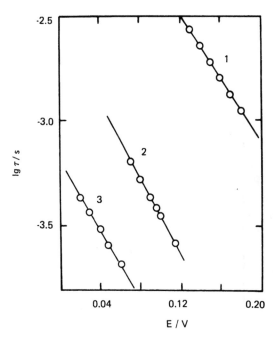

FIGURE 46 Relaxation times τ for charging the adsorption capacitance according to the reaction $H_2O \rightleftarrows (OH)_{ads} + H^+ + e^-$ as a function of the electrode potential at different pH values [175]. pH: 1, 4.64; 2, 4.0; 3, 3.0.

indicating that the rate constant $k^b a_{(H^+)}$ of the back reaction is much faster than the rate constant k^f of the forward reaction. Thus, in the electrode potential region considered, the relative coverage with OH is much smaller than one. The charges stored at equilibrium of reaction (43) are shown in Fig. 47. The potential dependence is described by $(\partial E/\partial \log Q) = 62 \pm 6$ mV and the pH dependence by $(\partial \log Q/\partial pH)_E = 1.2 \pm 0.1$. Within the limits of accuracy the experimental data are in agreement with $(\partial E/\partial \ln Q) = RT/F$ and $(\partial \log Q/\partial pH)_E = 1$ expected for the overall reaction (43). The authors [176] assumed a consecutive charge transfer with reaction (43) followed by the much slower step

$$Ni(OH)_{ads} \rightarrow NiOH^+ + e^- \qquad (44)$$

However, the question of whether reaction (43) proceeds in parallel with or consecutive to nickel ion transfer was not investigated. Data

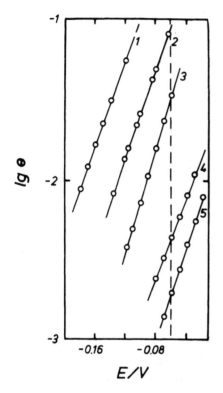

FIGURE 47 Steady-state coverages $\theta_{(OH)}$ related to an assumed saturation concentration of $\Gamma_{s,(OH)} = 10^{-9}$ mol cm^{-2} as a function of the electrode potential for different pH values [175]. pH; 1, 4.64; 2, 4.0; 3, 3.75; 4, 3.26; 5, 3.0.

for reaction (43) were taken at current densities much greater than the minimum current density where differences between the two mechanisms disappear.

The pH dependence of the rate of nickel dissolution can be reliably determined from steady-state polarization curves, but not from transient polarization curves, which can be measured starting from any steady state (see Sections 1.2 and 1.3).

If the corrosion potential at different pH values is selected as usually done, the resulting reaction order is fortuitous; e.g., the apparent reaction orders in deaerated and aerated solutions would be different. There are quite a few determinations of reaction orders from steady-state polarization curves. Kravtsov and Yan [169] observed a shift $(\partial E/\partial pH)i_{ss}^+ = -45$ mV in the region $1.75 \leq pH \leq 2.3$, yielding $y_{(OH^-)}^+ = 1.5$ with an experimental Tafel slope of $b_{ss}^+ =$

+30 mV. In the range $2.3 \leq pH \leq 3.25$ the reaction order $y_{(OH^-)}^+ = 1.7$ was slightly higher. In the acid sulfate solution $y_{(OH^-)}^+$ approached zero for $pH < 1.5$. Sato and Okamoto [170] measured steady-state polarization curves at three pH values, 0.45, 1.73, and 2.75. They state that $y_{(OH^-)}^+ = 1$, but closer inspection of their data yields $y_{(OH^-)}^+ = 1.1$ for the couple at lower pH values and $y_{(OH^-)}^+ = 1.7$ for the couple at higher pH values. In perchlorate solutions, Heusler and Gaiser [10,171] did not find a pH dependence of the reaction order but observed $y_{(OH^-)}^+ = 1.75 \pm 0.3$ down to pH 0.9. There was also no pH dependence of the reaction order in phosphate solutions [173], but the reaction order was not significantly higher than one.

The majority of the authors interpreted the observed kinetics in terms of a consecutive charge transfer mechanism with reaction (43) followed by reaction (44) as the rate-determining step and subsequent establishment of the homogeneous equilibrium

$$NiOH^+ + H^+ \rightleftarrows Ni_{sol}^{2+} + H_2O \qquad (45)$$

The main argument in favor of the consecutive mechanism is that Tafel slopes close to $b_{ss}^+ = 3RT/2F$ and reaction orders close to $y_{(OH^-)}^+ = 1$ are often observed. Even if more reliable values were available, they would not be sufficient to discriminate between the catalytic mechanism and the consecutive one. However, a number of experimental facts disagree with expectations based on a consecutive mechanism: Transient polarization curves would not cross the steady-state polarization curves at any steady state used to start the transient measurements, but would run at much higher current densities with a Tafel slope of about $b_t^+ = 2RT/F$. The currents after a potential step would decay toward the steady state in an exponential manner with a potential-dependent time constant. The behavior would be very similar to that actually observed for reaction (43) shown in Fig. 46.

However, a steady state with a Tafel slope of $b_{ss}^+ = 2RT/3F$ is not established after reaction (43) has come to equilibrium. Instead, a transient polarization curve with a much larger slope is then observed. Moreover, such transient polarization curves with Tafel lines intersecting at the equilibrium potential have similar slopes for dissolution and deposition [171]. From corresponding Tafel slopes one derives a charge number of the reaction $n = 2.0 \pm 0.14$, and transfer coefficients $\alpha^+ = 0.50 \pm 0.06$ and $\alpha^- = 0.49 \pm 0.05$ not compatible with consecutive charge transfer. From the small slopes of the steady-state anodic and cathodic Tafel lines, one calculates an apparent charge number $n_{app} = 4.1 \pm 0.4$, which is explained by the catalytic mechanism where the surface concentration of kinks grows exponentially

with increasing anodic and cathodic overvoltage. It is also not possible to argue that reactions (43) and (44) correspond to the transition from the transient polarization curve found for polarization times
of the order of 0.1 sec to the steady-state curves, because, among
other things, the characteristic times not only are potential-independent but also are orders of magnitude too long considering the charges
to be stored for the adsorption of OH. In order to remove this difficulty for the consecutive mechanism, Bockris and Kita [15] proposed
taking into account acidification of the solution next to the metal due
to the fast reaction (43) and subsequent decay of the pH difference
toward the bulk by diffusion. This theory does not explain why the
characteristic times are independent of pH and buffer capacity of the
solution, but depend on the nature of the metal. On the other hand,
all the arguments against the consecutive mechanism favor the catalytic mechanism with rate-determining transfer of nickel(II) in one
step at kinks and establishment of the steady-state surface concentration of kinks in a reaction sequence parallel to nickel ion transfer
involving chemisorbed OH, as described in the iron and cobalt dissolution mechanisms.

Halide ions increase the slope of the steady-state Tafel lines [10,
172] if their concentration exceeds a critical value. The critical values for the different halide ions are all larger than those for iron or
cobalt. The difference between transient and steady-state polarization curves disappears in a certain range of halide ion concentrations.
The reaction order with respect in hydroxyl ions drops to about one.
No decrease of the Tafel slope at high current densities and halide
ion concentration and no desorption potential were noticed with nickel
in solutions containing halide ions. In 2 M chloride solution, the dissolution rate at a given electrode potential was about two orders of
magnitude larger than that in perchlorate solution of the same pH,
according to Piatti et al. [174]. They found Tafel slopes close to
$b = RT/F$ at temperatures of 298 and 333 K for both anodic dissolution and cathodic deposition. Similar values were observed by Heusler
[10], who also noted that chloride can catalyze nickel deposition. In
the work of Kronenberg et al. [177], anodic Tafel slopes in chloride
solution were about $b^+ = +83$ mV and dissolution rates grew with the
chloride concentration. According to Bengali and Nobe [179], the
reaction order with respect to chloride changes from $y_{(Cl^-)}^+ = 0.5$
at low concentrations to $y_{(Cl^-)}^+ = 1$ at concentrations $c_{(Cl^-)} \geq$ M,
where the Tafel slopes were found to be $60 \leq b^+ \leq 75$ mV. The reaction order at constant chloride concentration was $y_{(OH^-)} = 0.5$ up
to 1 M acid and became negative in more acid solutions. The authors
[179] used the potentiodynamic method with sweep rates $0.3 \leq |dE/
dt| \leq 10$ mV min^{-1} and mentioned that they adjusted the sweep rates
in order to obtain optimally straight Tafel plots. This must be taken

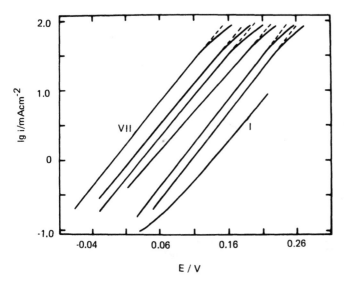

FIGURE 48 Tafel lines obtained potentiodynamically for anodic dissolution of nickel [180]. System: Ni(RDE)/x M HBr + y M HClO$_4$, x + y
= 1.0, ν_{rot} = 16 sec^{-1}. x: I, 0.02; II, 0.039; III, 0.057; IV, 0.167;
V, 0.286; VI, 0.412; VII, 0.80. Dashed lines were corrected for ohmic
drop.

as an indication that steady-state polarization curves deviated from
transient curves.

The same consideration applies to the experiments of Burstein and
Wright [180], who observed apparent reaction orders with respect to
fluoride $y_{(F^-)}^+$ = 0.3 ± 0.1, bromide $y_{(Br^-)}^+$ = 1.1 ± 0.1, and iodide
$y_{(J^-)}^+$ = 0.3 ± 0.1 for Tafel lines with slopes around b$^+$ = +105, 84,
and 82 mV, respectively. In solutions containing bromide or iodide
the potential was swept into the region of positive potentials where
pitting occurred. No significant pH dependence of the active dissolution rate was detected in the presence of bromide or iodide at concentrations 0.01 ≤ c_X^- ≤ 1.0 M. Figure 48 shows polarization curves
in the presence of bromide [180].

On the other hand, a reaction order of about $y_{(OH^-)}^+$ = 1 was found
by Heusler [10] as shown in Fig. 49. It should be noted that according to Kolotyrkin et al. [172], addition of 5 μM iodide to a given acid
solution inhibited dissolution of nickel. Coverages of adsorbed iodide
measured by a radiotracer technique did not significantly change with
electrode potential in the region of active dissolution, but decreased
with pH from about 5.5 × 10^{14} ions cm^{-2} at pH 0.4 to about 4 × 10^{14}
ions cm^{-2} at pH 2.2.

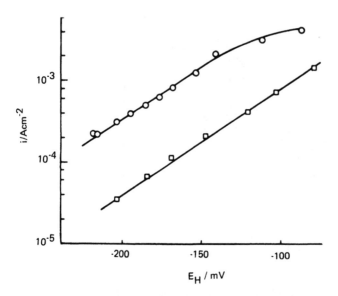

FIGURE 49 Steady-state polarization curves for the dissolution of nickel [10]. System: Ni/x M NaClO$_4$ + y M HClO$_4$ + 0.1 M BaCl$_2$, x + y = 0.1, pH: o, 3.7; □, 2.9. T = 338 K.

The effect of chemisorbed anions on the kinetics of nickel dissolution can be explained in terms of the mechanism already discussed for iron. In the case of nickel, the assumption of parallel reaction paths involving catalysis by halide ions and hydroxyl ions was also supported by Kolotyrkin et al. [181]. Rates of the path catalyzed by halide ions seem to be relatively high for nickel. Therefore, replacement of chemisorbed hydroxyl ions by halide ions does not inhibit the nickel dissolution significantly. Since the Tafel slope in the presence of halide ions is increased, their catalytic effect is more apparent at low overvoltages. The differences between steady-state and transient polarization curves disappear and the slope remains close to $b_{ss}^+ = RT/F$. Thus, one concludes that chemisorbed halide ions increase the surface concentration of kinks to the limit where their average distance becomes comparable to the distance of neighboring atoms in the metal lattice. The strong increase of the surface concentration of kinks is counteracted by a small rate constant for dissolution of nickel(II) from a kink. The product of the two factors yields dissolution rates comparable in magnitude to the dissolution rates in the absence of halide ions. The behavior of iron and cobalt is, in principle, the same, but in these cases dissolution via the path catalyzed by halide ions is always much slower than that via the path catalyzed by hydroxyl ions. The main reason seems to be the relatively small rate constant for ion transfer from kinks, while the roughening of the steps on an atomic scale will be similar for iron, cobalt, and nickel, as indicated by the high Tafel slopes at potentials more negative than the desorption potential. In the case of iron, changes of surface morphology also pointed in that direction [67].

1.5 SUMMARY

The kinetics of the anodic dissolution of iron, cobalt, and nickel are quite similar. A common mechanism was found to operate. This conclusion cannot be reached by considering only steady-state polarization curves; the time dependences of polarization as revealed by pulse techniques and ac impedance spectroscopy must also be taken into account. Chemisorption of species from the electrolyte modifies the kinetics. An important aspect is the correlation between the kinetics and the structure of the metal surface, which was studied down to the atomic level.

The general mechanism is the dissolution of divalent metal ions from kink sites in monoatomic steps:

$$\cdots MeMe_k \rightleftarrows \cdots Me_k + Me^{2+}_{solv} + 2e^- \tag{46}$$

The main reason for the rather complicated kinetic behavior of the iron group metals is the variability of the surface concentration of kinks, which depends on electrode potential, crystallographic orientation, lattice defects emerging at the metal surface, chemisorption of species from the electrolyte, etc. Such variability is possible because the equilibrium distance of kinks in steps is very long at metals with high specific edge energies of steps, surface energies, or melting points. There is no evidence for consecutive charge transfer via monovalent metal ions.

The influence of pH and surface-active anions such as halide ions on the dissolution kinetics is due to the electronic structure of the transition metals, the structural changes of the metal surface, and intermediate complex formation.

In the absence of surface-active anions one has dissolution of an intermediate complex $MeOH^+$. In the range of active metal dissolution, the surface concentration of kinks is proportional to the surface concentration of chemisorbed OH formed in a fast equilibrium with water and protons or hydroxyl ions. The metal ion dissolution is retarded at high relative coverages with 2-D oxidic species in the transition from prepassive ranges. There is a continuous transition from 2-D to 3-D oxidic layers in the passive layer formation range. The retarding effect is particularly strong with nickel.

In the presence of surface-active anions parallel pathways become possible by formation of additional intermediate complexes, but the structure of the surfaces is also altered. Inhibition is apparently due to stronger chemisorption of halide ions than OH to kink sites and a larger activation energy for dissolution from kink sites with chemisorbed halide. There is also the possibility of positive catalysis by halide ions, as with nickel under certain conditions.

Differences in the dissolution kinetics of iron, cobalt, and nickel can be explained within the same principal reaction mechanism by different values of the rate constants of the various reaction steps.

LIST OF SYMBOLS

a_S	activity of species S
ac	alternating current
b	Tafel slope
b^+, b^-	Tafel slopes of anodic and cathodic reactions, respectively
b_{ss}	steady-state Tafel slope
b_t	transient Tafel slope
c_S	concentration of species S
C	capacitance
C_{ads}	adsorption capacitance
C_{dl}	double-layer capacitance
dc	direct current
D	diffusion coefficient
D(S)	diffusion coefficient of species S
2-D	two-dimensional
3-D	three-dimensional
e	elementary charge of the electron
E	electrode potential
ΔE	potential difference
E^0	equilibrium potential or reference potential
E'	E for $Q/Q_p = 0.5$
E_{corr}	corrosion potential
E_H	electrode potential related to normal hydrogen electrode
E_{maxI}, E_{maxII}, E_{maxIII}	potentials of anodic current maxima I, II, and III, respectively
$E_{maxI'}$, $E_{maxII'}$, $E_{maxIII'}$	potentials of cathodic current maxima I', II', and III, respectively
E_{ss}	steady-state electrode potential

f	frequency
F	Faraday constant
g	constant in Eq. (33)
ΔH	enthalpy
ΔH^*	activation enthalpy
i	current density
i^+, i^-	current densities of anodic and cathodic reactions, respectively
i^0	exchange current density
i_{corr}	corrosion current density
i_F	faradaic current density
$\overset{\sim}{\cdot}$	initial current density
i_{maxI^+}, i_{maxII^+}, i_{maxIII^+}	anodic current maxima I, II, and III, respectively
i_{ss}	steady-state current density
i_t^0	transient exchange current density
k	rate constant or time constant
k^+, k^-	rate constants of anodic and cathodic reactions, respectively
k_{ss}	rate constant under steady-state conditions
k^f, k^b	rate constants of forward and backward reactions, respectively
K	kink site
K	equilibrium constant
n	charge number of electrode reaction
Q	charge
Q_p	passivation charge in Eq. (41)
R	gas constant
RDE	rotating disk electrode
$Re\{\overline{Z}_f\}$	real part of faradic impedance $\overline{Z}_f$
R_p	polarization resistance
s	Laplace variable [$s = j\omega$, with $j = (-1)^{1/2}$ and $\omega = 2\pi f$]
S^*	active center on surface of S
$S(0)$, $S(I)$, $S(II)$, $S(III)$	zero-, mono-, di-, and trivalent oxidation states of species S, respectively
S_{ads}	adsorbed species S
S_{des}	desorbed species S
S_{sol}, S_{aq}	solvated and hydrated species S, respectively
t	time
t_p	passivation time
T	temperature
x_s	step distance
X^-	halide anion
y	electrochemical reaction order

y^+, y^-	electrochemical reaction order in anodic and cathodic reactions, respectively
y_S	electrochemical reaction order related to species S
z	charge number of charge transfer step
α	charge transfer coefficient
α^+, α^-	charge transfer coefficients of anodic and cathodic reactions, respectively
β	coefficient
Γ	surface concentration
Γ^0	equilibrium surface concentration
$\Gamma_{a,s}$	saturation surface concentration at active sites
Γ_s	saturation surface concentration
Γ_S	surface concentration of S
ε	inhibition efficiency
ε_1, ε_2	inhibition efficiencies related to exposure times $t \to 0$ and $t \to \infty$, respectively
θ	Γ / Γ_s
ν_{rot}	rotational speed of disk electrode
τ	time constant
ψ	Γ_k / Γ_k^0

REFERENCES

1. W. A. Roiter, W. A. Juza, and E. S. Poluyan, *Acta Physio-chim. URSS, 10*: 389 (1939); W. A. Roiter, E. S. Poluyan, and W. A. Juza, *Acta Physiochim. URSS, 10*: 845 (1939).

2. B. N. Kabanov and D. I. Leikis, *Zh. Fiz. Khim., 20*: 995 (1946).

3. B. N. Kabanov and D. I. Leikis, *Dokl. Akad. Nauk SSSR, 58*: 1685 (1947).

4. B. N. Kabanov, R. Burstein, and A. N. Frumkin, *Discuss. Faraday Soc., 1*: 259 (1947).

5. B. N. Kabanov and D. I. Leikis, *Z. Elektrochem. Ber. Bunsenges. Phys. Chem., 62*: 660 (1958).

6. K. E. Heusler, Doctoral thesis, Univ. of Göttingen, 1957.

7. K. F. Bonhoeffer and K. E. Heusler, *Z. Phys. Chem. NF, 8*: 390 (1956).

8. K. F. Bonhoeffer and K. E. Heusler, *Z. Elektrochem. Ber. Bunsenges. Phys. Chem., 61*: 122 (1957).

9. K. E. Heusler, *Z. Elektrochem. Ber. Bunsenges. Phys. Chem., 62*: 582 (1958).

10. K. E. Heusler, Elektrochemische Auflösung und Abscheidung von Metallen der Eisengruppe, Habilitationsschrift, TH Stuttgart, 1966.

11. T. P. Hoar and T. Hurlen, *CITCE VIII 1956*, Butterworths, London, 1958, p. 445.
12. T. Hurlen, *Tek. Ukebl.*, *105*: 101, 119 (1958).
13. T. Hurlen, *Acta Chem. Scand.*, *14*: 1533, 1555, 1564 (1960).
14. J. O'M. Bockris, D. Drazic, and A. Despic, *Electrochim. Acta*, *4*: 325 (1961).
15. J. O'M. Bockris and H. Kita, *J. Electrochem. Soc.*, *108*: 676 (1961).
16. J. O'M. Bockris and D. Drazic, *Electrochim. Acta*, *7*: 293 (1962).
17. K. A. Christensen, H. Hoeg, K. Michelsen, G. B. Nielsen, and H. Nord, *Acta Chem. Scand.*, *15*: 300 (1961).
18. K. Schwabe and C. Voigt, *Werkst. Korros.*, *16*: 125 (1965).
19. K. Schwabe and C. Voigt, *J. Electrochem. Soc.*, *113*: 886 (1966).
20. C. Voigt, *Electrochim. Acta*, *13*: 2037 (1968).
21. K. Schwabe and C. Voigt, *Electrochim. Acta*, *14*: 853 (1969).
22. H. Fischer and H. Yamaoka, *Chem. Ber.*, *94*: 1477 (1961).
23. W. J. Lorenz, H. Yamaoka, and H. Fischer, *Ber. Bunsenges. Phys. Chem.*, *67*: 932 (1963).
24. H. Kaesche, *Z. Elektrochem. Ber. Bunsenges. Phys. Chem.*, *63*: 492 (1959).
25. E. J. Kelly, *J. Electrochem. Soc.*, *112*: 124 (1965).
26. J. J. Podesta and A. J. Arvia, *Electrochim. Acta*, *10*: 159, 171 (1965).
27. K. E. Heusler, in *Encyclopedia of Electrochemistry of the Elements* (A. J. Bard, ed.), Dekker, New York, 1982, vol. 9, pt. A, p. 230.
28. G. Eichkorn and W. J. Lorenz, *Naturwissenschaften*, *52*: 618 (1965).
29. W. J. Lorenz and G. Eichkorn, *J. Electrochem. Soc.*, *112*: 1255 (1965).
30. W. J. Lorenz and G. Eichkorn, *Ber. Bunsenges. Phys. Chem.*, *70*: 99 (1966).
31. G. Eichkorn, W. J. Lorenz, L. Albert, and H. Fischer, *Electrochim. Acta*, *13*: 183 (1968).
32. G. Eichkorn and W. J. Lorenz, *Z. Metalloberfläche*, *22*: 102 (1968).
33. W. J. Lorenz, Das elektrochemische Verhalten des aktiven Reineisens in sauren Lösungen, Habilitationsschrift, Universität Karlsruhe, 1968.
34. F. Hilbert, Y. Miyoshi, G. Eichkorn, and W. J. Lorenz, *J. Electrochem. Soc.*, *118*: 1919, 1927 (1971).
35. H. Rosswag, G. Eichkorn, and W. J. Lorenz, *Werkst. Korros.*, *25*: 86 (1974).
36. A. Akiyama, R. E. Patterson, and K. H. Nobe, *Corrosion*, *26*: 51 (1970).
37. G. Bech-Nielsen, *Electrochim. Acta*, *19*: 821 (1974).
38. Ph. Morel, thesis, Univ. of Paris, 1968.

39. M. Keddam, thesis, Univ. of Paris, 1968.
40. I. Epelboin and M. Keddam, *J. Electrochem. Soc.*, *117*: 1052 (1970).
41. H. Takenouti, thesis, Univ. of Paris, 1971.
42. I. Epelboin, Ph. Morel, and H. Takenouti, *J. Electrochem. Soc.*, *118*: 1282 (1971).
43. I. Epelboin and M. Keddam, *Electrochim. Acta, 17*: 177 (1972).
44. I. Epelboin, M. Keddam, and J. C. Lestrade, *Faraday Discuss. Chem. Soc.*, *56*: 264 (1974).
45. I. Epelboin, C. Gabrielli, M. Keddam, and H. Takenouti, *Electrochim. Acta, 20*: 913 (1975).
46. C. Gabrielli, M. Keddam, E. Stupnisek-Lisac, and H. Takenouti, *Electrochim. Acta, 21*: 757 (1976).
47. B. Bechet, I. Epelboin, and M. Keddam, *J. Electroanal. Chem.*, *76*: 129 (1977).
48. M. Keddam, O. R. Mattos, and H. Takenouti, *J. Electrochem. Soc.*, *128*: 257, 266 (1981).
49. D. Geana, A. A. El Miligy, and W. J. Lorenz, *Corros. Sci.*, *13*: 505 (1973).
50. D. Geana, A. A. El Miligy, and W. J. Lorenz, *Corros. Sci.*, *14*: 657 (1974).
51. A. A. El Miligy, D. Geana, and W. J. Lorenz, *Electrochim. Acta, 20*: 273 (1975).
52. J. Bessone, L. Karakaya, P. Lorbeer, and W. J. Lorenz, *Electrochim. Acta, 22*: 1147 (1977).
53. P. Lorbeer and W. J. Lorenz, *Corros. Sci.*, *20*: 405 (1980).
54. P. Lorbeer and W. J. Lorenz, *Electrochim. Acta, 25*: 375 (1980).
55. H. Nord and G. Bech-Nielsen, *Electrochim. Acta, 16*: 849 (1971).
56. G. Bech-Nielsen and J. C. Reeve, in *Proceedings of the 6th Scandinavian Corrosion Congress*, Gothenburg, 1971.
57. G. Bech-Nielsen, *Electrochim. Acta, 18*: 671 (1973).
58. G. Bech-Nielsen, *Electrochim. Acta, 19*: 821 (1974).
59. G. Bech-Nielsen, *Electrochim. Acta, 20*: 619 (1975).
60. G. Bech-Nielsen, *Electrochim. Acta, 21*: 627 (1976).
61. G. Bech-Nielsen, *Electrochim. Acta, 23*: 425 (1978).
62. G. Bech-Nielsen, *Acta Chem. Scand. Ser. A, 32*: 781 (1978).
63. M. Mogensen, G. Bech-Nielsen, and E. Maahn, *Electrochim. Acta, 25*: 919 (1980).
64. G. Bech-Nielsen, in *Passivity of Metals* (R. P. Frankenthal and J. Kruger, eds.), The Electrochemical Society, Princeton, N. J., 1978, p. 614.
65. G. Bech-Nielsen, *Electrochim. Acta, 27*: 1383 (1982).
66. G. Bech-Nielsen, paper at the International Conference on Corrosion Inhibition, Dallas, 1983.
67. W. Allgaier, Doctoral thesis, Univ. of Clausthal, 1975.
68. W. Allgaier and K. E. Heusler, *Z. Phys. Chem. NF, 98*: 161 (1975).
69. G. T. Burstein and G. W. Ashley, *Corrosion, 39*: 241 (1983).

70. P. Lorbeer and W. J. Lorenz, *Corros. Sci.*, *21*: 79 (1981).
71. P. Lorbeer, K. Jüttner, and W. J. Lorenz, *Werkst. Korros.*, *34*: 290 (1983).
72. W. Allgaier and K. E. Heusler, *Z. Metallkd.*, *67*: 766 (1976).
73. W. Allgaier and K. E. Heusler, *J. Appl. Electrochem.*, *9*: 155 (1979).
74. K. E. Heusler, *DECHEMA Monogr.*, *93*: 193 (1983).
75. G. J. Bignold and M. Fleischmann, *Electrochim. Acta*, *19*: 363 (1974).
76. J. A. Harrison and W. J. Lorenz, *Electrochim. Acta*, *22*: 205 (1977).
77. H. Schweickert, W. J. Lorenz, and H. Friedburg, *J. Electrochem. Soc.*, *127*: 1693 (1980); *128*: 1295 (1981).
78. H. Schweickert, Doctoral thesis, Univ. of Karlsruhe, 1978.
79. K. Jüttner and W. J. Lorenz, in *Proceedings of the 8th International Congress on Metallic Corrosion*, Mainz, 1981, vol. I, p. 87, DECHEMA, Frankfurt/M., 1981.
80. W. J. Lorenz, G. Eichkorn and C. Mayer, *Corros. Sci.*, *7*: 357 (1967).
81. P. Lorbeer, Doctoral thesis, Univ. of Karlsruhe, 1978.
82. G. M. Florianovich, A. Sokolova, and Ya. M. Kolotyrkin, *Electrochim. Acta*, *12*: 879 (1967).
83. M. Pourbaix, *Atlas d'Equilibres Electrochimiques a 25°C*, Gauthier-Villars, Paris, 1963, p. 307.
84. W. J. Lorenz and A. A. El Miligy, *J. Electrochem. Soc.*, *120*: 1698 (1973).
85. J. R. Vilche and W. J. Lorenz, *Corros. Sci.*, *12*: 785 (1972).
86. B. Folleher and K. E. Heusler, *J. Electroanal. Chem.*, *180*: 77 (1984).
87. A. A. El Miligy, F. Hilbert, and W. J. Lorenz, *J. Electrochem. Soc.*, *120*: 247 (1973).
88. D. M. Drazic and C. S. Hao, 33rd ISE Meeting, Extended Abstract IC 19, p. 222, Lyon, 1982.
89. J. Hitzig, J. Titz, K. Jüttner, W. J. Lorenz, and E. Schmidt, *Electrochim. Acta*, *29*: 287 (1984); K. Jüttner and W. J. Lorenz, paper 188, 166th Meeting of the Electrochemical Society, 1984, in *Proceedings of the Symposium on Computer Aided Acquisition and Analysis of Corrosion Data*, The Electrochemical Society, Pennington, N. J., 84-3: 144 (1985); E. Schmidt, J. Hitzig, J. Titz, K. Jüttner, and W. J. Lorenz, *Electrochim. Acta*, *31*: No. 8 (1986), in press.
90. U. F. Franck, *Werkst. Korros.*, *11*: 401 (1960).
91. J. Osterwald, *Z. Elektrochem. Ber. Bunsenges. Phys. Chem.*, *66*: 401 (1962).
92. D. Gilroy and B. E. Conway, *J. Phys. Chem.*, *69*: 1259 (1965).
93. J. L. Ord and J. H. Bartlett, *J. Electrochem. Soc.*, *112*: 160 (1965).

94. K. E. Heusler, *Ber. Bunsenges. Phys. Chem.*, *72*: 1197 (1968).

95. P. Lorbeer and W. J. Lorenz, in *Passivity of Metals* (R. P. Frankenthal and J. Kruger, eds.), The Electrochemical Society, Princeton, N. J., 1978, p. 607.

96. W. J. Lorenz, unpublished results.

97. P. P. Russell and J. Newman, *J. Electrochem. Soc.*, *130*: 547 (1983).

98. I. Epelboin, C. Gabrielli, M. Keddam, J.-C. Lestrade, and H. Takenouti, *J. Electrochem. Soc.*, *119*: 632 (1972).

99. I. Epelboin, C. Gabrielli, and M. Keddam, *Corros. Sci.*, *15*: 155 (1975).

100. M. Baddi, C. Gabrielli, M. Keddam, and H. Takenouti, in *Passivity of Metals* (R. P. Frankenthal and J. Kruger, eds.), The Electrochemical Society, Princeton, N. J., 1978, p. 625.

101. B. D. Cahan and P. J. Pearson, Abstract 484, 164th Meeting of The Electrochemical Society, Washington, D. C., 1983.

102. O. Faust, *Z. Elektrochem.*, *13*: 161 (1907).

103. F. Förster, *Z. Elektrochem.*, *14*: 295 (1908).

104. F. Förster and P. Herold, *Z. Elektrochem.*, *16*: 461 (1910).

105. K. G. Weil, *Z. Elektrochem.*, *62*: 638 (1958).

106. T. Hurlen, *Electrochim. Acta*, *8*: 609 (1963).

107. A. M. Sukhotin and K. M. Kartashova, *Corros. Sci.*, *5*: 393 (1965).

108. V. N. Flerov and L. I. Pavlova, *Elektrokhimiya*, *3*: 621 (1967).

109. C. M. Sheppard and S. Schuldiner, *J. Electrochem. Soc.*, *115*: 1124 (1968).

110. S. Asakura and K. Nobe, *J. Electrochem. Soc.*, *118*: 536 (1971).

111. R. D. Armstrong and I. Baurhoo, *J. Electroanal. Chem.*, *34*: 41 (1972).

112. G. J. Bignold, *Corros. Sci.*, *12*: 145 (1972).

113. D. D. McDonald and D. Owen, *J. Electrochem. Soc.*, *120*: 317 (1973).

114. H. Cnobloch, D. Göppel, W. Nippe, and F. v. Sturm, *Chem. Ing. Tech.*, *45*: 203 (1973).

115. A. P. Plyankova and Z. A. Jofa, *Elektrokhimiya*, *10*: 1344 (1974).

116. N. A. Hampson, R. J. Latham, A. Marshall, and R. D. Giles, *Electrochim. Acta*, *19*: 374 (1974).

117. D. Geana, A. A. El Miligy, and W. J. Lorenz, *J. Appl. Electrochem.*, *4*: 337 (1974).

118. P. Doig and P. E. J. Flewitt, *Corros. Sci.*, *17*: 369 (1977).

119. R. S. Schrebler Guzman, J. R. Vilche, and A. J. Arvia, *Electrochim. Acta*, *24*: 395 (1979).

120. D. M. Drazic and C. S. Hao, *Electrochim. Acta*, *27*: 1409 (1982).

121. A. J. Salkind, C. J. Venuto, and S. U. Falk, *J. Electrochem. Soc.*, *111*: 493 (1964).

122. T. K. Teplinskaya, N. N. Federova, and S. A. Rozentsveig, *Zh. Fiz. Khim.*, *38*: 2176 (1964).
123. C. L. Foley, J. Kruger, and C. J. Bechtold, *J. Electrochem. Soc.*, *114*: 994 (1967).
124. H. G. Silver and E. Leaks, *J. Electrochem. Soc.*, *117*: 5 (1970).
125. I. Geronov, T. Tomov, and S. Georgiev, *J. Appl. Electrochem.*, *5*: 351 (1975).
126. L. Öjefors, *J. Electrochem. Soc.*, *123*: 1691 (1976).
127. T. P. Hoar and T. W. Farrer, *Corros. Sci.*, *1*: 49 (1961).
128. Z. A. Foroulis, *Proc. 3rd Eur. Symp. Corros. Inhibitors*, Ann. Univ. Ferrara, N.S., Sez. V, Suppl. 5, 723 (1970).
129. S. Asakura and K. Nobe, *J. Electrochem. Soc.*, *118*: 13 (1971).
130. M. Erbil and W. J. Lorenz, *Werkst. Korros.*, *29*: 505 (1978).
131. M. A. Morsi, Y. A. Elewady, P. Lorbeer, and W. J. Lorenz, *Werkst. Korros.*, *31*: 108 (1980).
132. F. Mansfeld, M. W. Kendig, and S. Tsai, *Corros. Sci.*, *22*: 455 (1982).
133. J. O'M. Bockris, M. A. Genshaw, V. Brusic, and H. Wroblowa, *Electrochim. Acta*, *16*: 1859 (1971).
134. W. J. Lorenz and F. Mansfeld, in *Proceedings of the Symposium on Fundamental Aspects of Corrosion Protection by Surface Modification*, The Electrochemical Society, Pennington, N.J., 1984, vol. 84-3, p. 144.
135. W. J. Lorenz and F. Mansfeld, paper at the International Conference on Corrosion Inhibition, Dallas, 1983; W. J. Lorenz and F. Mansfeld, *Electrochim. Acta*, *31*: 467 (1986).
136. G. Trabanelli and V. Carrassiti, in *Advances in Corrosion Science and Technology* (M. G. Fontana and R. W. Staehle, eds.), Plenum, New York, 1970, vol. 1, p. 147.
137. H. Fischer, *Werkst. Korros.*, *24*: 525, 575 (1973).
138. F. C. Raducanu and W. J. Lorenz, *Electrochim. Acta*, *16*: 995, 1143 (1971).
139. W. J. Lorenz and F. Mansfeld, *Corros. Sci.*, *21*: 647 (1981); K. Jüttner, K. Manandhar, U. Seifert—Kraus, W. J. Lorenz, and E. Schmidt, *Werkst. Korros.*, *37*: No. 7 (1986), in press.
140. P. Lorbeer and W. J. Lorenz, *Proc. 5th Eur. Symp. Corros. Inhibitors*, Univ. of Ferrara, 1980, vol. 1, p. 377.
141. K. E. Heusler and G. H. Cartledge, *J. Electrochem. Soc.*, *108*: 732 (1961).
142. W. J. Lorenz, *Corros. Sci.*, *5*: 121 (1965).
143. W. J. Lorenz and H. Fischer, *Ber. Bunsenges. Phys. Chem.*, *69*: 689 (1965).
144. N. A. Darwish, F. Hilbert, W. J. Lorenz, and H. Rosswag, *Electrochim. Acta*, *18*: 421 (1973).

145. L. L.. Cavallaro, L. Felloni, G. Trabanelli, and F. Pulidori, *Electrochim. Acta, 9*: 485 (1964).
146. K. Schwabe and C. Voigt, *Electrochim. Acta, 14*: 869 (1969).
147. E. McCafferty and N. Hackerman, *J. Electrochem. Soc., 119*: 999 (1972).
148. W. J. Lorenz, C. Mayer, and H. Fischer, *Z. Phys. Chem. NF, 52*: 180, 193 (1967).
149. L. Horner, *Chem. Ztg., 100*: 247 (1976).
150. L. Horner and D. Schödel, *Werkst. Korros., 25*: 711 (1974).
151. V. I. Kravtsov, *Acta Chim. Acad. Sci. Hung., 18*: 321 (1959).
152. V. I. Kravtsov and N. Kh. Pikov, *Vestn. Leningr. Univ. Fiz. Khim., 4*: 70 (1961).
153. V. I. Kravtsov and O. G. Lokshtanova, *Zh. Fiz. Khim., 36*: 2362 (1962).
154. K. E. Heusler, *Z. Elektrochem. Ber. Bunsenges. Phys. Chem., 66*: 177 (1962).
155. K. E. Heusler, *Ber. Bunsenges. Phys. Chem., 7*: 620 (1967).
156. Z. A. Jofa and Wei Pao-ming, *Zh. Fiz. Khim., 36*: 2558 (1962).
157. J. Bjerrum, G. Schwarzenbach, and L. G. Sillen, *Stability Constants*, Part II: *Inorganic Ligands*, Spec. Publ. 7, The Chemical Society, London, 1958, p. 12.
158. K. E. Heusler and R. Knoedler, *Ber. Bunsenges. Phys. Chem., 7*: 1085 (1967).
159. K. E. Heusler, *Corros. Sci., 6*: 183 (1966).
160. I. Epelboin, M. Micinic, and Ph. Morel, *Mem. Sci. Rev. Metall., 68*: 727 (1971).
161. K. E. Heusler, 14th CITCE Meeting, Moscow, 1963; *Abstract in Electrochim. Acta, 8*: 30 (1963); *Osnowie voprosi sovremenoi teoreticheskoi elektrokhimii* (Fundamental problems of modern theoretical electrocyemistry), MIR, Moscow, 1965, p. 453.
162. Z. A. Jofa and Wei Pao-ming, *Zh. Fiz. Khim., 37*: 2300 (1963).
163. Z. A. Jofa, V. V. Batrakov, and Cho-Ngok-Ba, *Electrochim. Acta, 9*: 1645 (1964).
164. V. V. Batrakov, K. Kh. Avad, and Z. A. Jofa, *Elektrokhimiya, 8*: 603 (1972).
165. T. P. Hoar and D. Havenhand, *J. Iron Steel Inst., 133*: 239P (1936).
166. G. T. Burstein and G. A. Wright, *Electrochim. Acta, 20*: 95 (1975); *21*:311 (1976).
167. V. I. Kravtsov and Chzhan Chzhi-bin, *Vestn. Leningr. Univ., 22*: 81 (1959); *Russ. J. Phys. Chem., 34*: 2205 (1960).
168. Chzhan Chzhi-bin, V. I. Kravtsov, and Ya. V. Durdin, *Russ. J. Phys. Chem., 34*: 2041 (1960).
169. V. I. Kravtsov and Yan Pien-chzhao, *Vestn. Leningr. Univ., 10*: 107 (1962).

170. N. Sato and G. Okamoto, *J. Electrochem. Soc.*, *111*: 897 (1964).
171. K. E. Heusler and L. Gaiser, *Electrochim. Acta*, *13*: 59 (1968).
172. Ya. M. Kolotyrkin, G. G. Lopovok, and L. A. Medvedeva, *Zashch. Met.*, *2*: 527 (1966); *5*: 3 (1969).
173. R. M. Dvorkina, L. K. Ilina, A. L. Lvov, and L. V. Tyurina, *Elektrokhimiya*, *19*: 957 (1983).
174. R. C. V. Piatti, A. J. Arvia, and J. J. Podesta, *Electrochim. Acta*, *14*: 541 (1969).
175. M. Turner, G. E. Thompson, and P. A. Brook, *Corros. Sci.*, *13*: 985 (1973).
176. T. R. Agladze, O. O. Sushkova, and H. Sasaki, *Elektrokhimiya*, *16*: 1459 (1980).
177. M. L. Kronenberg, J. C. Banter, E. Yeager, and F. Hovorka, *J. Electrochem. Soc.*, *110*: 1007 (1963).
178. K. E. Heusler and T. Ohtsuka, *J. Electroanal. Chem.*, *100*: 319 (1979).
179. A. Bengali and K. Nobe, *J. Electrochem. Soc.*, *126*: 1118 (1979).
180. G. T. Burstein and G. A. Wright, *Electrochim. Acta*, *20*: 95 (1975); *21*: 311 (1976).
181. Ya. M. Kolotyrkin, Yu. A. Popov, and Yu. V. Alekseev, *Elektrokhimiya*, *9*: 624, 629 (1973).

2

ALLOY DISSOLUTION

HERMANN KAISER

Institute for Materials Science, Friedrich-Alexander-University Erlangen-Nürnberg, Erlangen, Federal Republic of Germany

2.1 THERMODYNAMIC CONSIDERATIONS

Considering a hypothetical alloy AB in contact with a solution containing monovalent ions of both components with activities a_A^+ and a_B^+, the equilibrium potentials $E_{A,B}$ of the components may be expressed by

$$E_A = E_A{}^0 + \frac{RT}{F} \ln \frac{a_A^+}{f_A X_A} \tag{1}$$

and

$$E_B = E_B{}^0 + \frac{RT}{F} \ln \frac{a_B^+}{f_B X_B} \tag{2}$$

where $E_{A,B}^0$ are the standard electrode potentials of the components, $f_{A,B}$ and $X_{A,B}$ are the activity coefficients and the atomic fractions of the components in the alloy phase, respectively, and the other quantities have their usual meaning.

Equilibrium between the alloy phase and the solution is established if $E_A = E_B$. Introducing the above expressions for E_A and E_B and rearranging, one finds as the equilibrium condition

$$\frac{a_B^+ f_A (1 - X_B)}{a_A^+ f_B X_B} = \exp\left[(E_A{}^0 - E_B{}^0) \frac{F}{RT}\right] \tag{3}$$

Equation (3) indicates that, for a given atomic fraction of the alloy, equilibrium requires a distinct activity ratio a_{B^+}/a_{A^+} in the electrolyte. For all other ratios, both the electrolyte concentration and the alloy composition must vary by preferential dissolution of the less noble and deposition of the more noble alloy component. Since any change of the alloy composition necessitates the occurrence of volume diffusion in that phase, it is obvious that, at room temperature, the establishment of equilibrium will be difficult for solid alloy electrodes. In liquid alloy electrodes, the diffusivity is known to be much higher [1], and their equilibrium potential is, therefore, more easily established. If, in addition, the difference between the standard potentials of their components $\Delta E^0 = E_A^0 - E_B^0$ is marked (e.g., for Zn-amalgam electrodes), it can be deduced from Eq. (3) that the equilibrium potential is mainly determined by the equilibrium potential of the less noble component [2].

2.2 PHENOMENOLOGICAL ASPECTS

In the field of corrosion, the behavior of solid alloy electrodes at anodic polarization is of primary interest. In many cases their steady-state dissolution mode is virtually simultaneous; i.e., anodic dissolution apparently proceeds without preferential dissolution processes. This holds, in particular, if alloys with similar electrochemical behavior of the components are polarized to potentials that are anodic to the equilibrium potential of the more noble component.* If, on the other hand, the nobilities of the alloy components differ considerably, and the electrode potential does not exceed the reversible potential of the more noble component significantly, steady-state anodic dissolution may occur selectively, i.e., by preferential dissolution of less noble components.

For a binary alloy AB, with B being the less noble component, this process may be described by the overall reaction

$$AB_{cryst} \rightarrow A'_{cryst} + B^{z+} + ze^- \tag{4}$$

or, with consideration of the formation of an insoluble reaction product BY_z with the anions Y^- of the electrolyte, by

$$AB_{cryst} + zY^- \rightarrow A'_{cryst} + BY_z + ze^- \tag{5}$$

---

*The simultaneous dissolution in the steady state may be preceded, however, by a transient period of preferential dissolution of less noble components (see Section 2.3).

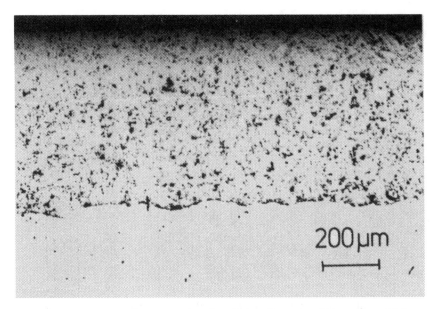

FIGURE 1 Dezincification of a Cu-30Zn alloy resulting from 96 hr anodic polarization in acidified 0.5 M NaCl at 65°C and E_H = 0 mV. (a) Porous reaction layer; (b) reaction front; (c) unattacked alloy.

In both cases, there is agglomeration of the pure, or almost pure, component A' at the electrode surface. In the English literature the effect is, therefore, called *dealloying*.

The best-known dealloying phenomenon appears to be the selective dissolution of brasses [3,4]. As expected from the standard electrode potentials ($E^0_{Zn/Zn^{2+}}$ = -0.763 V; $E^0_{Cu/Cu^{2+}}$ = +0.377 V) it results in a "dezincification." This process is known to occur either uniformly (*layer-type dezincification*) or locally (*plug-type dezincification*). As shown by Fig. 1, uniform dezincification proceeds by the advance of a fairly plane reaction front (alloy/electrolyte interface), leaving behind a porous reaction layer of pure, or almost pure, copper. As a consequence of its porosity, the reaction layer is of poor mechanical strength. The effect of dezincification on the mechanical properties of Cu-Zn alloys is, therefore, very detrimental. Factors that stimulate the dezincification tendency are

Increasing Zn concentration of the alloy (the β-phase of α/β-
brasses being more intensely dezincified than the α-phase)
Stagnant electrolyte solutions
The presence of both cuprous and chloride ions
Increased temperatures

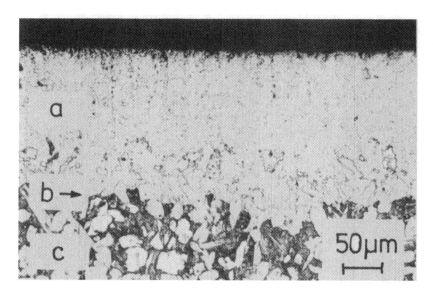

FIGURE 2 Dealuminization of a complex aluminum bronze (Cu-10Al-4Fe-3Mn-0.5Ni) resulting from 48 hr anodic polarization in 1 N H_2SO_4 at 25°C and E_H = 240 mV. (a) Porous reaction layer containing crystals of undissolved α-phase; (b) reaction front; (c) unattacked alloy, essentially consisting of α-phase (bright) as well as β'- and γ_2-phases (dark).

On the other hand, the dezincification of brasses is known to be suppressed by alloying additions of

Sn (admiralty brass)
Al (aluminum brass)
As and/or P (typical concentrations ranging from 0.02 to 0.06%)

It should be noted that the beneficial effects of both As and P are not observed with the β-phase in duplex α/β-brasses. In order to explain the effect of As, different theories have been proposed in the literature [5--7], which necessarily reflect the authors' opinions concerning the mechanism of dezincification. Regarding the beneficial effect of Al additions, it was shown by X-ray photoelectron spectroscopy that in $FeSO_4$-containing seawater solutions, a protective film of hydrotalcite, $Mg_6Al_2(OH)_{16}CO_3 \cdot 4H_2O$ is formed [8]. The corrosion behavior of Cu-Al alloys (aluminum bronzes) is, in many respects, similar to the behavior of brasses, since they may suffer from selective dissolution of aluminum (Fig. 2). This is especially true of

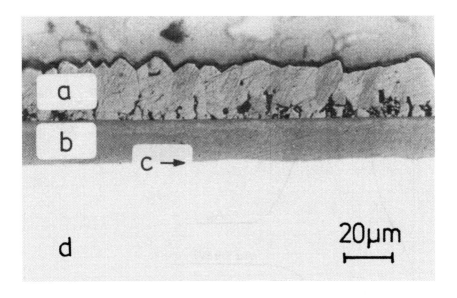

FIGURE 3 Selective sulfidation of a Cu-13Au alloy resulting from 300 hr anodic polarization at E_H = 0 mV in a sulfide-containing buffer solution of pH = 5 and pS^{2-} = 15. (a) Cu_2S layer; (b) porous reaction layer consisting of gold-rich metal and Cu_2S; (c) reaction front; (d) unattacked alloy.

the martensitic decomposition products of the β-phase, whereas the face-centered cubic (fcc) α-phase appears not to be dealuminized, most likely as a consequence of the formation of complex protective layers [9—11]. Other well-known examples of selective dissolution phenomena are found in the Cu-Ni, Cu-Mn, and Cu-Cd systems [12—14].

In addition to the copper alloys, there are numerous other systems that suffer from dealloying [15]. Among these, noble metal systems such as Cu-Au and Ag-Au have been of particular interest with respect to a process of dealloying as described by Eq. (4) [16,17]. In the presence of sulfide ions they tend, in addition, to preferential sulfidation of components such as Cu and Ag according to Eq. (5). As shown by Fig. 3, the microscopic examination of an electrochemically sulfidized Cu-Au alloy reveals the formation of a two-layer structure, which consists of an outer layer of Cu_2S and an inner reaction layer, most probably consisting of porous gold-rich metal with Cu_2S inside its pores [18]. It is believed that similar processes of preferential sulfidation are responsible for the tarnishing of noble metal

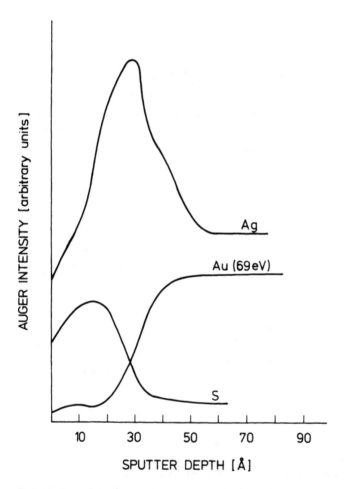

FIGURE 4 AES depth profiles of Ag-70Au after 21 days exposure in
a test climate containing 1 ppm H_2S at 75% relative humidity and 25°C.
(From Ref. [20].)

dental alloys in human saliva [19] and for the degradation of electrical contact materials in sulfur-containing atmospheres. The thickness of the sulfide layers in these cases is known, however, to be
much smaller than that in Fig. 3. This is illustrated by Fig. 4,
showing the enrichment of both sulfur and silver on the surface of
an Ag-70% Au alloy as revealed by Auger electron spectroscopy
(AES) after exposure to an H_2S-containing test climate [20].

2.3 ELECTROCHEMICAL ASPECTS

As pointed out by Gerischer [21], the composition of a liquid alloy AB, upon anodic polarization, will change continuously by the selective dissolution of the less noble component B. The variation of the mole fraction X_B may then be expressed by

$$\frac{dX_B}{dt} = -\frac{1}{M_A + M_B}\left[(1 - X_B)\frac{I_B}{z_B F} - X_B \frac{I_A}{z_A F}\right] \tag{6}$$

where M_A and M_B are the numbers of moles of the components in the alloy, I_A and I_B are the anodic partial currents of these components, and z_A and z_B denote the valences of the ions of A and B, respectively. Setting $dX_B/dt = 0$, the following steady-state condition with respect to the alloy composition may be derived from Eq. (6).

$$Z \equiv \frac{z_A I_B X_A}{z_B I_A X_B} = 1 \tag{7}$$

That is, selective dissolution of B will occur only if $Z > 1$. Equation (7) may also be used to describe the selectivity of the dissolution of solid alloys. This concept was introduced by Marshakov and Bogdanov [22] to describe the dezincification tendency of brasses. In that case Z may be called the *dezincification factor*, whereas in the general case it is called a *selectivity coefficient*.

It is obvious from Eq. (7) that the selectivity coefficient depends on the ratio of the partial rates of anodic dissolution of the alloy components. For an evaluation of the partial dissolution rate of an individual alloy component, it should be considered that, even in the absence of cathodic reactions, current meter readings as obtained with standard polarization techniques always give the sum of the rates of all possible anodic reactions. Therefore, they represent the partial current of a single component only under very specific conditions. This difficulty may be overcome by:

1. Application of the rotating ring disk electrode (RRDE), which is based on convective transport of ions from a dissolving disk to a surrounding ring electrode. There, by suitable choice of the ring potential, only the electropositive component will be deposited. In the case of a binary alloy AB, the partial current of the dissolution of the electropositive component A from the disk will be given by

$$I_A \equiv I_A{}^{disk} = \frac{I^{ring} - I_0{}^{ring}}{N_N} \tag{8}$$

where I^{ring} and $I_0{}^{ring}$ are the current of deposition of A and the background current at the ring electrode, respectively, and N_N denotes the collection efficiency [23]. If cathodic reactions at the disk electrode are absent, the partial current of the electronegative component I_B may also be calculated from the difference between the total disk current $I_{tot}{}^{disk}$ and I_A as determined by Eq. (8) [24]. Alternatively, I_B may be measured by use of a split ring-disk electrode [25].

2. Determination of the variation of the electrolyte concentration c with time by other methods of chemical analysis. Using Faraday's law, the current of the nth component, I_n, follows from

$$I_N = zFv \frac{dc_n}{dt} \tag{9}$$

where v is the volume of the electrolyte solution. Among the various possible methods of chemical analysis, γ-spectroscopy appears to provide the best combination of sensitivity, selectivity, and quick response [26,27]. As shown by Fig. 5, it is based on simultaneous and continuous detection of the radioactivity of γ-isotopes that are contained in specially prepared alloys and enter the electrolyte solution during anodic polarization.

By using the methods outlined above, partial anodic dissolution rates and/or selectivity coefficients have been obtained for various alloys as a function of the electrode potential. As an example, Fig. 6 shows quasi-stationary anodic polarization curves for the dissolution of Cu from binary Cu-Au alloys in deaerated 0.1 N Na_2SO_4/0.01 N H_2SO_4 solutions [28,29]. It is obvious from these curves that Cu dissolution from all alloys starts at potentials that are anodic to the reversible potential of the Cu/Cu^{2+} electrode ($E_{Cu/Cu^{2+}} = 0.163$ V for $a_{Cu^{2+}} = 10^{-6}$ M). However, the higher the Au concentration of the alloy, the more the partial rate of Cu dissolution differs from that of pure Cu. A detailed examination of the partial polarization curves reveals that they may be divided into four characteristic ranges. After the first increase of the current density, an anodic limiting current density is observed, which decreases from a poorly resolved value for the Cu-8Au alloy to about 2.10^{-8} A/cm^2 for the Cu-25Au alloy. In other words, alloys of high Au concentration are quasi-immune with respect to the selective dissolution of Cu and therefore may be used in practice. After reaching some sort of critical potential E_c, however, the region of quasi-immunity is terminated and the

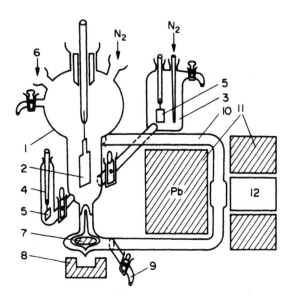

FIGURE 5 Cell for electrochemical and radiotracer measurements: 1, main cell; 2, working electrode; 3, 4, side cells; 5, auxiliary electrode; 6, electrolytic connection with reference electrode; 7, centrifugal pump; 8, magnetic stirrer; 9, stopcock for sampling; 10, circulation tube with cuvette; 11, lead shield; 12, scintillation detector. (From Ref. [26], with permission.)

current density of Cu dissolution increases by several orders of magnitude. The critical potential E_c itself is shifted in the anodic direction by increasing the noble metal concentration. For practical purposes, therefore, the noble metal concentration must be high enough to ensure that E_c remains anodic with respect to the free corrosion potential, E_{corr}. Finally, at electrode potentials $E > E_c$, a second plateau of the polarization curve is observed which may be associated with the formation of insoluble corrosion products [30].

Similar polarization curves were reported for the preferential dissolution of Zn from various Cu-Zn phases (Fig. 7), as well as for the dissolution of Ag and Mg from Ag-Au [17,32] and Mg-Cd alloys [33], respectively. It appears, therefore, that for the preferential dissolution of less noble components in the active state, the shape of the anodic polarization curve as discussed above is the general one.

It is also evident from Fig. 7 that, as a consequence of the high Cu concentration, the critical potential of the α-Cu-Zn phase in acid sulfate solutions is shifted to values that are close to the dissolution potential of Cu from that phase. At $E > E_c$, Zn and Cu therefore tend

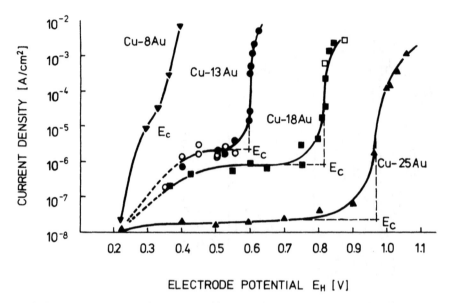

FIGURE 6 Quasi-stationary anodic polarization curves and critical potentials E_C for the dissolution of Cu from Cu-Au alloys in 0.1 N Na_2SO_4/ 0.01 N H_2SO_4 solutions. (Data for Cu-13Au and Cu-18Au from Ref. [28], for Cu-8Au and Cu-25Au from Ref. [29].) Open symbols, chemical analysis; closed symbols, microammeter readings.

to dissolve in the proportion in which they exist in the alloy. A similar result was derived from the anodic polarization of Cu-rich brasses in chloride solutions: plotting the dezincification factor Z against the electrode potential, a maximum was observed close to the free corrosion potential of pure Cu, which was followed by a decrease of Z to virtually unity at increased anodic polarization [34]. Both results thus indicate a transition of the steady-state dissolution mode of α-brass from preferential to simultaneous. This transition in the dissolution mode may occur with any alloy system if the equilibrium potential of the electropositive component is exceeded. From an analysis of the polarization data for numerous alloy/electrolyte systems, it was shown, in particular, that this situation is more likely to occur as the separation of the standard potentials of the alloy components, ΔE^0, decreases and as the concentration of the electropositive component in the alloy increases [35].

It is an important feature of the simultaneous mode of alloy dissolution that, for $\Delta E^0 > 0$ and under steady-state conditions, the dissolution rate of the less noble component is controlled by the codissolving noble component. There is considerable evidence that this

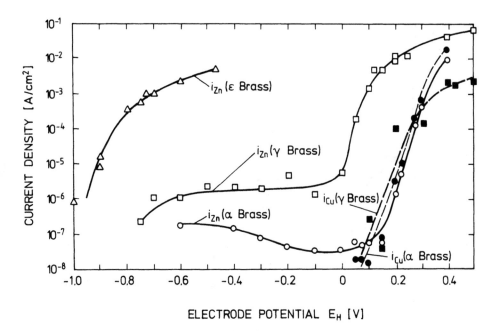

FIGURE 7 Partial anodic polarization curves for the dissolution of Zn (solid lines) and Cu (dashed lines) from single-phase Cu-Zn alloys in buffered Na_2SO_4 solutions of pH 5. Open symbols, chemical analysis; closed symbols, microammeter readings. (From Ref. [31].)

control is established during a preceding transient period of preferential dissolution, during which the alloy surface is enriched in the electropositive component. Experimental proof of the latter effect has been obtained, in particular, from a soft X-ray study of the near-surface composition of brass electrodes that were polarized at appropriate potentials in acidic sulfate solutions [36]. It appears, therefore, that the simultaneous mode of alloy dissolution is only a special case of the more general phenomenon of preferential alloy dissolution. This view has been corroborated by galvanostatic pulse experiments with α-brass electrodes that were initially kept in acidified NaCl solutions at a potential cathodic to the reversible potential of the zinc electrode [37]. As shown by Fig. 8, the subsequent anodic current pulse results in a first plateau of the electrode potential at values close to the reversible potential of the zinc electrode ($E_{Zn/Zn^{2+}} = -0.937$ V for $a_{Zn^{2+}} = 10^{-6}$ M), indicating preferential dissolution of Zn at the beginning of the polarization. Due to the depletion of the alloy surface with that component, the electrode potential is subsequently shifted in the anodic direction until a second

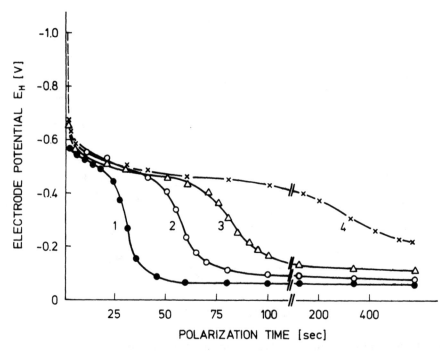

FIGURE 8 Potential transients for galvanostatic anodic polarization of
Cu-30Zn electrodes in 1 N NaCl/0.01 N HCl solutions at various current densities (in amperes per square centimeter): 1, 10^{-4}; 2, 3 10^{-5};
3, 10^{-5}; 4, open circuit. (From Ref. [37].)

plateau appears close to the reversible potential of the copper electrode,* indicating codissolution of Cu in the steady state. Similar
conclusions may be drawn from an analysis of the partial anodic current transients as obtained by γ-spectroscopy for the dissolution of
Zn and Cu from the same alloy: as shown by Fig. 9, i_{Zn} decreases
from rather high values immediately after the beginning of the anodic
polarization, whereas i_{Cu} only increases after about 2 min. Steady-
state values, corresponding to $Z_{Zn} = 1$, are reached after about 6
min [39].

---

*In chloride solutions, the dissolution of Cu proceeds to the monoval-
ent state via its chloride complex. For a detailed discussion of the
reversible potential of the Cu/Cu^{+} electrode under these conditions
see [38].

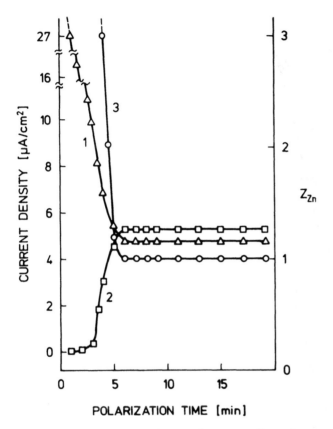

FIGURE 9 Time dependence of the partial anodic current densities of Zn (1) and Cu (2) and of the dezincification factor Z_{Zn} (3) during galvanostatic anodic polarization of a Cu-30Zn electrode in 1 N NaCl/ 0.01 N HCl at 10^{-5} A/cm^2. (From Ref. [37].)

Few data have been presented for the partial dissolution rates of noble alloy components. From Fig. 7 it may be seen that the partial rates of dissolution of Cu from α- and γ-brass are almost equal. In addition, they correspond well to the dissolution rate of pure Cu under the same conditions [40]. It might be concluded, therefore, that the dissolution kinetics of noble components are not changed significantly by less noble components. By the use of γ-spectroscopy, a more detailed analysis of this problem was carried out for the dissolution of Fe and Cr from binary Fe-Cr alloys in 1 N H$_2$SO$_4$ [27]. As shown by Fig. 10, the partial dissolution rate of the electropositive component, Fe, in the active state is slightly increased by increasing

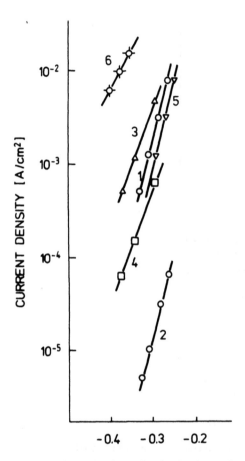

FIGURE 10 Partial anodic polarization curves of the dissolution of Fe
(1, 3) and Cr (2, 4) from Fe-Cr alloys with 0.85% Cr (1, 2) and 13.2%
Cr (3, 4) as well as of pure Fe (5) and pure Cr (6) in 1 N H_2SO_4 at
50°C. (From Ref. [27].)

the Cr concentration of the alloy. The slopes of the corresponding
Tafel lines decrease, however, from the value characteristic of pure
Fe to that of pure Cr, indicating a change in the dissolution mechanism by a coupling of the anodic partial reactions. For similar reasons,
the slopes of the Tafel lines for the dissolution of Cr decrease on increasing the Fe concentration of the alloy. Moreover, by comparison
with the polarization curve of pure Cr, it follows from Fig. 10 that

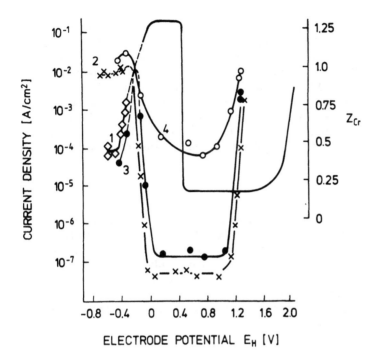

FIGURE 11 Stationary anodic polarization curves of Fe (1), Cr (2), and Fe-28Cr (3) as well as dependence of the selectivity coefficient of Fe-28Cr on the electrode potential in 1 N H_2SO_4 at 40°C (4). (From Ref. [27].)

the dissolution rate of Cr from the Fe-Cr alloys is decreased to an extent that, in accordance with the above principles, the steady-state dissolution mode of the alloys is virtually simultaneous.

The latter situation is changed, however, if the Fe-Cr alloys are passivated. This may be seen, in particular, from Fig. 11, which shows a plot of the steady-state values of the selectivity coefficient Z_{Cr} against the electrode potential for an Fe-28% Cr alloy together with the anodic polarization curves of the alloy and its components.

It is obvious that only in the active state of the alloy does the value of Z_{Cr} exceed unity. On the transition to the passive state, Z_{Cr} decreases considerably and reaches a minimum value close to 0.4 at a potential of about 0.7 V before it again increases due to the transpassivity of Cr [27]. In other words, in the passive region of the alloy there is a preferential dissolution of the electropositive component, Fe, whereas Cr must be accumulated in the metallic phase of the electrode surface and/or in the passive film. Experimental evidence for both possibilities was obtained by X-ray photoelectron spectroscopy and by Auger electron spectroscopy of passivated Fe-18Cr surfaces [41]. Similar results have been reported for Ni-Mo alloys [42] as well as for austenitic stainless steels. In particular, it has been suggested that the beneficial effects of Ni and Mo on the passivation of austenitic stainless steels are due to enrichment of these elements in the metallic phase below the passive film [43]. On the basis of these findings, it appears, therefore, that selective dissolution phenomena are essential for the passivation of such alloys.

2.4 THEORY

Any theory of the anodic dissolution of homogeneous alloys should be able to predict the rate of this process. Early approaches to this problem were made on the basis of an independent superposition of the anodic partial reactions: assuming that a solid solution may be described by the model of a heterogeneous dispersion of atomic dimensions with an area fraction f_i for the ith component, the anodic current density i_{AB} of a binary alloy AB at a given electrode potential would be given by

$$i_{AB} = f_A i_A^* + f_B i_B^* \tag{10}$$

where i_A^* and i_B^* are the current densities of the pure components at the same potential [44]. Even if these considerations are restricted to the simultaneous mode of alloy dissolution, however, they fail to predict the shape of the anodic polarization curve of alloy electrodes without arbitrary manipulations. This is due to the fact that even for the simultaneous dissolution mode, the concept of independent anodic partial reactions is not applicable to alloy electrodes with different electrochemical behavior of the components.

Keeping in mind that preferential dissolution of less noble components is the fundamental process of alloy dissolution, it appears necessary to determine the sequence of individual steps of the overall electrode reaction as given by Eqs. (4) and (5) and, in particular, to evaluate the rate-determining one. Unfortunately, there is no general agreement on this question at present. It is generally accepted, however, that at least one of the following three "mechanisms" will be operative during the preferential dissolution of alloy components from solid solutions.

2.4.1 Ionization-Redeposition Mechanism

For a binary alloy AB, this mechanism assumes an initial simultaneous ionization of both components. As shown by Fig. 12, this step

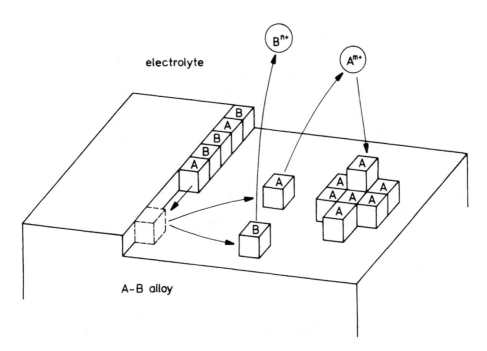

FIGURE 12 Selective dissolution of a binary alloy AB by the ionization-redeposition mechanism, assuming the intermediate formation of
adatoms (schematic).

may involve intermediate formation of adatoms of the less noble component, B. Alternatively, direct transfer of B ions from kink sites
to the electrolyte solution must be considered [45, 46]. Irrespective
of these details of the initial charge transfer step, the electropositive component A is thought to redeposit on the alloy surface in a
consecutive step.

By thermodynamic reasoning, it can be shown that this process
is impossible unless, in accordance with the principles of irreversible thermodynamics, a coupling between the anodic partial reactions occurs, which may reflect an average activity of adsorbed adatoms of the electropositive component A that exceeds unity [16].
If this situation prevails, A ions may be dissolved from the alloy
at underpotentials that are cathodic with respect to the reversible
potential E_A' for the dissolution of the pure, or almost pure, component A' that agglomerates on the electrode surface. They have,
therefore, the character of an unstable intermediate species and
will be plated out on the surface of A' crystals at the same electrode potential.

Experimental evidence for the existence of A ions at E < $E_{A'}$ and hence for the ionization-redeposition mechanism may be obtained from RRDE measurements. Due to the outward flow of electrolyte solution at the rotating disk electrode, a fraction of possible intermediate A ions will be carried toward the outer ring electrode, where, by suitable choice of the ring potential, they may be detected by the occurrence of a deposition current. If, on the other hand, the potential of the disk electrode is anodic to the reversible potential E_A of the more noble component (e.g., at galvanostatic conditions and small separation ΔE^0 of the standard potentials of the alloy components), the existence of A ions tells nothing about the mechanism. Nevertheless, the ionization-redeposition mechanism would necessitate the occurrence of a reduced collection efficiency N, which follows from the N_0 as calculated from the electrode geometry by

$$N = \frac{N_0}{1 + k\delta_N/D} \tag{11}$$

Here k denotes the rate constant of the redeposition reaction at the disk electrode, δ_N is the thickness of the Nernst diffusion boundary layer, and D represents the diffusion coefficient of the ions of the electropositive component [47].

Based on the above RRDE techniques, no evidence for the ionization-redeposition mechanism was found for the anodic dissolution of a Cu-10% Au alloy in an acidified $CuSO_4$ solution [16]. This indicates that the above mechanism is generally impossible if it requires the ionization of an electropositive component A at potentials far below its equilibrium potential E_A (e.g., at an underpotential of about 0.6 V for Cu-13Au in Fig. 6). On the other hand, it should be checked thoroughly for small underpotential conditions, particularly for alloys with a small separation of E_C and E_A, such as the Cu-rich brasses. The experimental situation with the latter is, however, still open to question. In acidified sulfate solutions a reduced collection efficiency of Cu, and hence a probable indication of a redeposition reaction, was observed for α/β-brasses [48], but not for α-brasses [16, 48]. Moreover, a very complex behavior of α-brass in acidified chloride solutions has been inferred from a plot of the dezincification factor Z_{Zn} versus the polarization time [26]. As shown by Fig. 13, Z_{Zn} quickly decreases from an initially high value to virtually unity, indicating the usual change from the preferential mode of dissolution to the simultaneous dissolution of both Zn and Cu at galvanostatic anodic polarization. However, after about 1 hr, Z_{Zn} increases considerably. This effect, which has been called *pseudopreferential dissolution*, may be delayed or accelerated by replacement of the electrolyte or by the introduction of a 1.5×10^{-4} M CuCl solution,

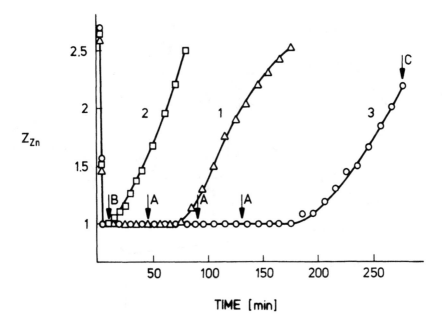

FIGURE 13 Time dependence of the dezincification factor Z_{Zn} of Cu-30Zn in air-saturated 1 N NaCl/0.01 N HCl at open circuit potential and 20°C. 1, Continuous accumulation of Cu ions in the solution, 2, replacement of solution at point B with a solution containing 1.5×10^{-4} M CuCl; 3, replacement of original electrolyte at points A with a new a new one. (From Ref. [26].)

respectively. From this observation it is obvious that the increase of Z_{Zn} requires a critical concentration of dissolved Cu ions of about 10^{-4} M and may therefore be related to the redeposition of Cu from its chlorocomplex.

2.4.2 Volume Diffusion Mechanism

The volume diffusion mechanism, which was introduced by Pickering and Wagner [16], does not consider an "underpotential dissolution" of the electropositive component. Instead, the atoms of the latter are assumed to accumulate as mobile adatoms on the electrode surface. It may be expected, therefore, that these adatoms will have an increasing tendency to move back to the kink sites and steps and to block any further removal of the less noble component from these positions by the usual mechanism of metal dissolution. Provided that the anodic overvoltage is high enough, there is, however, an alternative possibility that the preferential dissolution will proceed from

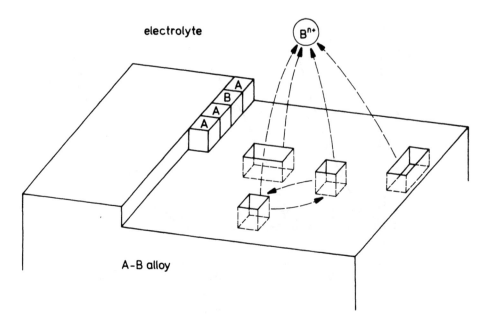

FIGURE 14 Formation of surface (di)vacancies during the selective dissolution of a binary alloy AB by the volume diffusion mechanism (schematic).

terrace sites (which may be in the vicinity of the kink sites [49]). As shown by Fig. 14 for a binary Cu-Au alloy, this process results in the formation of surface vacancies. The major concept of the volume diffusion mechanism is that by the injection of these surface vacancies into the bulk of the alloy, a large vacancy supersaturation is developed adjacent to the alloy/electrolyte interface which enhances the diffusivity beyond its thermal value. Repopulation of the electrode surface with Cu atoms via volume diffusion is therefore considered to be possible even at room temperature.

   A detailed analysis of this diffusion problem reveals that it is characterized by the interdiffusion of Cu and Au atoms and by a moving reaction front (alloy/electrolyte interface). For a Cu-20% Au alloy, this situation is illustrated by Fig. 15, assuming that, within an interdiffusion zone of thickness δ, the mole fraction of Cu falls from its bulk value X_{Cu}^{0} to virtually zero, whereas the mole fraction of Au is increased from its bulk value to virtually one. The concentration profile $X_{Cu}(\xi,t)$, as shown schematically by Fig. 15, has been calculated, and by the additional application of Fick's first law, the flux of Cu atoms to the interface, j_{Cu}, was estimated to be

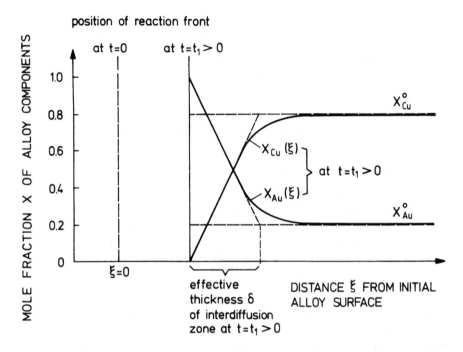

FIGURE 15 Recession of alloy/electrolyte interface and compositional changes in a Cu-20Au alloy due to the volume diffusion mechanism of selective alloy dissolution (schematic).

$$j_{Cu} = \frac{X^0_{Cu}}{V_m}\left[\frac{D}{2(1 - X^0_{Cu})t}\right]^{1/2} \tag{12}$$

where X^0_{Cu} again is the bulk mole fraction of Cu in the Cu-Au alloy, V_m is the molar volume, and D is the interdiffusion coefficient, which was shown to be approximately constant within the interdiffusion zone [16]. Assuming further that the volume diffusion of Cu is the rate-determining step of the dissolution process, the current density i_{Cu} of the formation of divalent Cu ions is given by

$$i_{Cu} = 2j_{Cu}F = \frac{2FX^0_{Cu}}{V_m}\left[\frac{D}{2(1 - X^0_{Cu})t}\right]^{1/2} \tag{13}$$

where F is Faraday's constant. The most crucial point for the volume diffusion mechanism is the evaluation of the interdiffusion coefficient

D. From the extremely low diffusivity D_{1V} of monovacancies in Cu at 25°C, it follows that the diffusion of Cu atoms via monovacancies cannot account for the selective dissolution of this component by the above mechanism. If it is considered, however, that in fcc lattices the diffusivity of divacancies is much higher ($D_{2V} = 1.3 \times 10^{-12}$ cm²/sec vs. $D_{1V} = 3 \times 10^{-19}$ cm²/sec in Cu at 25°C [50]), it may be assumed that the interdiffusion proceeds via a divacancy mechanism. In this case, the effective interdiffusion coefficient will be given by

$$D = D_{2V} X_{2V} \tag{14}$$

where X_{2V} is the increased mole fraction of divacancies in the vicinity of the electrode surface. The problem is then to calculate X_{2V} from a model that describes the process of vacancy injection into the alloy with due consideration of vacancy annihilation at the interface and vacancy losses at internal sinks and by void formation. No solution of this problem is yet available. It has been assumed, however, that $X_{2V} = 10^{-2}$ might be a reasonable value, i.e., that annihilation and vacancy loss processes may be neglected and a rather high supersaturation of divacancies prevails [16,51]. By substitution of Eq. (14) in Eq. (13), the true current density i_{Cu} of a plane electrode surface may then be calculated. For $X_{Cu}^0 = 0.9$, $t = 10^3$ sec, and the above value for X_{2V}, i_{Cu} was shown to be of the order of 2×10^{-4} A/cm², indicating that the volume diffusion of Cu via divacancies may be fast enough to be instrumental in the preferential dissolution of Cu from Cu-Au alloys at room temperature [16]. This conclusion may be valid even if the above estimate of X_{2V} is too large: it has been shown that a plane electrode surface will be unstable under volume diffusion control of the selective dissolution process [52, 53]. As a consequence, there will be an appreciable surface roughening of the electrode, which, according to Fig. 16, finally results in the formation of the porous reaction zone. The apparent current density, therefore, will be higher than the true current density that follows from Eq. (13).

One possible way to demonstrate the occurrence of the volume diffusion mechanism experimentally is to prove the existence of the interdiffusion zone. Since for a Cu-10% Au alloy the thickness δ of this zone after 10^3 sec was estimated to be of the order of 10^{-2} μm [16], it may not be resolved by conventional electron beam microanalysis. However, as indicated by Fig. 16, the thickness d over which compositional differences exist may far exceed the effective thickness of the interdiffusion zone due to the increasing surface roughening. The interdiffusion zone, therefore, should contribute to an X-ray or electron diffraction pattern of the bulk alloy. As a consequence of the variation of the lattice constant within this zone, one would expect, in particular, broadened diffraction lines to occur at Bragg angles

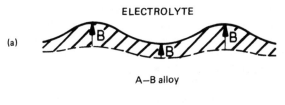

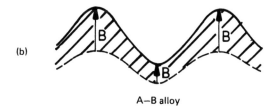

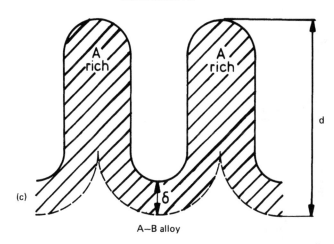

FIGURE 16 Onset of surface instability during selective removal of less noble component B under rate-limiting diffusion in the alloy ((a), (b)), as well as pore formation at a later stage of the process (c). For (c) the thickness, d, over which compositional differences exist far exceeds the thickness δ of the interdiffusion zone. (From Ref. [51].)

varying between the values for the bulk alloy and that for pure Au. Moreover, with increasing surface roughening, the intensity of these lines should increase at the expense of the intensity of the line pattern of the bulk. Simultaneously, their peak positions should be shifted to lower Bragg angles that correspond to higher Au concentrations.

Virtually all of the above predictions have been observed experimentally. A typical example is given by Fig. 17, showing X-ray diffraction patterns of a polycrystalline Cu-3% Au alloy at successive stages of anodic dissolution in 1 M H_2SO_4 [54]. It has been argued, however, that similar observations might be made if the reaction zone consisted of an agglomerate of small individual crystals formed by a

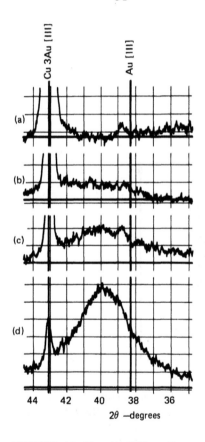

FIGURE 17 X-ray diffraction patterns for a Cu-3Au specimen after increasing amounts of anodic dissolution at 5 mA/cm^2 in 1 M H_2SO_4, showing the occurrence of a broadened diffraction line of increasing intensity at a Bragg angle corresponding to a gold-rich alloy. The $2\theta_{\{111\}}$ position of pure Au is indicated by the vertical line at $2\theta = 38.2°$. (a) Prior to anodic dissolution; (b) 3.0 C/cm^2 passed; (c) 6.0 C/cm^2 passed; (d) 24 C/cm^2 passed. (From Ref/ [54], with permission.)

nucleation and growth process [55]. In this case the line broadening
in essence would be a particle size effect and the occurrence of dif-
fracted intensity at Bragg angles below the value for pure Au could
be due to incorporation of Cu atoms into growing Au particles. In
addition, the effect of an increased stacking fault probability on the
peak positions in the diffraction pattern for the Au particles should
be considered [56]. Analogous X-ray investigations were carried out
with Zn-rich ε- and γ-Cu-Zn alloys. In addition to the effects described above, the selective dissolution of Zn from these intermetallic
phases resulted in the formation of the more Cu-rich γ- and α-phases,
respectively [49,57]. Similar results were reported for the anodic
dissolution of Cd from ε-Ag-Cd alloys [58]. Even though there were
no data available on the diffusivity of divacancies in these lattices,
the results were taken as a strong affirmation of the volume diffusion mechanism. It should be noted, however, that the formation of
new phases might be understood as well in terms of incorporation of
the less noble component into growing particles of the more noble
one. In particular, it has been claimed that the ionization-redeposition mechanism may be operative [59]. In addition to the X-ray diffraction work, electron diffraction experiments were carried out in
order to establish the existence of the interdiffusion zone. In contrast to the results obtained from reflection electron diffraction patterns of Cu-10Au electrodes [54], selected area diffraction patterns
from thin films of corroded Cu_3Au and Ni-Au alloys did not show any
evidence for a continuous change in the lattice parameter that might
be due to an interdiffusion zone [60,61]. On the other hand, it was
revealed by Auger electron spectroscopy that the surfaces of binary
Cu-Pd alloys were enriched with Pd to a depth of about 20 nm after
the alloys were subjected to anodic polarization below their critical
potential in an acidified Na_2SO_4 solution [62].

Other attempts to confirm the validity of the volume diffusion concept included (a) a study of the creep rate of stressed α-brass electrodes, which was shown to be increased by anodic polarization; from
this observation it was concluded that divacancies were generated at
the electrode surface in agreement with the volume diffusion hypothesis [63]; and (b) an analysis of current-time transients. Disregard-
ing surface roughening effects, it follows from Eq. (13) that, under
potentiostatic control, the partial anodic current density i_B of the
less noble component of a binary alloy AB should decrease with $1/\sqrt{t}$.
(For galvanostatic conditions, i.e., at a constant rate of recession of
the alloy/electrolyte interface, the calculation of the partial current
transient is, in general, more complicated [26,36].). In accordance
with the above prediction, experimental plots of i_B vs. $1\sqrt{t}$ were reported to be linear over various time spans for several alloy/electro-
lyte systems [24,32,58,64]. A typical example, obtained for the anodic
dissolution of Ag from an Ag-Au alloy, is shown in Fig. 18. It remains

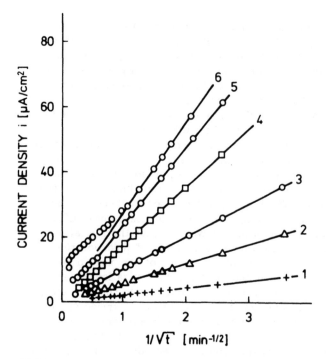

FIGURE 18 Partial dissolution rate of Ag from an Ag-15Au alloy plotted against $1/\sqrt{t}$ at electrode potentials E_H of 1, 0.60 V; 2, 0.70 V;
3, 0.75 V; 4, 0.80 V; 5, 0.85 V; and 6, 0.90 V in 0.1 M KNO_3. (From
Ref. [32].)

unclear in this figure whether i_{Ag} decreases continuously to zero or
attains a small, but finite steady-state value as it is the case with Cu-
Au alloys (see Fig. 6). Apart from this problem, it is evident that the
slope of the i_{Ag} vs. $1/\sqrt{t}$ plots increases with increasing anodic polarization. On the basis of Eq. (13), this result has been interpreted in terms of a potential dependence of the interdiffusion coefficient
D [32], which may be understood by assuming a potential dependence
of the divacancy mole fraction X_{2V} according to Eq. (14). An alternative explanation was given by Schwitzgebel et al. [58]. Ignoring
the problem of a moving alloy/electrolyte interface, the authors calculated the anodic partial current density i_B by the simplified expression

$$i_B = zF(c_B^0 - c_B^s)\left(\frac{D}{\pi t}\right)^{1/2}$$

(15)

where $c_B{}^0$ and $c_B{}^s$ are the bulk and surface concentrations of component B, respectively. Following this treatment, the potential dependence of the slope of the i_B vs. $1/\sqrt{t}$ plots may as well reflect a potential dependence of the surface concentration $c_B{}^s$, whereas in the derivation of Eq. (13), $c_B{}^s$ was assumed to be zero throughout. In work by the same authors, a possible contribution of grain boundary diffusion to the current density transient was considered by the second term in square brackets in the expression

$$i_B = zF(c_B{}^0 - c_B{}^s)\left[\left(\frac{D}{\pi t}\right)^{1/2} + \frac{const}{d}\,(\delta_{gb}D_{gb})^{1/2}\left(\frac{D}{t}\right)^{1/4}\right] \qquad (16)$$

where d is the grain diameter, δ_{gb} is the grain boundary width, and D_{gb} is the grain boundary diffusion coefficient [64]. It should be noted that defect-enhanced diffusion has also been considered to account for the occurrence of steady-state values of the rate of Cu dissolution from Cu-Au alloys at $E < E_c$ by the volume diffusion mechanism [28].

2.4.3 Surface Diffusion Mechanism

In addition to the redeposition and volume diffusion mechanisms, there is a third conceivable mechanism for the selective dissolution of alloys which involves the nucleation and growth of crystals of the pure, or almost pure, noble component via a surface diffusion process. Following earlier suggestions by Gerischer [21], this mechanism may be described in detail by a model which, for a binary AB, includes

1. Removal of both components from steps and kink sites and formation of adatoms of the more noble component A at electrode potentials $E < E_c$. The adatoms may subsequently (a) crystallize via surface diffusion or (b) accumulate at the steps and kink positions, where they are thought to block the removal of the less noble species. However, in contrast to the assumptions that were made with the volume diffusion process, a small but finite steady-state dissolution rate of B atoms is now considered to be possible by exchange of places between A and B atoms.
2. Removal of B atoms from terrace sites in order to account for the increased dissolution rate at $E > E_c$. Since the remaining A atoms are expected to form a porous layer of small individual crystallites, this process is thought to operate without transport of B atoms to the electrode surface via volume diffusion.

On the basis of this model it should be possible to calculate the partial anodic polarization curve for the dissolution of a less noble component B by superposition of the rates of the two processes above, i.e., by

$$i_B(E) = i_B{}^k(E) + i_B{}^t(E) \tag{17}$$

where $i_B{}^k$ and $i_B{}^t$ denote the anodic partial current densities of the dissolution of B atoms from kink and terrace sites, respectively. Assuming that B atoms are ionized by direct, rate-limiting transfer from the above lattice positions, Kaesche [30] expressed $i_B{}^k(E)$ and $i_B{}^t(E)$ by the usual approximation for high anodic polarization, i.e.,

$$i_B{}^k = k_1 \theta_k \exp\left(\frac{\alpha zF}{RT} E\right) \tag{18}$$

$$i_B{}^t = k_2 \theta_t \exp\left(\frac{\alpha zF}{RT} E\right) \tag{19}$$

Here k_1 and k_2 are the forward reaction rate constants of the two dissolution processes, while θ_k and θ_t are the coverage with kink sites and the fraction of terrace sites occupied by B atoms, respectively. Taking into account further that the initial coverage with kink sites, $\theta_k{}^0$, may decrease upon anodic polarization according to $\theta_k = \theta_k{}^0(1 - i_B{}^k/i_{limit})$, where i_{limit} is the anodic limiting current density as described in Section 2.3, and making empirical assumptions about the dependence of i_{limit} and θ_t on the bulk mole fraction of the more noble component, $X_A{}^0$, the following expression was finally derived [30].*

$$i_B = \frac{k_1 \theta_k{}^0 \exp[(\alpha zF/RT)E]}{1 + [\theta_k{}^0/(1 - X_A{}^0)^m] \exp[(\alpha zF/RT)E]}$$

$$+ k_2 \theta_t{}^0 (1 - X_A{}^0)^n \exp\left(\frac{\alpha zF}{RT} E\right) \tag{20}$$

With a suitable choice of the empirical parameters m and n, and provided that $k_1 \theta_k{}^0 \gg k_2 \theta_t{}^0$, Eq. (20) describes the characteristic shape of the partial anodic polarization curve of a less noble alloy component up to current densities of several milliamperes per square centimeter.

---

*A slightly modified expression is obtained if the dissolution of the less noble component from terrace sites is described by the model of a layer-by-layer two-dimensional growth of surface cavities [65].

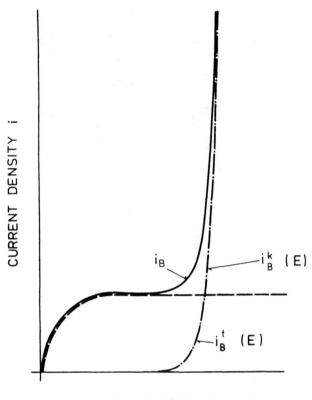

FIGURE 19 Partial anodic polarization curve for the dissolution of B from an alloy AB as obtained by the superposition of the dissolution rates of B from kink sites, i_B^k, and from terrace sites, i_B^t (schematic).

As shown schematically by Fig. 19, the main attraction of the above concept is that the existence of a critical potential E_c follows without constraint from the intercept of $i_B^k(E)$ and $i_B^t(E)$. The volume diffusion hypothesis, on the other hand, must arbitrarily assume that E_c is related to the onset of gross surface roughening. Considerable support for the surface diffusion/surface recrystallization mechanism came from a study of the micromorphology of the selective attack of nitric acid on thin-film Ag-Au alloys on a pure Au substrate by transmission electron microscopy. With this technique it was possible to detect Au-rich nuclei at the early stages of selective dissolution of

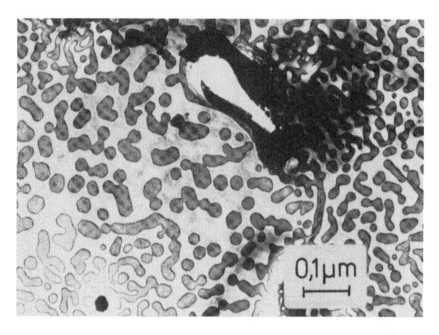

FIGURE 20 Transmission electron micrograph showing the formation
of an island and channel structure by the growth of Au-rich nuclei
(dark) on the surface of an Ag-50Au alloy at an early stage of attack
by 35% HNO_3. (From Ref. [66], with permission.)

Ag [67]. As shown by Fig. 20, these nuclei overlap to form characteristic islands that spread as corrosion proceeds. At the same time,
the channels between the islands shrink but deepen. From the simi-
larity between these findings and those for the deposition of thin films
by electrocrystallization as well as by vapor deposition [67], it was
concluded that the growth of the islands is accomplished by reorder-
ing of the alloy surface via surface diffusion. On the basis of this
result it has been proposed that, at a later stage of the dissolution
process, the selective dissolution of Ag from Ag-Au alloys will include the formation of deep pits or tunnels. This concept is illustrated by Fig. 21, showing successive stages of selective dissolution
of an Ag-Au alloy by the above mechanism. By a theoretical analysis of that model, it was concluded that there is a time dependence
of the partial current density of Ag. For atomic fractions $X_{Au} < 1/2$,

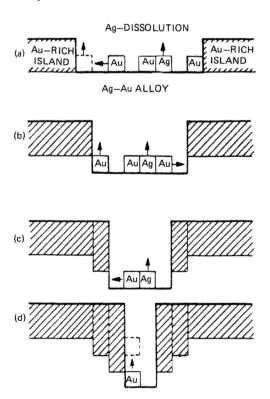

FIGURE 21 Successive stages (a-d) of island growth and pit (tunnel) formation by the selective dissolution of Ag-Au alloys via a surface diffusion model. (From Ref. [68], with permission.)

the current density transient has been assumed to be given by [68]

$$i_{Ag}(t) \propto \left(\frac{1}{t}\right)^{X_{Au}/(1 - 2X_{Au})} \tag{21}$$

The formation of a porous structure and the decrease of the partial current density of a less noble component during its selective dissolution, in other words, do not appear to require a volume diffusion mechanism. It should be noted, however, that for Ag-Au alloys, the rapid growth of islands via the surface diffusion of Au may involve the formation of a gold oxide by a very specific mechanism [69]. It is open to question, therefore, whether the above conclusions may be generalized.

REFERENCES

1. H. W. Schadler and R. F. Grace, *Trans. AIME, 215*: 559 (1959).
2. G. Kortüm, *Lehrbuch der Elektrochemie*, Verlag Chemie, Weinheim, 1966.
3. U. R. Evans, *The Corrosion and Oxidation of Metals*, Arnold, London, 1971.
4. R. H. Heidersbach, Jr., and E. D. Verink, Jr., *Corros. NACE, 28*: 397 (1972).
5. V. F. Lucey, *Br. Corros. J., 1*: 9 (1965); *1*: 53 (1965).
6. H. G. Feller, *Z. Metallkd., 58*: 875 (1967).
7. M. J. Pryor and K.-K. Giam, *J. Electrochem. Soc., 129*: 2157 (1982).
8. D. C. Epler and J. E. Castle, *Corros. NACE, 35*: 451 (1979).
9. P. Süry, thesis, Univ. of Zürich, 1970.
10. P. Süry and H. R. Oswald, *Corros. Sci., 12*: 77 (1972).
11. R. Langer, H. Kaiser, and H. Kaesche, *Werkst. Korros., 29*: 409 (1978).
12. R. G. Blundy and M. J. Pryor, *Corros. Sci., 12*: 65 (1972).
13. D. S. Kier and M. J. Pryor, *J. Electrochem. Soc., 127*: 2138 (1980).
14. Y. Hori, M. Ikawa, and T. Mukaibo, *Electrochim. Acta, 19*: 569 (1973).
15. V. V. Losev and A. P. Pchelnikov, *Itogi Nauki Tech. Serija Elektrokhim., 15*: 62 (1978).
16. H. W. Pickering and C. Wagner, *J. Electrochem. Soc., 114*: 698 (1967).
17. R. P. Tischer and H. Gerischer, *Z. Elektrochem., 62*: 50 (1958).
18. H. Stocklossa and H. Kaiser, unpublished results.
19. W. Popp, H. Kaiser, H. Kaesche, W. Brämer, and F. Sperner, *Proc. 8th Int. Contr. Met. Corros.*, Dechema, Frankfurt/Main, 1981.
20. K.-L. Schiff, N. Harmsen, and R. Schnabl, in *Proceedings of the 21st Holm Seminar "Electrical Contacts,"* Illinois Inst. of Technology, Chicago, 1975, p. 37.
21. H. Gerischer, in *Korrosion XIV (Korrosionsschutz durch Legieren*, Verlag Chemie, Weinheim, 1962.
22. I. K. Marshakov and V. Bogdanov, *Zh. Fiz. Khim., 37*: 2767 (1963).
23. W. J. Albery and M. L. Hitchman, *Ring-Disc Electrodes*, Oxford Univ. Press, London, 1971.
24. N. V. Vyazovikina, I. K. Marshakov, and N. M. Tutukina, *Elektrokhimija, 17*: 838 (1981).
25. B. Miller, *J. Clectrochem. Soc., 116*: 1117 (1969).
26. A. P. Pchelnikov, A. D. Sitnikov, I. K. Marshakov, and V. V. Losev, *Electrochim. Acta, 26*: 591 (1981).
27. Ya. M. Kolotyrkin, *Electrochim. Acta, 25*: 89 (1980).

28. H. W. Pickering and P. J. Byrne, *J. Electrochem. Soc.*, *118*: 209 (1971).
29. W. Popp and H. Kaiser, unpublished results.
30. H. Kaesche, *Die Korrosion der Metalle*, 2nd ed., Springer, Berlin, 1979, *Metallic Corrosion* (English translation by R. A. Rapp), NACE, Houston, 1985.
31. H. W. Pickering and P. J. Byrne, *J. Electrochem. Soc.*, *116*: 1492 (1969).
32. N. V. Vyazovikina and I. K. Marshakov, *Zashch. Metall.*, *15*: 656 (1979).
33. J. I. Gardiazábal and J. R. Galvele, *J. Electrochem. Soc.*, *127*: 255 (1980).
34. E. Maahn and R. Blum, *Br. Corros. J.*, *9*: 39 (1974).
35. H. W. Pickering, *Corros. Sci.*, *23*: 1107 (1983).
36. H. W. Pickering and J. Holliday, *J. Electrochem. Soc.*, *120*: 470 (1973).
37. V. V. Losev, A. P. Pchelnikov, and I. N. Marshakov, *Elektrokhimija*, *15*: 837 (1979).
38. J. O'M. Bockris, B. T. Rubin, A. Despic, and B. Lovrecek, *Electrochim. Acta*, *17*: 973 (1972).
39. A. D. Sitnikov, A. P. Pchelnikov, I. K. Marshakov, and V. V. Losev, *Zashch. Metall.*, *14*: 258 (1978).
40. H. Kaiser, unpublished results.
41. C. Leygraf, G. Hultquist, I. Olefjord, and B.-O. Elström, *Corros. Sci.*, *19*: 343 (1979).
42. Ya. M. Kolotyrkin, *Sächs. Akad. Wiss. Math. Naturwiss. Kl.*, *53*: 131 (1975).
43. I. Olefjord and B.-E. Elfstrom, *Corros. NACE*, *38*: 46 (1982).
44. R. F. Steigerwald and N. D. Greene, *J. Electrochem. Soc.*, *109*: 1026 (1962).
45. K. J. Vetter, *Electrochemical Kinetics*, Academic Press, New York, 1967.
46. T. Vitanov, A. Popov, and E. Budevski, *J. Electrochem. Soc.*, *121*: 207 (1974).
47. A. C. Riddiford, in *Advances in Electrochemistry and Electrochemical Engineering* (P. Delahay, ed.), vol. 4, Interscience, New York, 1966.
48. H. G. Feller, *Corros. Sci.*, *8*: 259 (1968).
49. H. W. Pickering, *J. Electrochem. Soc.*, *117*: 8 (1970).
50. R. Ramstetter, G. Lampert, A. Seeger, and W. Schüle, *Phys. Status Solidi*, *8*: 863 (1965).
51. H. W. Pickering and Y. S. Kim, *Corros. Sci.*, *22*: 621 (1982).
52. C. Wagner, *J. Electrochem. Soc.*, *103*: 571 (1956).
53. J. D. Harrison and C. Wagner, *Acta Met.*, *7*: 722 (1956).
54. H. W. Pickering, *J. Electrochem. Soc.*, *115*: 143 (1968).
55. H. Kaiser and H. Kaesche, *Werkst. u. Korros.*, *31*: 347 (1980).

56. M. S. Paterson, *J. Appl. Phys.*, *23*: 805 (1952).
57. C. W. Stillwell and E. S. Turnipseed, *Ind. Engng. Chem.*, *26*: 740 (1934).
58. G. Schwitzgebel, Y. Zohdi, and P. Michael, *Acta Metall.*, *23*: 1551 (1975).
59. L. Piatti and R. Grauer, *Werkst. Korros.*, *14*: 551 (1963).
60. P. R. Swann, *Corros. NACE*, *25*: 147 (1969).
61. P. R. Swann and W. R. Duff, *Metall. Trans.*, *1*: 69 (1970).
62. J. Gniewek, J. Pezy, B. G. Baker, and J. O'M. Bockris, *J. Electrochem. Soc.*, *125*: 17 (1978).
63. R. W. Revie and H. H. Uhlig, *Corros. Sci.*, *12*: 669 (1972).
64. G. Schwitzgebel, P. Michael, and J. Lang, *Werkst. Korros.*, *33*: 448 (1982).
65. H. Kaiser, unpublished results.
66. A. J. Forty and P. Durkin, *Philos. Mag. A*, *42*: 295 (1980).
67. D. W. Pashley, *Philos. Mag.*, *4*: 316, 324 (1959).
68. A. J. Forty and G. Rowlands, *Philos. Mag. A*, *43*: 171 (1981).
69. P. Durkin and A. J. Forty, *Philos. Mag. A*, *45*: 95 (1981).

3

CORROSION INHIBITORS

GIORDANO TRABANELLI

Corrosion Study Center "Aldo Dacco," University of Ferrara, Ferrara, Italy

3.1 INTRODUCTION

Inhibition is a preventive measure against corrosive attack on metallic materials. It consists of the use of chemical compounds which, when added in small concentrations to an aggressive environment, are able to decrease corrosion of the exposed metal.

By considering the electrochemical nature of corrosion processes, constituted by at least two electrochemical partial reactions, inhibition may also be defined on an electrochemical basis. Inhibitors will reduce the rates of either or both of these partial reactions (anodic oxidation and/or cathodic reduction). As a consequence we could have anodic, cathodic, and mixed inhibitors. Other tentative classifications of inhibitors have been made by taking into consideration their chemical nature (organic or inorganic substances), their characteristics (oxidizing or nonoxidizing compounds), or their technological field of application (pickling, descaling, acid cleaning, cooling water systems, etc.).

Inhibitors can be used in electrolytes at different pH values, from acid to near-neutral or alkaline solutions. Because of the very different situations created by changing various factors such as medium and inhibitor in the system metal/aggressive medium/inhibitor, various inhibition mechanisms must be considered [1–6].

An accurate analysis of the different modes of inhibiting electrode reactions including corrosion was carried out by Fischer [7]. He distinguished among various mechanisms of action, such as:

Interface inhibition
Electrolyte layer inhibition

Membrane inhibition
Passivation

Subsequently, Lorenz and Mansfeld [8] proposed a clear distinction between *interface* and *interphase* inhibition, representing two different types of retardation mechanisms of electrode reactions including corrosion. Interface inhibition presumes a strong interaction between the inhibitor and the corroding surface of the metal [1,7,9]. In this case the inhibitor adsorbs as a potential-dependent two-dimensional layer. This layer can affect the basic corrosion reactions in different ways:

By a geometric blocking effect of the electrode surface due to the adsorption of a stable inhibitor at a relatively high degree of coverage of the metal surface.

By a blocking effect of active surface sites due to the adsorption of a stable inhibitor at a relatively low degree of coverage.

By a reactive coverage of the metal surface. In this case the adsorption process is followed by electrochemical or chemical reactions of the inhibitor at the interface.

According to Lorenz and Mansfeld [8], interface inhibition occurs in corroding systems exhibiting a bare metal surface in contact with the corrosive medium. This condition is often realized for active metal dissolution in acid solutions.

Interphase inhibition presumes a three-dimensional layer between the corroding substrate and the electrolyte [7,10,11]. Such layers generally consist of weakly soluble corrosion products and/or inhibitors. Interphase inhibition is mainly observed in neutral media, with the formation of porous or nonporous layers. Clearly, the inhibition efficiency strongly depends on the properties of the formed three-dimensional layer.

In the following chapters we discuss inhibition mechanisms, which vary according to the different conditions considered.

3.2 ACID SOLUTIONS

Usually, corrosion of metals and alloys in aqueous acid solutions is very severe; nevertheless, this kind of attack can be inhibited by a large number of organic substances. These include triple-bonded hydrocarbons, acetylenic alcohols, sulfoxides, sulfides and mercaptans, aliphatic, aromatic, or heterocyclic compounds containing nitrogen, and many other families of simple organic compounds or of condensation products formed by the reaction between two different species such as aldehydes and amines.

Generally, it is assumed that the first stage in the action mechanism of the inhibitors in aggressive acid media is adsorption of the inhibitors onto the metal surface. The processes of adsorption of inhibitors are influenced by the nature and surface charge of the metal, by the chemical structure of the organic inhibitor, and by the type of aggressive electrolyte. Physical (or electrostatic) adsorption and chemisorption are the principal types of interaction between an organic inhibitor and a metal surface.

In the adsorption of organic inhibitors the water molecules adsorbed at the metal surface in contact with the aqueous solution are involved. As a consequence, the adsorption of an organic substance at the metal/ solution interface may be written [12] according to the following displacement reaction:

$$Org_{(sol)} + nH_2O_{(ads)} = Org_{(ads)} = nH_2O_{(sol)}$$

where n is the number of water molecules removed from the metal surface for each molecule of inhibitor adsorbed. According to Bockris and Swinkels [12], n is assumed to be independent of coverage or charge of the electrode.

Clearly, the value of n will depend on the cross-sectional area of the organic molecule with respect to that of the water molecule. Adsorption of the organic molecule occurs because the interaction energy between the inhibitor and the metal surface is higher than the interaction energy between the water molecules and the metal surface [5].

In the following, the various adsorption phenomena and the influencing parameters are discussed.

3.2.1 Physical Adsorption

Physical adsorption is the result of electrostatic attractive forces between inhibiting organic ions or dipoles and the electrically charged surface of the metal. The surface charge of the metal is due to the electric field at the outer Helmholtz plane of the electrical double layer existing at the metal/solution interface. The surface charge can be defined by the potential of the metal (E_{corr}) vs. its zero-charge potential (ZCP) ($E_{q=0}$) [13]. When the difference $E_{corr} - E_{q=0} = \phi$ is negative, cation adsorption is favored. Adsorption of anions is favored when ϕ becomes positive. This behavior is related not only to compounds with formal positive or negative charge, but also to dipoles whose orientation is determined by the value of the ϕ potential.

According to Antropov [13], at equal values of ϕ for different metals, similar behavior of a given inhibiting species should be expected in the same environment. This has been verified for adsorption of organic charged species on mercury and iron electrodes, at the same ϕ potential for both metals.

In studying the adsorption of ions at the metal/solution interface, it was first assumed that ions maintained their total charge during the adsorption, giving rise in this way to a pure electrostatic bond. Lorenz [14–16] suggested that a partial charge is present in the adsorption of ions; in this case a certain amount of covalent bond in the adsorption process must be considered. The partial charge concept was studied by Vetter and Schulze [17–20], who defined as *electrosorption valency* the coefficient for the potential dependence and charge flow of electrosorption processes. The term electrosorption valency was chosen because of its analogy with the *electrode reaction valency* which enters into Faraday's law as well as the Nernst equation.

Considering the concepts discussed above in relation to corrosion inhibition, when an inhibited solution contains adsorbable anions, such as halide ions, these adsorb on the metal surface by creating oriented dipoles and consequently increase the adsorption of the organic cations on the dipoles. In these cases a positive synergistic effect arises; thus, the degree of inhibition in the presence of both adsorbable anions and inhibitor cations is higher than the sum of the individual effects. This could explain the higher inhibition efficiency of various organic inhibitors in hydrochloric acid solutions compared to sulfuric acid solutions [21] (Table 1). A similar interpretation has been given [21] for the increase in inhibition by quaternary ammonium ions in sulfuric acid solution when the solution contains bromide ions (Fig. 1).

TABLE 1 Inhibition Efficiency of Some Pyridinium Derivatives at the Same Molar Concentration (1×10^{-4} M) on Armco Iron in Hydrochloric and Sulfuric Acid Solutions at 25°C [21]

Additive	Inhibition efficiency (%)	
	1 N HCl	1 N H_2SO_4
n-Decylpyridinium bromide	87.6	20.0
n-Decyl-3-hydroxypyridinium bromide	94.8	57.5
n-Decyl-3-carboxypyridinium bromide	92.7	76.5
n-Decyl-3,5-dimethylpyridinium bromide	92.5	30.2

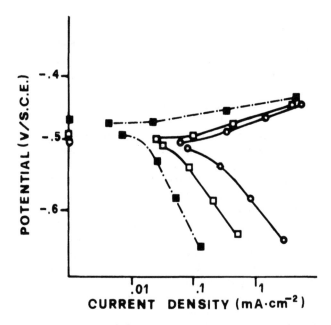

FIGURE 1 Polarization curves of Armco iron electrodes in 1 N H_2SO_4 at 25°C in the presence of 2×10^{-3} M n-decylpyridinium (DP) bromide or hydroxide [21]. o—o = 1 N H_2SO_4; □—□ = +2 × 10^{-3} M DP hydroxide; ■—■ = +2 × 10^{-3} M DP bromide.

A very detailed discussion of electrostatic adsorption has been given by Foroulis [2], who also considered the importance of structural parameters, such as hydrocarbon chain length and the nature and position of substituents in aromatic rings, in influencing the electrical charge of the organic ions, since these factors could change the degree of inhibition.

The inhibiting species whose action is to be attributed to electrostatic adsorption interact rapidly with the electrode surface, but they are also easily removed from the surface. The electrostatic adsorption process has a low activation energy, and it proves to be relatively independent of temperature [22]. On the other hand, electrostatic adsorption appears to depend on:

The electrical characteristics of the organic inhibitors
The position of the corrosion potential with respect to the zero-charge potential
The type of adsorbable anions present in the aggressive solution

3.2.2 Chemisorption

Another type of metal/inhibitor interaction is chemisorption. This process involves charge sharing or charge transfer from the inhibitor molecules to the metal surface in order to form a coordinate type of bond.

The chemisorption process takes place more slowly than electrostatic adsorption and with a higher activation energy. It depends on the temperature; higher degrees of inhibition should be expected at higher temperatures. Chemisorption is specific for certain metals and is not completely reversible [22]. The bonding occurring with electron transfer clearly depends on the nature of the metal and the nature of the organic inhibitor. In fact, electron transfer is typical for transition metals having vacant, low-energy electron orbitals. Concerning inhibitors, electron transfer can be expected with compounds having relatively loosely bound electrons. This situation may arise because of the presence in the adsorbed inhibitor of multiple bonds or aromatic rings, whose electrons have a π character. Clearly, even the presence of heteroatoms with lone-pair electrons in the adsorbed molecule will favor electron transfer. Most organic inhibitors are substances with at least one functional group regarded as the reaction center for the chemisorption process. In this case, the strength of the adsorption bond is related to the heteroatom electron density and to the functional group polarizability. For example, the inhibition efficiency of homologous series of organic substances differing only in the heteroatom is usually in the following sequence:

$$P > Se > S > N > O$$

An interpretation may be found in the easier polarizability and lower electronegativity of the elements on the left in the above sequence. On this basis, a surface bond of the Lewis acid-base type, normally with the inhibitor as electron donor and the metal as electron acceptor, has been postulated [22].

The principle of soft and hard acids and bases (SHAB) [23] has also been applied to explain adsorption bonds and inhibition effects [24]. Softness and hardness are usually associated with high or low polarizability. The SHAB principle states that hard acids prefer to coordinate with hard bases and soft acids prefer to coordinate with soft bases. Metal atoms $M°$ on oxide-free surfaces are considered soft acids which in acid solutions are able to form strong bonds with soft bases, such as sulfur-containing organic inhibitors. By comparison, nitrogen-containing or oxygen-containing organic compounds are considered hard bases and may establish weaker bonds with metal surfaces in acid solutions. From these considerations, the importance of the concepts of functional group electron density, polarizability, and electronegativity with respect to inhibition efficiency is confirmed.

The structural characteristics of the rest of the molecule influence the electron density on the heteroatom and, as a consequence, the strength of the chemisorption bond. Regular and systematic changes in the molecular structure, such as the introduction of substituents in various positions on the aromatic and heterocyclic compounds, may influence the electron density and the ability of compounds to inhibit corrosion.

Relationships between structural characteristics and inhibition efficiencies of organic compounds have been given on the basis of different parameters [1]. As an example, the ionization potential may constitute a measure of the difference in heteroatom electron density in aliphatic amines or in nitrogen-containing heterocyclic rings, assuming that for these inhibitors the electron involved in the first ionization belongs to the lone pair [25]. Higher inhibition efficiencies are to be expected corresponding to the lower values of the ionization potential.

Nuclear magnetic resonance (NMR) has proved to be useful in determinations of the electron density distribution in organic molecules. Cox et al. [26] examined the high-resolution NMR spectra of several substituted anilines, comparing the NMR frequency of amine protons with the corrosion inhibition efficiency of the various aniline derivatives. The NMR frequency of amine protons should indicate the electron density on the nitrogen atom. A linear relation between the resonance frequency of amine protons of substituted anilines and the efficiency of these compounds in inhibiting steel corrosion in hydrochloric acid was found [26]. The authors concluded that all derivatives having a greater amine electron density than anilines were better inhibitors.

The values of Hammett's constant (σ) or Taft's constant (σ^*) [27] constitute a measure of electron density at an atom. Hammett's σ is defined as the *substituent constant* in the Hammett equation concerning kinetic processes involving parent or substituted compounds. In the equation, $\log \text{th } k/k_0 = \rho \sigma$, k_0 is the rate constant of the parent compound, k the rate constant of the derivative considered, and ρ the reaction constant. In considering the reactivity of a family of organic compounds, H in the parent compound has a σ of 0.00, whereas substituents which withdraw electrons from the reaction center, by decreasing its electronic charge density, have positive values of σ. Negative values of σ are associated with electron-providing substituents, which increase the electronic charge density of the reaction center. Values of σ have frequently been used to correlate electron density and inhibiting characteristics of substituted organic additives [28--33]. Applications may be found for aromatic nitriles [34] (Fig. 2), thiophene derivatives [35], and aromatic amines.

In general, regular expected variations of the inhibition efficiencies are obtained by replacing a hydrogen atom with nucleophilic ($\sigma < 0$) or

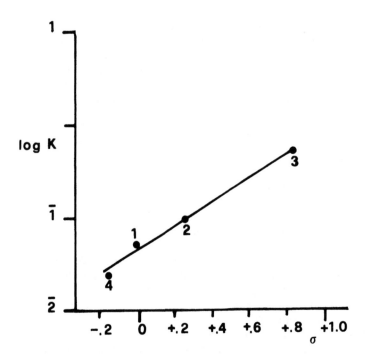

FIGURE 2 Logarithm of corrosion rate (K in mg cm^{-2} hr^{-1}) as a function of the value of Hammett's constant (σ) for different substituents of benzonitrile. Measurements of K were made in 1 N sulfuric acid at 25°C on Armco iron electrodes. Additive concentration: 1×10^{-2} M [34]. 1, Benzonitrile; 2, p-chloro; 3, p-nitro; 4, p-methyl.

electrophilic ($\sigma > 0$) substituents, which respectively increase or decrease the electron density. Nevertheless, in some cases the substitution of a hydrogen atom with either nucleophilic or electrophilic substituents leads to an increase in the inhibition efficiency. The increase of inhibition proves to be proportional to the absolute value of σ, but independent of the sign. Smialowska and Kaminsky [35] showed this behavior for thiophene derivatives, giving the following interpretation: (a) the introduction of nucleophilic substituents, by increasing the electron density and enhancing the chemical adsorption interaction, determines the ability of the compounds to efficiently inhibit corrosion; and (b) the increase of the inhibitor efficiency, even in the presence of electrophilic substituents which diminish electron density, is rationalized by the increase in the dipole moment of the substituted molecules. As a consequence, an increase in the adsorption ability of the molecule can be expected.

Relationships between electronic structure and efficiency of various classes of inhibitors have been deduced from quantum mechanical calculations [32–36]. In this way, the π-electron density, the order of the π bond along the bond line, the index of free valence and the charges of atoms in the free and adsorbed states have been calculated.

Other structural parameters influencing the inhibiting efficiency may be mentioned. Thus, the projected molecular area [25], molecular weight [37], and molecular configuration [38] of various series of organic compounds have been correlated with the variations in inhibiting efficiencies.

For example, by studying the behavior of a series of pyridine derivatives, Hackerman and co-workers [39,40] found a correlation between the inhibition and the projected areas of the compounds arranged parallel to the metallic surface. In this case the interaction would take place through the π-electron sextet of the heterocyclic ring.

The influence of molecular weight on inhibition has been evaluated by studying the behavior of homologous series of organic compounds. In work with nitriles [34], amines, or mercaptans [37] the inhibiting efficiency was shown to increase with increased length of the hydrocarbon chain (Fig. 3).

The importance of molecular configuration is clearly emphasized by examining the behavior of various organic sulfides [38]. A comparison may be made among the di-*n*-butyl, di-*s*-butyl, and di-*t*-butyl sulfides. The tertiary derivative shows no inhibition, although its sulfur atom has the highest electron density. This occurs as a result of the steric configuration of the molecule, because the sulfur is screened by the *t*-butyl groups and cannot adsorb strongly to the metal surface. As a consequence of the steric effect, the inhibition efficiency increases from tertiary to secondary and primary derivatives (Table 2).

3.2.3 Interactions Between Adsorbed Inhibitors

When the coverage of the metal surface by adsorbed inhibitor species increases, lateral interactions between inhibitor molecules may arise, influencing the inhibition efficiency.

Attractive lateral interactions usually give rise to stronger adsorption and higher inhibition efficiency. This effect has been shown in the case of compounds containing long hydrocarbon chains, due to attractive van der Waals forces [41,42]. In the presence of ions or molecules containing dipoles, repulsive interactions may occur, weakening the adsorption and diminishing the inhibiting efficiency.

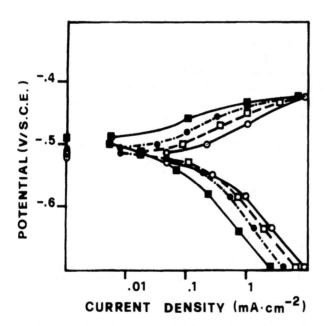

FIGURE 3 Polarization curves in 1 N sulfuric acid solutions at 25°C for Armco iron electrodes in the presence of some organic nitriles at 1×10^{-2} M concentration. C_2-CN, propionitrile; C_3-CN, butyronitrile; C_5-CN, capronitrile [34]. o—o = 1 N H_2SO_4; □—□ = 10^{-2} M C_2^-—CN; ●---● = $+10^{-2}$ M C_3-CN; ■—■ = $+10^{-2}$ M C_5-CN.

3.2.4 Relationships Between Inhibitor Reactivity and Efficiency

The nature of the inhibitor initially present in acid solutions may change with time and/or the electrode potential as a consequence of reduction reactions, polymerization reactions, or the formation of surface products. The inhibition due to the reaction products is usually called *secondary inhibition*, whereas *primary inhibition* is attributed to the compound initially added to the solution. Secondary inhibition may be higher or lower than primary inhibition, depending on the effectiveness of the reaction products [43].

An example of inhibitors undergoing electrochemical reduction is that of sulfoxides, the most important being dibenzyl sulfoxide, whose reduction gives rise to a sulfide which is more effective than the primary compound [44].

On the contrary, the reduction of thiourea and its alkyl derivatives give rise to HS^- ions, whose accelerating effect is known [45]. In some cases the reduction reaction may be followed by polymerization reactions at the metal/electrolyte interface. This mechanism of

TABLE 2 Inhibition Efficiency of Various Organic Sulfides at the Same Concentration (10^{-4} M for Armco Iron Corrosion in 1 N H_2SO_4 at 25°C) [37]

Additive	Inhibiting efficiency (%)
Di-n-butyl sulfide	70
Di-s-butyl sulfide	30
Di-t-butyl sulfide	0

action is generally accepted for acetylenic derivatives [46,47]. Electrochemical measurements on iron electrodes in sulfuric acid solutions inhibited by alkynes [48] showed that acetylenic compounds act as cathodic inhibitors, giving rise to a surface barrier phenomenon. Duwell et al. [46] found hydrogenation and dehydration reaction products in heptane extracts of acid/iron powder/ethynylcyclohexan-1-ol. According to Duwell et al., the efficiency of ethynylcyclohexan-1-ol as a corrosion inhibitor apparently depends on the properties and rates of formation of the reaction products. Putilova et al. [47] used a similar extraction technique to show that acetylene reacted at the iron/hydrochloric acid interface to form thick polymolecular films on the iron. Podobaev et al. [49], analyzing the results obtained for a large number of acetylene derivatives, concluded that these compounds act according to an adsorption-polymerization mechanism; the adsorption on iron takes place mainly via a triple bond. According to Podobaev et al. [49], the presence in the molecule of polar groups weakens the stability of the triple bond and increases the probability of adsorption and polymerization. The effect of the polar groups on the triple bond reactivity is maximum when the substituent is in position 3 relative to a triple bond.

Poling [50], using a multiple reflectance technique, recorded the infrared spectra of surface films formed on iron and steel mirrors by acetylene, 1-propyne-3-ol, and 1-ethynylcyclohexane-1-ol acid corrosion inhibitors. These spectra showed that the acetylenic molecules reacted at the metal surfaces in hydrochloric acid solutions to produce protective polymer film coatings. Corrosion protection increased markedly as the polymer film grew from nearly two-dimensional chemiadsorbed layers to films up to several hundred angstroms thick. These films are mainly composed of saturated hydrocarbon materials. Hydrogen evolved by the acid corrosion reaction probably participated in hydrogenating adsorbed acetylenic species. The polymer also contained several polar species, including hydroxyl and carbonyl

groups. The latter probably resulted from the following type of catalyzed hydration reaction:

$$HC \equiv CH + HOH \xrightarrow[\text{HCl}]{\text{Fe}} H_2C = CHOH \longrightarrow H_3C - CHO$$

Condensation products successively formed probably contribute to
polymer films having high inhibitor efficiencies.

The formation of surface products was proposed by Balezin and co-
workers [51] to explain the effectiveness of compounds such as quinoline and toluidines which are able to form organometallic-type complexes. According to these workers, the inhibiting action of the
compounds is due to a reaction between inhibitor, metallic ions, and
anions of the aggressive solution, which form a layer of reaction product on the metal surface. This surface film must be adherent, have a
very low solubility, and prevent access of the solution components to
the metal in order to guarantee optimum protection. If the complexes
are soluble, stimulation of the corrosion process may be expected.

3.3. MECHANISMS OF INHIBITOR ACTION IN ACID CORROSION PROCESSES

The effect of adsorbed inhibitor on the acid corrosion of metals is to
retard either the anodic dissolution reaction of the metal, the cathodic hydrogen evolution reaction, or both. This action may occur
by means of different mechanisms involving:

Changes in the electrical double layer
Formation of a physical barrier
Reduction of metal reactivity
Participation of the inhibitor in partial electrochemical reactions

Modern instrumental techniques [52,53] and surface analysis methods [54–57] can contribute to the understanding of the inhibition
mechanism involved in the system metal/acid/inhibitor. In particular, the recording of electrochemical characteristics, such as steady-
state corrosion potential, anodic and cathodic polarization curves, polarization resistance [58], and electrochemical impedance [52,53,59–
65] in the presence and absence of the additive, may contribute to
the identification of the prevailing mechanism of inhibitor action.

Clearly, the formulation of such different action mechanisms, as
previously mentioned, is not absolute. According to Foroulis [2], it
indicates the prevailing mechanism involved in any given instance, always keeping in mind that in most cases the adsorbed inhibitor acts to
some degree as a physical barrier.

3.3.1 Changes in the Electrical Double Layer

Corrosion inhibition related to changes in the structure of the electrical double layer at the metal/solution interface takes place as a consequence of the electrostatic adsorption of ionized inhibiting species. The modification of the electrical double layer is represented in the kinetic equations corresponding to the partial corrosion processes by the appearance of the adsorption potential jump ψ [66,67]. In the case of a corrosion process in which hydrogen evolution constitutes the cathodic reaction, the ψ value can be calculated from the equation [68,69]:

$$\psi = (\psi_2 - \psi_1) = \frac{b_0(b_a + b_c)}{b_0(b_a + b_c) - b_a b_c} \, \text{th} \, \Delta E_c$$

where ψ_2 and ψ_1 represent the potentials in the plane of closest approach to the metal surface in the presence and absence of the inhibitor, respectively, b_0 is equal to $2.3RT/F$, and b_a and b_c represent the anodic and cathodic Tafel slopes. No variation of the Tafel slopes is assumed for addition of the inhibitor [68,69]. In this case, the presence of the inhibitor determines only a reduction of the active area of the metal surface, by creating an additional energy barrier for the proton discharge. In the equation, $\Delta E_c = E_2 - E_1$ is the difference between the corrosion potential values in the presence (E_2) and absence (E_1) of the inhibitor.

Adsorption of organic cations, such as quaternary ammonium ions [69] or pyridinium ions, to the iron surface in acid solution results in a positive adsorption potential jump. This positive shift in the potential shows that the rate of the hydrogen evolution reaction is reduced. This is realized in completely deaerated acid solutions. Conversely, in aerated acid solutions in the presence of organic cations which inhibit hydrogen evolution, the contribution of the oxygen reduction reaction may become important. This is to be attributed to the fact that the diffusion current of oxygen reduction should not be influenced by the presence of selective inhibitors which retard hydrogen evolution much more effectively than the oxygen reduction reaction. Adsorption of anions on iron surfaces in acid solutions is observed to stimulate the hydrogen evolution reaction, which results in a negative adsorption potential jump. Measurements of ΔE_c as a function of the additive concentration permit calculation of the adsorption isotherms [70].

3.3.2 Formation of a Physical Barrier

Some classes of inhibitors, such as sulfoxides, acetylene derivatives, or substances with a high number of carbon atoms in the hydrocarbon

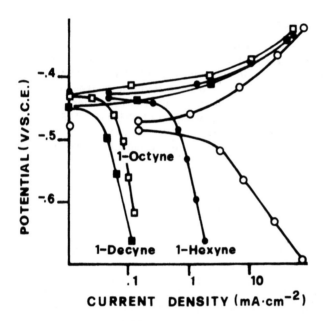

FIGURE 4 Polarization curves in 5 N sulfuric acid solutions at 25°C for Armco iron electrodes in the presence of various alkynes (saturated solutions) [48]. ○--○--○ = 5 N H_2SO_4.

chain, are able to form multimolecular layers on the metal surface. The resulting barrier action is quite independent of the nature of adsorption forces between the inhibitor molecules and the metal surface. Chemisorption bonds, π-electron interactions, hydrogen bonding, and attractive lateral interactions may be involved. The formed layer interferes with the diffusion of ions to or from the metal surface. The hindering of mass transport causes inhibition of the corrosion reaction. Analysis of the corresponding polarization curves shows concentration polarization and resistance polarization on the cathodic branches (Fig. 4).

3.3.3. Reduction of Metal Reactivity

The inhibition mechanism related to a reduction of the reactivity of the metal does not necessarily involve complete coverage of the metal surface by the adsorbed inhibitor. The type of interacting forces is important, and higher efficiencies are to be expected when stronger bonds, such as chemisorption bonds, are established.

According to this mechanism, the inhibitor adsorbs on sites active with respect to the partial electrochemical reactions. A reduction of either the anodic or the cathodic reaction or both arises from the blockage of the corresponding active sites. Clearly, the reaction rates will be reduced in proportion to the extent to which the active sites are covered by the adsorbed inhibitor. This type of surface coverage does not change the reaction mechanism, as demonstrated by the analysis of polarization curves recorded in the presence of inhibitors, such as quinoline derivatives [71]. The polarization curves are shifted toward lower current density values without modifications in the Tafel slope values.

3.3.4 Participation of the Inhibitor in Partial Electrochemical Reactions

Both the anodic reaction of metal dissolution and the cathodic reaction of hydrogen evolution proceed by steps with the formation of adsorbed intermediates on the metal surface. According to this mechanism of action, the adsorbed additive may participate in the intermediate formation, promoting either a decrease in or a stimulation of the electrode reaction depending on the stability of the adsorbed surface complex. The stimulating action will be discussed later (Section 3.4.1). When the presence of the organic compound provokes a decrease in the corrosion rate, we may hypothesize the formation of a stable surface complex containing the inhibitor.

Most research work has focused on the anodic dissolution of metals. As an example, we may consider the anodic process of iron dissolution, in which the formation of intermediates such as adsorbed (FeOH) is generally assumed [72]. In the presence of organic inhibitors (Inh) the formation of a stable chelate $[(FeOH) \cdot Inh_n]$ adsorbed to the iron surface is assumed [28,30]. In the presence of the surface complex the rate of anodic dissolution of iron is reduced by changing the reaction mechanism. As a consequence, a variation in the anodic Tafel slope is observed. The anodic Tafel slope for uninhibited zone-refined iron has been found by Donahue et al. to be 47 ± 2 mV, while in the presence of 0.3 M aniline slopes of 55 ± 3 mV were obtained. According to the authors [29], this increase in Tafel slope suggests a mode of inhibition involving an interposition of organic into the charge transfer process for the anodic reaction. This is consistent with a low surface concentration of free FeOH and the presence of adsorbed $(FeOH) \cdot Inh_n$ [29].

3.4 INHIBITION PROBLEMS IN ACID SOLUTIONS

The main problems arising from incorrect choice or use of organic inhibitors in acid solutions may be corrosion stimulation and/or hydrogen penetration into the metal.

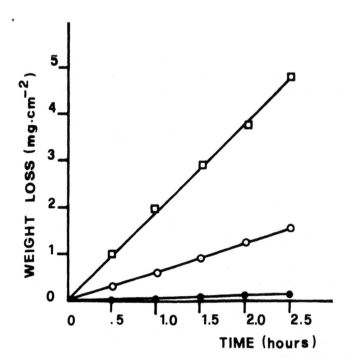

FIGURE 5 Weight loss as a function of time for Armco iron specimens
in 1 M sulfuric acid solutions in the presence of various amounts of
2-mercaptobenzoxazole (2-MBO) [45]. o—o = 1 M H_2SO_4; •—• = +10^{-3}
M 2-MBO; □—□ = +10^{-4} M 2-MBO.

3.4.1 Corrosion Acceleration in the Presence of Inhibitors

Acceleration of acid corrosion of iron often occurs at low inhibitor
concentrations. This adverse action depends on the type of acid,
as clearly shown by Pevneva et al. [73] in a study of the inhibiting
properties of bis(4-dimethylaminophenyl)antipyrilcarbinol and its derivatives. These compounds at 10^{-4} M concentration inhibited corrosion of steel in hydrochloric acid solutions, but stimulated attack in
sulfuric acid solutions, although the latter effect could be suppressed
by further addition of 10^{-4} M potassium iodide.

Generally, stimulation is not related to the type and structure of
the organic molecule. Stimulation of acid corrosion of iron was found
with mercaptans, sulfoxides, azole and triazole derivatives, nitriles,
and quinoline (Fig. 5).

Similar phenomena were also observed for nonferrous metals.
Foroulis [74] has shown that certain nitrogen-containing organic

compounds (amine type) stimulate corrosion of copper in sulfuric and perchloric acids mainly due to depolarization of the cathodic reaction. According to Foroulis [74], this behavior may be attributed to a catalytic action exerted by these substances in the form of the free amine on the copper surface.

According to Rosenfeld et al. [75], who used amines (i.e., triethylamine) as acid corrosion inhibitors for steel, special attention must be paid to the microstructure of the metal and to the fact that such substances could accelerate both the cathodic and anodic reactions, and consequently stimulate corrosion.

The causes of adverse stimulating effects [45] of the organic inhibitor may be classified as follows:

1. Stimulation caused by inhibitor decomposition products. This process seems to be basically related to the use of critical concentrations of organic inhibitors containing sulfur—thiourea and its derivatives, thiocyanates, etc.

2. Stimulation through preferential paths of partial electrochemical reactions in corrosion processes. When an additive is present in an aggressive environment which can provide a catalytic path of lowered activation energy for an electrochemical reaction, it is noticed that a stimulation of the partial process ensues, or even a stimulation of the global electrochemical process. It is principally in connection with the behavior of amines that this concept can be applied. This is based on an observation by Hoar and Khera [41] that cathodic stimulation is caused by the ability of amines to produce a catalytic path of lowered activation energy for proton discharge.

3. Stimulation caused by inhibitor participation in the metal dissolution process. According to this mechanism, the acceleration of corrosion in the presence of organic compounds is related to the oxidative propensity of the surface chelates mentioned above (Section 3.3.4). Inhibition persists until the chelate is adsorbed. If charge transfer comes about with desorption of the complex ion according to the reaction:

$$\left[(FeOH) \cdot Inh_n\right]_{ads} = \left[(FeOH) \cdot Inh_n\right]_{sol}^{+} + e^-$$

the additive will undoubtedly act as a stimulator [29].

Many other results on the stimulation of corrosion processes are reported in the literature, particularly for cases involving low additive concentrations [40,76—78]. Since, in practice, one can find situations where the inhibitor concentration might inadvertently fall below the critical value, care should be taken in the use of organic inhibitors.

3.4.2 Hydrogen Penetration into the Metal

The problem of hydrogen penetration into the metal may assume re-markable importance in the application of organic inhibitors for pick-ling, acid cleaning, etc. While any additive which covers part of the metallic surface is likely to decrease the total amount of hydrogen pro-duced, the proportion of hydrogen evolving in molecular form may be decreased even further, in which case hydrogen penetration into the metal may actually be increased.

It was shown [37,79,80] that, although some classes of organic compounds containing $=C=S$ or $\equiv C-SH$ bonds, such as thiourea and its derivatives and mercaptans, effectively act as corrosion in-hibitors for ferrous metals, they nevertheless stimulate hydrogen penetration, creating the prerequisites for embrittlement of the metal. It was assumed that such compounds are partially reduced in the cath-odic zones, thus forming hydrogen sulfide, which acts as a promoter for hydrogen penetration. In agreement with the data mentioned above, Saito and Nobe [81] showed that both hydrogen sulfide and organic compounds forming hydrogen sulfide in solution accelerate the hydrogen penetration rate in 4130 and 4330 high-strength steels in acidic solutions.

These authors also found that some acetylenic compounds may sub-stantially diminish the stimulating effect of hydrogen sulfide. Usually, in the presence of organic substances such as amines, aldehydes, ni-triles, sulfides, and sulfoxides, which do not form penetration pro-moters, the adsorption bond between organic molecules and the sur-face atoms of the metal promotes corrosion inhibition and simulta-neously considerably retards hydrogen penetration (Table 3).

The hydrogen penetration into the metal is sometimes attributed to the different modes of organic compound adsorption to the metal sur-face. Thus, Bockris and co-workers [82] showed that naphthonitrile, benzonitrile, and valeronitrile decrease the hydrogen penetration rate because of their vertical adsorption to the surface, promoting inhibi-tion of proton discharge and a decrease in the hydrogen surface cov-erage. On the other hand, the same authors showed that naphthalene increases the rate of hydrogen penetration into iron due to the inter-action between the π-electron system of naphthalene aromatic rings and the d-orbitals of iron, with an ensuing decrease in the strength of the Me-H_{ads} bond.

Results are quoted in the literature [83] which show that the ac-tion of the inhibitor against hydrogen penetration is dependent on the carbon content of the steel. Thus, some water-soluble sulfates or higher pyridine bases, which decrease the corrosion rate and hy-drogen uptake of 0.23-0.45% carbon steel in sulfuric acid solutions, have little effect on 0.66% carbon steel and stimulate both processes at higher carbon content in steel.

TABLE 3 Inhibition of Hydrogen Embrittlement by Organic
Sulfoxides, Sulfides, and Mercaptans [37] at the Same
Concentration (10^{-5} M) on Armco Iron Electrodes
(Cathodically Charged) in 1 N H_2SO_4[a]

Additive	Hydrogen embrittlement inhibition (%)
Di-*n*-butyl sulfoxide	+92
Dibenzyl sulfoxide	+100
Di-*n*-butyl sulfide	+100
Dibenzyl sulfide	+100
n-Butyl mercaptan	-24
Benzyl mercaptan	-96

[a]All the organic additives cited do inhibit overall general
corrosion of Armco iron in 1 N H_2SO_4.

Data have also been reported [84] showing an increase in ductility
of steel after pickling in HCl solution inhibited with 10^{-2} M tribenzyl-
trihydro-*sym*-triazine, suggesting a reduction in the hydrogen penetration. This effect is attributed to the improved surface conditions
(lower surface roughness and fewer stress raisers) and to the surface plasticizing action of the adsorbed inhibitor molecules. The increase in ductility appears to be of remarkable practical interest.

3.5 NEAR-NEUTRAL SOLUTIONS

Generally, inhibitors efficient in acid solutions have little or no effect
in near-neutral aqueous solutions. This specific behavior is due to
differences in the mechanism of the corrosion processes. In acid solutions the inhibitor action is due to adsorption on oxide-free metal
surfaces. In these media the main cathodic process is hydrogen evolution.

The corrosion processes of metals in contact with almost neutral
aqueous solutions result in the formation of sparingly soluble surface
products such as oxides, hydroxides, or salts; the cathodic partial
reaction is oxygen reduction. The interaction of inhibitors on oxide-
covered metal surfaces has been demonstrated [1,4-6]. In these

cases, the inhibitor action will be exerted on the oxide-covered surface by increasing or maintaining the protective characteristics of the oxide or of the surface layers in the aggressive solutions. The displacement of preadsorbed water molecules by adsorbing inhibitor molecules may be usually considered the fundamental step of inhibition. Chemical or electrochemical reactions of the inhibitor at the surface may also be assumed in order to explain the inhibitor efficiency. Because of these reactions, additional inhibitor uptake may take place.

As a result of the adsorption or adsorption-reaction of the inhibitor at the oxide-covered metal surface, there may be different inhibitor mechanisms. Thick surface layers having poor electronic-conductive properties are found in the presence of inhibitors that restrict diffusion of oxygen; these additives interfere with the oxygen reduction reaction and are referred to as cathodic inhibitors. Additives giving rise to thin passivating films usually inhibit the anodic metal dissolution reaction; as a consequence, these types of inhibitors are considered anodic inhibitors.

According to Thomas [4], the mechanism of action of both inorganic and organic inhibitive anions on the corrosion of various metals such as Fe, Al, and Zn in near-neutral solutions involves the following:

Stabilization of the passivating oxide film by reducing its dissolution rate

Repassivation of the surface due to repair of the oxide film by promoting re-formation of the oxide

Repair of the oxide film by formation of insoluble surface compounds and consequent plugging of pores

Prevention of the adsorption of aggressive anions because of the competitive adsorption of inhibitive anions

Corrosion inhibition in neutral solutions is also obtained by using oxygen scavengers, such as hydrazine and its derivatives or sodium sulfite. In this case, corrosion prevention is the result of removal of the dissolved oxygen through a chemical reaction of the oxygen with the scavenger. This effect is not an actual inhibitor action, but rather a modification of the aggressive environment.

3.5.1 Inorganic Inhibitors

Inhibition of the corrosion of metals and alloys in near-neutral aqueous solutions has been achieved in many cases by using inorganic compounds [85]. An attempt to classify inorganic inhibitors according to their mechanisms of action is the following:

1. Ca^{2+} and Mg^{2+} ions, usually present as constituents in industrial waters. Their action can be related to the precipitation, due to

the local alkalinity produced, of corresponding carbonates on metal surface sites, active for the cathodic oxygen reduction.

2. Ni^{2+}, Co^{2+}, Zn^{2+}, Fe^{2+} intentionally added to water and intended to modify surface film protective properties. Different action mechanisms may be supported. In the case of corrosion inhibition of zinc in 3% NaCl solution by the action of diluted cobalt chloride, Leidheiser and Suzuki [86] attributed the inhibiting efficiency to the introduction of Co atoms into the zinc surface oxide, which led to inhibition of cathodic oxygen reduction. On the other hand, the inhibiting efficiency of Fe^{2+} against corrosion of Cu-Ni alloys in water is attributed to the formation of a Y-FeOOH protective layer.

3. Inorganic anions such as polyphosphates, phosphates, silicates, and borates. All of these contribute to the formation and maintenance of protective films according to various mechanisms [87]. It is generally assumed that they affect the cathodic reaction by restricting dissolved oxygen diffusion to the metal surface. Some of them are reported even to affect the anodic reaction.

4. Oxidizing inhibitors such as chromates and nitrites. These compounds are useful in reducing the corrosion rate on metals and alloys with active-passive anodic behavior. They function by causing self-passivation of the metallic material. The disadvantage in adopting oxidizing inhibitors is related to the need to maintain a sufficient concentration to ensure spontaneous passivation conditions. If this "safe" condition is not achieved, "dangerous" corrosion attack may occur. Below a critical concentration oxidizers tend to increase pitting and other forms of localized attack. Therefore, oxidizing inhibitors should be used with caution in the presence of chloride ions or other ions associated with localized attacks [88].

3.5.2 Organic Inhibitors

The use of sodium salts of organic acids such as benzoate, salicylate, cinnamate, tartrate, and azelate [88] has been suggested as an alternative to the inorganic compounds mentioned above, particularly with ferrous metals. The action of these substances may be related to adsorption of the anion on the oxide surface [4] in a manner similar to that previously discussed for adsorption of inhibitors on oxide-free surfaces in acid media. Another possible action in neutral solutions is an ion-exchange process which takes place when adsorbed inhibiting anions substitute for oxide ions leaving the oxide lattice.

In some cases even complexes formed between anions and metal ions have been found in surface corrosion products. In inhibited solutions, the presence of certain anions, such as chlorides or sulfates, may accelerate the rate of film dissolution or break down the oxide film.

Consequently, the inhibitor concentration necessary for effective protection will depend on the concentration of aggressive anions [89].
This relationship seems to support the idea that some form of adsorption competition exists between inhibitive and aggressive anions. One
can hypothesize that the action of the inhibitive anion will prevail
when the surface concentration of the aggressive anion is reduced
below a critical level, as in the inhibition by aliphatic (acetate, pro-
pionate, and azelate) and aromatic (benzoate, phthalate, and cinnamate) anions [90—94]. In this case, rather than the critical concentration, the critical pH value for inhibition must be considered.
The oxygen dissolved in air-saturated, near-neutral solutions en-
sures the critical degree of oxidizing power of the solution. It has
been shown that by increasing the dissolved oxygen concentration,
both the critical pH value for inhibition and the critical concentration of inhibitive anions may be lowered.

Other inhibitor formulations for ferrous metals include organic
phosphorus-containing compounds, often in conjunction with zinc
ions. Salts of aminomethylenephosphonic acid [95], hydroxyethyl-
idenediphosphonic acid, and phosphinocarboxylic acid [96] have been
suggested. Marshall [96] has shown that the inhibition of ferrous
metal corrosion by organophosphorus compounds/zinc formulations is
due to the formation of a film on the metal surface which hinders the
diffusion of species in the vicinity of the surface and retards the rate
of both the anodic dissolution reaction and the cathodic oxygen reduction reaction. By a surface analysis technique, the presence of both
phosphorus and zinc in the film was detected. Other filming compounds, such as polyacrylate and polymethacrylate, have been tested
as scaling and corrosion inhibitors in waters [97].

Inhibition of zinc corrosion in near-neutral aqueous solutions has
been achieved with inhibitive anions such as borate and nitrocinna-
mate [98,99]. The action of these anions on zinc is quite similar to
the action on iron. As in the case of iron, the presence of dissolved
oxygen in the aqueous solution is necessary to achieve protection by
inhibitive anions.

The corrosion rate of aluminum in near-neutral solutions is reduced by inhibitive anions such as acetate or benzoate [100,101]. Because of the wide interval of stability of the aluminum oxide, the presence of dissolved oxygen in the solution is not necessary for the action of inhibitive anions.

In some cases inhibitors have been formulated for the protection of
specific metals. Examples are heterocyclic compounds [102] such as
benzotriazole and its derivatives [102—108], 2-mercaptobenzothiazole
[109,110], and 2-mercaptobenzimidazole [111,112], suggested for corrosion inhibition for copper and copper base alloys in neutral aqueous
solutions (Fig. 6). Some of the inhibitors mentioned maintain their
efficiency even in solutions polluted with sulfide ions.

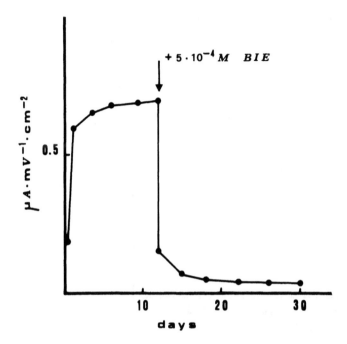

FIGURE 6 CuZn29Sn1 in flowing synthetic seawater (2.81 m sec^{-1}). Influence of the addition of 2-mercaptobenzimidazole (BIE) on the polarization conductance of the alloy [111].

Prefilming treatment with organic inhibitors has been suggested [111,113–116], since the corrosion resistance of copper and copper base alloys in neutral aqueous solutions is clearly dependent on the nature and properties of the surface films. These treatments are intended to protect the surface temporarily during transport and storage and to ensure a low corrosion rate during the initial service period of the metal by means of the formation of a regular protective surface film. Prefilming treatment of the metal surface with an inhibitor also permits reduction of the inhibitor concentration in the aggressive solution (Fig. 7).

The mechanism of action of some of these inhibitors has been studied by ellipsometry [104,115,117], ultraviolet, visible, and infrared reflection spectroscopy [118,119], X-ray photoelectron spectroscopy (XPS) [120] and Auger electron spectroscopy (AES) [121], and electrochemical techniques [102,111]. It has been assumed that in neutral media benzotriazole is chemisorbed on the copper surface [104]. Suggestions have also been made about the successive formation of Cu^{I}-

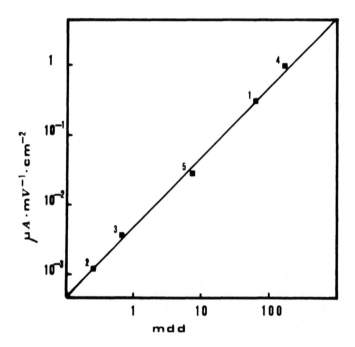

FIGURE 7 CuZn29Sn1 in flowing synthetic seawater (2.81 m sec^{-1}).
Correlation between the gravimetric corrosion rate (mdd) and the
average values of the polarization conductance. Test duration, 30
days; T, 25°C; BIE, 2-mercaptobenzimidazole [111]. 1, Seawater;
2, $+10^{-3}$ M BIE; 3, $+10^{-5}$ M BIE, BIE prefilmed surface; 4, polluted
seawater (10 ppm S^{2-}); 5, $+10^{-3}$ M BIE.

benzotriazolate on the metal surface [118]. The protective efficiency
of benzotriazole has also been attributed to the formation of a surface
polymeric complex [103]. Such a complex should be able to reinforce
the Cu$_2$O film which is usually present on the metal surface.

3.5.3 Chelating Agents as Inhibitors

Chelating agents are organic molecules which contain at least two functional polar groups able to form coordinate bonds with metal cations.
Both basic (i.e., -NH$_2$ or heterocyclic nitrogen) and acidic (i.e.,
-COOH or -SH) groups are involved in the chelation reaction. As a
consequence of the chelation reaction stable ring structures are established. The most stable structures consist of five-membered rings
containing the metal ion [122,123].

Surface-active chelating agents act as efficient corrosion inhibitors when insoluble surface chelates are formed. In contrast, the formation of soluble chelates may provoke stimulation of the corrosive attack.

Different mechanisms of action may be suggested for chelating agents as inhibitors. In some cases, the thickness and the physical characteristics of the surface layer formed in solutions inhibited by chelating agents can support a surface precipitation reaction of the chelate formed between the organic additive and the dissolved metal ion. The thick layer formed creates a physical barrier, hindering the contact between the electrolyte and the metal surface.

In the presence of a very thin surface layer, a different action mechanism can be hypothesized. Initially, chemisorption of the chelating agent can be assumed. Subsequently, this agent interacts either with metal ions still bound in the crystal lattice or with metal ions already in association with the surface oxide film.

Various surface-active chelating agents are suggested as corrosion inhibitors for different metals:

Alkyl-catechol derivatives and sarcosine derivatives [122], carboxymethylated fatty amines, and α-mercaptocarboxylic acids [123] for steel in industrial cooling systems
Azo compounds [124], cupferron, and rubeanic acid [125] for aluminum alloys
Azole derivatives and alkyl esters of thioglycolic acid for zinc [126] and galvanized steel
Oximes and quinoline derivatives for copper [127]
Cresolphthalexon and thymolphthalexon derivatives for titanium in sulfuric acid solutions [128]

3.6 ALKALINE SOLUTIONS

Information about corrosive attack on metallic materials by aqueous solutions of alkalis can be found in the literature. In particular, all metals whose hydroxides are amphoteric and metals covered by protective oxides which are broken in the presence of alkalis are subject to caustic attack. Localized attack on metals may also occur because of pitting and crevice formation.

The inhibition data reported [1,51,129] mainly concern aluminum, zinc, copper, and iron corrosion prevention in various alkaline solutions. Organic substances such as tannins, gelatin, saponin, and agar-agar are often suggested. Their action may consist of broadening the pH stability range of amphoteric oxide and hydroxide layers [130], repairing pores in oxide and hydroxide films, decreasing the rate of diffusion of reactant to the surface, and removing corrosion products from the surface [129].

Simpler substances such as thiourea, substituted phenols and naphthols, beta-diketones, 8-hydroxyquinoline, and quinalizarin are also reported to be efficient inhibitors in alkalis. To interpret the inhibiting action various mechanisms involving adsorption, increase in overvoltage, and formation of chelate compounds covering the metallic surface have been proposed.

3.7 INHIBITORS FOR TEMPORARY PROTECTION

Temporary protection against atmospheric corrosion, particularly in the case of finished metallic materials or of machinery parts, can be achieved by means of surface treatments with corrosion inhibitors dissolved in adequate solvents or in an oily phase [131–133]. At the end of the transportation or of the storage period, the protective layer can easily be removed.

Efficient protection can also be obtained by controlling the environmental aggressiveness, either by eliminating the moisture and the aggressive gases or by introducing a vapor phase inhibitor. This last proposal is obviously possible only in closed environments such as internal car parts, packing containers, and museum showcases. With respect to other methods, the treatment with vapor phase inhibitors offers the advantage of preserving the objects independently of their form and their surface state.

Among organic compounds studied as contact or vapor phase inhibitors are substances [134] belonging to the following classes: (a) aliphatic, cycloaliphatic, aromatic, and heterocyclic amines; (b) amine salts with carbonic, carbamic, acetic, benzoic, nitrous, and chromic acids; (c) organic esters; (d) nitro derivatives; and (e) acetylenic alcohols.

The inhibiting action of these substances presupposes the physical presence of their molecules or their active parts on the metallic surface. This problem does not exist for compounds sprayed on the metal surface in an oily or solvent phase. For vapor phase inhibitors whose action is related to their vapor pressure, different mechanism of transport to the metallic surface is assumed. In the case of organic compounds, such as amines, the inhibitor vaporizes in a nondissociated molecular form. Successive reaction with water and/or a dissociation reaction can take place when the inhibitor is already on the metallic surface. Such a transport mechanism was considered by Rosenfeld et al. [135] even for dicyclohexylamine nitrite; they hypothesized a nondissociative evaporation in the form of a molecular amine complex having two hydrogen bonds in the molecule. Only after reaching the metallic surface by diffusion and convection would the compound react with water, liberating the protective groups.

In contrast, a dissociative transport mechanism was proposed by Baker [136] for dicyclohexylamine nitrite. This mechanism was based on the fact that, since dicyclohexylamine nitrite is the salt of a weak volatile acid and base, hydrolysis is competitive with dissociation. As a result, there is liberation of the amine and the nitrous acid, which arrive independently at the metallic surface.

The action of the inhibitors, both the original molecules and their decomposition products, can be attributed, at least in the first stage, to adsorption [137] on the metallic surface, as shown experimentally by using labeled amine [138] (Fig. 8). This adsorption process can be of a physical as well as a chemical nature.

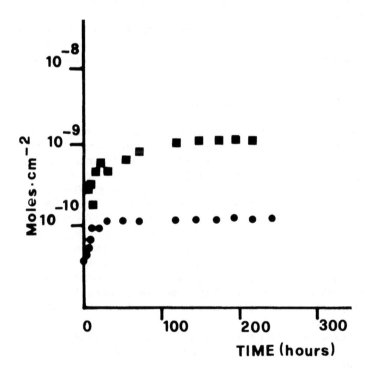

FIGURE 8 Adsorption of *n*-dodecylamine (tritium-labeled) on rusted and on bright Armco iron specimens at 25°C [138]. (Roughness factor ratio about 10:1.) ■ = Rusted Armco iron; ● = bright Armco iron.

An attempt to correlate inhibiting efficiency with molecular characteristics may be carried out in order to show whether the adsorption
phenomena mentioned above involve the electronic availability of the
atom or atoms considered as reaction centers. In the case of amines,
keeping in mind that normally it is the nitrogen atom which acts as
reaction center, the electronic charge density of this element is important. As an example, it was shown that aliphatic and cycloaliphatic amines have inhibiting efficiencies superior to those of the corresponding aromatic substances with similar vapor pressure [139].

A problem which can be linked to inhibitor structural characteristics is that of their specificity of action on different metals. Generally, compounds that are efficient in preventing atmospheric attack
on ferrous metals do not have an analogous effect on nonferrous materials, sometimes even accelerating the corrosion process, as in the
case of some amines acting on copper. Among the inhibitors able to
preserve different metals from atmospheric corrosion, Rosenfeld et
al. [135] suggested the hexamethyleneimine benzoate series and, in
particular, the 3,5-dinitrobenzoate.

The various electrochemical techniques adopted [140–143] for studying and monitoring atmospheric corrosion phenomena have also contributed to the interpretation of the mechanisms of action of the inhibitors.

3.8 INHIBITION OF LOCALIZED CORROSION

In modern industry, on the basis of weight of lost metal, general corrosion causes the greatest destruction of material. However, from the
technical point of view, uniform attack is not the most important form,
since the life span of the industrial apparatus can be predicted and
the attack can be prevented by using corrosion inhibitors. On the
other hand, an evaluation by various companies of metal failure frequency in chemical plants demonstrated that about 70% of all corrosion
failures were due to stress corrosion cracking, corrosion fatigue, pitting, and erosion-corrosion, and only 30% to general corrosion [144].

As mentioned above, corrosion inhibitors are usually able to prevent general corrosion, but their effect on localized corrosion processes is limited [145]. Generally, a higher inhibitor concentration
is required to prevent localized corrosion processes than is necessary
to inhibit general corrosion.

Staehle [146] examined the possibility of inhibiting localized attack
which does not depend on metallurgical structures. He considered
phenomena related to:

Geometric effects, such as galvanic corrosion and crevice corrosion
Simple localization of the attack, such as pitting and dezincification
Effects of relative motion, such as erosion and cavitation

The application of organic inhibitors to control galvanic corrosion presents several problems. It is well known that an inhibitive treatment efficient on a single metal, because of the specific action of the inhibitor, may fail to control corrosion if dissimilar metals are in contact. The behavior of a zinc-steel couple in sodium benzoate solution was studied by Brasher and Mercer [147]. Steel is protected in benzoate solution, but usually it corrodes when coupled to zinc, as in galvanized iron, in that solution. An interpretation of the phenomenon was given [147], emphasizing that rusting of steel coupled to zinc in benzoate solution can be prevented by bubbling air through the solution for 1 to 2 days immediately after immersion. After this time the system should have become stabilized and no corrosion of the steel should take place when the air stream is discontinued. Venczel and Wranglen [148] studied the steel-copper combination in a hot-water system. The best inhibitive efficiency was obtained by using a mixture of benzoate and nitrite.

Brunoro et al. [149] studied the behavior of steel-aluminum brass and steel-aluminum bronze couples in flowing aqueous solutions in the absence or presence of various inhibitors. Among them, benzotriazole and benzimidazole-2-thiol were found to be efficient in reducing the coupling effect.

A galvanic effect was observed [150] on coupling admiralty brass with other copper alloys (Al-brass, Al-bronze, Cu-Ni) in flowing synthetic seawater. The galvanic effect was highly reduced in the presence of 10^{-3} M benzimidazole-2-thiol. Prefilming treatment of the metallic surfaces with this compound was reliable in reducing the galvanic attack, even when the seawater contained only a low concentration (10^{-5} M) of the inhibitor.

Inhibition of automotive cooling systems [151] and, most recently, of solar energy systems [152–154] is usually achieved with complex mixtures of inorganic and organic inhibitors.

Crevice corrosion and pitting corrosion are often found in neutral solutions. Inorganic inhibitors are used to protect equipment against severe crevice corrosion [155]. Addition of organic inhibitors to the bulk solution in order to inhibit the anodic reaction within the occluded cells has also been proposed. It appears uncertain, however, whether such inhibitors will enter the crevices in sufficient amount [156]. The use of inhibited sealing materials has been suggested for filling crevices to prevent the entry of moisture. Inhibitive paint systems have been proposed by Godard [157] to protect aluminum structures from crevice corrosion.

Considerations similar to those in occluded cell chemistry may be applied to pitting corrosion and pitting corrosion inhibition [156].

Various theories of pitting corrosion, in order to explain the initiation of pit, assume the presence and accumulation of aggressive anions (e.g., Cl$^-$) on or in the passive film, creating the conditions for local

breakdown of the film. The early stages of pit formation on iron and
nickel in buffer solutions with the addition of chloride ions have been
studied by Strehblow [158]. By working with potentiostatically polarized electrodes, Strehblow and Ives [159] found a large current
density flow within corroding pits at relatively high applied anodic
potentials. This fact suggested that the pit surface is not in the active region of the polarization curve. According to Strehblow and
Ives [159], in the early stages of pit formation the potential within
the pits is nearly the same as that over the unpitted surface and the
current density within the pits is controlled by activated processes
similar to those associated with anodic polarization in the absence of
a passivating film. According to these authors [159], it is possible
that some other product film is formed at the pit surface in which the
metal ion is much faster than in a passive film. Some types of chloride-containing adsorption layers [159] or the presence of a poreless
salt film [160] could be sufficient to prevent the development of a
passive film. During the propagation of the pits the most important
reaction is the anodic dissolution of the metal in the occluded cell
within the pit. An interesting electrochemical method for the determination of pit growth kinetics has been proposed by Hunkeler and
Böhni [161]. The results obtained on aluminum showed that the pit
growth rate is time-dependent. It is also markedly influenced by the
applied potential and the chloride concentration of the electrolyte.
Limiting potential values have also been determined below which the
pit growth is not possible. Inhibition of pitting corrosion may be
achieved at both the initiation stage and the propagation stage.

It is well known [162] that pitting depends not only on the concentration of aggressive anions in the solution, but also on the concentration of nonaggressive anions. For this reason, special attention was paid to the effect of inhibitive anions in the aggressive solution. The presence of nonaggressive anions produces different effects [162]: (a) shifting the pitting potential to more positive values;
(b) increasing the induction period for pitting, and (c) reducing the
number of pits.

In studying the behavior of various anions in pitting, Strehblow
[158] found that pit nucleation on iron electrodes in phthalate buffer
containing chloride ions was prevented by picrate ions. Strehblow
and Titze [163] emphasized that the pitting potential of iron in borate buffer containing chloride ions was ennobled by the presence of
capronate. On the other hand, Vetter and Strehblow [164] showed
that sulfate ions can reduce the current density at the pit bottom
area, inhibiting the propagation stage of pitting corrosion. The minimum anion activity necessary to inhibit pitting of aluminum [165],
iron [166], and austenitic stainless steel [167] in solutions with different chloride ion activities was determined by Uhlig and co-workers.
The general equation obtained under potentiostatic conditions was

$$\text{logth } a_{Cl^-} = B\text{th logth } a_{An^-} + C$$

where a_{An^-} represents the activity coefficient of inhibiting anions such as nitrate, perchlorate, benzoate, and acetate; the constants B and C are determined experimentally. The anions mentioned, in sufficient concentration, act as pitting inhibitors, shifting the critical potential to more noble values. This behavior was interpreted in terms of competitive adsorption of chloride ions and inhibitive anions on the metal surface.

Adsorption of organic molecules may also produce inhibitive effects. Some results have been obtained in our laboratory [168] on the effect exerted by 0.2% *n*-decylamine, cyclohexylamine, and piperidine against pitting corrosion of AISI 304 stainless steel in 4 N NaCl solution boiling at 107°C. The recording of the anodic polarization curves showed an increase in pitting critical potential and a decrease in the passive current in the presence of the substances mentioned. According to Horvath et al. [169], the synergism between organic cations and previously adsorbed anions should be effective in retarding the pitting process. In research on the inhibitive effect of various amines (butylamines, benzylamine, cyclohexylamine) against pitting of AISI 304 in 0.1 N NaCl at 25°C, Horvath et al. showed that the greatest effect was produced by amines with a greater tendency to onium salt formation. These substances, by synergistic action with the chloride ions, are able to retard the initiation or propagation of pitting. They increase the pH value within the pit and may prevent the migration of aggressive anions. Grigoriev et al. [170] investigated the effect of some aniline derivatives, benzaldehyde, pyrillium salts, etc. on the pitting corrosion of 18Cr-8Ni stainless steel in sulfuric, acetic, formic, and propionic acid solutions containing sodium chloride. According to Grigoriev et al. [170], the additives tested did not affect the pitting potential values, but they reduced the rate of attack by a factor varying between 2 and 10. These results appear to demonstrate that adsorption-type organic substances might be of interest in preventing or reducing pitting corrosion.

Dezincification is the most common example of selective leaching. It is usually prevented by using less susceptible alloys. The phenomenon can also be minimized by reducing the aggressiveness of the environment with corrosion inhibitors. Kravaeva et al. [171] demonstrated that the addition of surface-active substances such as saponin, dextrin, or benzotriazole can inhibit dezincification of single-phase and diphasic brass in 0.5 N NaCl and HCl solutions. The highest inhibiting efficiency was obtained with benzotriazole. Zucchi et al. [172] investigated the effect of *o*-hydroxyquinoline, benzoinoxime, benzimidazole-2-thiol, and benzotriazole against dezincification of Cu58-Zn42 brass in 0.1 N NaCl at pH 4 at 50°C. Benzimidazole-2-thiol and

benzotriazole added to the aggressive solution or employed in prefilming treatment were very efficient in inhibiting dezincification. The
inhibitive effect was also maintained at anodic imposed potentials during accelerated tests.

In a broad review of the principles and practice of corrosion inhibition, Mercer [85] pointed out that corrosion can often be initiated or
intensified by the conjoint action of mechanical factors. Typical examples include erosion or cavitation effects. Inhibitors that are efficient in the absence of these factors may not be effective in their
presence. According to Fontana and Greene [173], the effective use
of inhibitors to decrease erosion-corrosion depends, in many cases,
on the nature and type of film formed on the metal as a result of reaction between the metal and the inhibitor. The performance of cavitation inhibitors depends on their ability to rebuild the surface layer
destroyed by the implosion of cavitation bubbles caused by sudden
pressure changes in the circulating liquid. Pini and Weber [174], in
selecting cavitation inhibitors to be applied in cooling water systems,
showed that good protection against cavitation was provided by organic inhibitors based on 2-aminopyridine. A positive synergistic effect between borate and benzothiazole-2-thiol was also found [174].
Cavitation inhibitors have been proposed to reduce attack in diesel
cooling systems. Mixtures of nitrite with benzoate and triethanol-
amine phosphate reduce the extent of the attack. In general, inhibitors have proved more valuable in service than might have been expected on the basis of cavitation laboratory tests, possibly because
they prevent, during the static period encountered in service, the
repeated heavy initial corrosion of freshly exposed cavitation-dam-
aged surfaces [175].

Research has been performed on inhibiting structurally dependent
localized forms of attack, such as intergranular corrosion or intergranular stress corrosion cracking [176]. The behavior of sensitized
AISI 304 stainless steel in dilute sulfuric or sulfamic acid solutions at
70°C, which are normally used for acid cleaning, was studied in both
the absence and presence of organic inhibitors. It was found [177]
that several organic substances containing at least one sulfur atom
with free electronic lone pairs were able to block the intergranular
attack (Table 4). If we assume that, as a consequence of sensitization treatment and chromium depletion, local galvanic cells may be established between grain boundaries and matrix, the efficiency of the
inhibitors may be related to their ability to inhibit local galvanic effects [178].

Inhibition of stress corrosion cracking (SCC) remains one of the
main objectives of scientists. New metal/environment combinations
producing stress corrosion cracking are continually being found. The
spectrum of situations involving stress corrosion cracking can only be

TABLE 4 Inhibition of Intergranular Corrosion of Sensitized AISI 304 Stainless Steel in 1 N NH_2SO_3H at 70°C [177] after 24 hr Testing

Additive	Concentration (mM/dm^3)	Corrosion morphology[a]
None	—	I.G.
Acetylenic derivative	16	I.G.
Benzonitrile	10	I.G.
Benzimidazole	1	I.G.
Benzotriazole	1	I.G.
N-containing commercial inhibitor	0.3% v/v	I.G.
Dibenzyl sulfoxide	1	No I.G.
Phenylthiourea	1	No I.G.
Benzothiazole	1	No I.G.
3-Mercaptobenzimidazole	1	No I.G.
2,5-Dihydroxy-1,4-dithian	1	No I.G.

[a]I.G., intergranular attack.

explained on the basis of various mechanisms [179,180]. Thus, different explanations were found even for the action of inhibitors capable of blocking stress corrosion cracking under very different conditions.

Additives able to control stress corrosion cracking may act in the following ways:

1. They may destroy or prevent the formation of compounds promoting stress corrosion cracking.
2. They may shift the potential of the metallic material to a safe range.
3. They may reduce the penetration of hydrogen to the metal, thus preventing failures caused by entry of the hydrogen which arises from corrosion reactions.
4. They may contribute, i.e., by adsorption, to blocking one or more of the reactions responsible for stress corrosion cracking.

TABLE 5 Inhibition of Stress Corrosion Cracking of AISI 304
Stainless Steel in $MgCl_2$ at 130°C [183][a]

Additive	Concentration (mM/dm^3)	T_f (min)
None	—	130
1-Hexylamine	2.5	574
Di-*n*-hexylamine	0.85	2,974
Dicyclohexylamine	11	2,093
1-Decylamine	3.5	13,000
1-Dodecylamine	10	1,561
Benzylamine	6.1	1,390
Piperidine	20	8,134
Acridine	1	97
8-Hydroxyquinoline	1	307
Tetrahydroquinoline	1	735
Nitroguanidine	1	1,606
N-Decyl-3-hydroxypyridinium bromide	10	3,500

[a] AISI 304 wires, 1 mm in diameter; applied stress, 10 kg mm^{-2};
T_f, time to fracture.

    The prevailing mechanism of inhibitor action depends on the nature
of the metals and the environments. Positive results in inhibiting
stress corrosion cracking have been obtained in various aggressive
systems.
    Mild steels are susceptible to SCC in nitrate, caustic, and carbon-
ate solutions. It has been demonstrated that the presence of inorganic
anions such as nitrate or chromate [179] or organic chemicals (acetate,
benzoate) can inhibit the localized attack.
    Research on SCC inhibition with austenitic stainless steels in mag-
nesium chloride solutions has been performed by Lee and Uhlig [181]
and Pinard [182]. In the presence of small concentrations of piperidine, *n*-decylamine, and other nitrogen-containing substances [183]
(Table 5), it was possible to notably lengthen the failure time of AISI

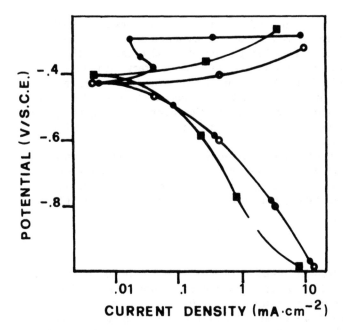

FIGURE 9 Polarization curves in boiling (130°C) $MgCl_2$ solutions for AISI 304 stressed wire electrodes in the presence of some organic additives [183]. o—o = $MgCl_2$; ●—● = +20 mM piperidine; ■—■ = +1 mM acridine.

304 wires under constant load in boiling magnesium chloride solutions. The polarization curves showed that the efficient additives exerted an inhibitive effect mainly on the anodic process (Fig. 9). The trend of the potential-time curves showed that organic additives interfere with the formation of the surface film, whereas the penetration of the cracks is slowed down by inhibition of the anodic process through competitive adsorption with the chloride ions.

The influence of some organic substances on the stress corrosion cracking of austenitic stainless steels in sulfuric acid solutions containing chlorides or in dilute hydrochloric acid solutions at room temperature has also been studied [184]. Very promising results have been obtained with benzonitrile, 2-mercaptobenzimidazole, 2-mercaptobenzothiazole, and thiourea derivatives, depending on the type of aggressive medium. Measurements have been carried out by recording potential-time curves and polarization curves and by the scratching electrode technique. The slow strain rate technique has also been applied [185]. The overall results support the idea that SCC of

austenitic stainless steels in acid chloride media occurs on the alloys
in the active state and is related to the presence of an adsorbed layer
of chloride ions. The stress brings "bare" metal areas, which are
more active than the surrounding areas, into contact with the solution. A very localized attack can thus occur with the consequent formation of a crack. In this way, only organic substances able to rapidly adsorb on the surface zones where the layer is destroyed by the
mechanical stress can block the fissuring attack by inhibiting the anodic process of metal dissolution.

In regard to corrosion fatigue, there are two principal ways to
mitigate the phenomenon: by lowering the stress amplitude or reducing the corrosivity of the medium. The latter can be achieved
by varying pH, decreasing temperature, or adding inhibitors [186].

3.9 OUTLOOK

An overview of corrosion inhibition mechanisms has been given. This
chapter does not include all cases of corrosion inhibition studied in
the laboratory or found in industrial applications. In addition, there
are always new situations created by new systems: new combinations
of metallic materials and aggressive environments lead to new trends
in research on corrosion inhibition.

Among the problems of interest from both the scientific and the
technological point of view are the following:

   Evaluation of corrosion inhibitors to be introduced in paints and
      in coating formulations. Satisfactory results in inhibition
      should be obtained by introducing as inhibitors organic sub-
      stances compatible with the coating formulation. This might
      be achieved by comparing the structure of the suggested in-
      hibitors (e.g., guanidine derivatives, heterocyclic compounds,
      benzoates) with that of the polymeric coating in order to avoid
      interferences or secondary reactions.
   Evaluation of inhibitors efficient in fighting localized corrosion pro-
      cesses such as crevice and pitting corrosion, intergranular at-
      tack and stress corrosion cracking, corrosion fatigue, and de-
      alloying. The best action is to be expected from substances
      which are able to block the corrosive attack in the incubation
      period. Only very rapidly adsorbing substances could inhibit
      localized corrosion processes during the propagation stage.
   Improvement of prefilming treatments with inhibitors on metallic
      surfaces in order to achieve temporary protection against cor-
      rosion during transport and storage. This type of treatment
      should also control the corrosion rate during the initial service

of the industrial apparatus, such as condensers or heat exchangers. Prefilming treatments should also offer the possibility of maintaining a long-term satisfactory corrosion inhibition in service by adding very low inhibitor concentrations in the aggressive solution.

Extension of research and experience in corrosion inhibition to surfaces other than "bare" metal surfaces. In several cases corrosion inhibitors become effective through the interaction with surface layers or with corrosion products (e.g., surface chelate or salt layer formation, stabilization of oxide films by inhibitors, interface vs. interphase inhibitors). This corresponds to broadening the concept of "adsorption inhibition" by including the formation of new phases, whose contribution in retarding corrosion processes becomes fundamental.

As in the past, electrochemical techniques should continue to contribute to the interpretation of corrosion and inhibition mechanisms. Nevertheless, new comtributions to the understanding of inhibition phenomena will certainly be made by the new methods of surface analysis.

ACKNOWLEDGMENT

The author thanks A. Frignani and G. Brunoro for their assistance with this manuscript.

REFERENCES

1. G. Trabanelli and V. Carassiti, in *Advances in Corrosion Science and Technology* (M. G. Fontana and R. W. Staehle, eds.), vol. 1, Plenum, New York, 1970, p. 147.
2. Z. A. Foroulis, in *Proceedings of Symposium on Basic and Applied Corrosion Research*, National Association of Corrosion Engineers, Houston, 1969.
3. O. L. Riggs, Jr., in *Corrosion Inhibitors* (C. C. Nathan, ed.), National Association of Corrosion Engineers, Houston, 1973, p. 7.
4. J. G. Thomas, in *Corrosion* (L. L. Shreir, ed.), vol. 2, Newnes-Butterworths, London, 1976, p. 18.3.
5. E. McCafferty, in *Corrosion Control by Coatings* (H. Leidheiser, Jr., ed.), Science Press, Princeton, 1979, p. 279.
6. I. L. Rosenfeld, in *Corrosion Inhibitors*, McGraw-Hill, New York, 1981.

7. H. Fischer, *Werkst. Korros.*, *23*: 445 (1972).
8. W. J. Lorenz and F. Mansfeld, *International Conference on Corrosion Inhibition*, National Association of Corrosion Engineers, Dallas, 1983, paper 2.
9. W. J. Lorenz and F. Mansfeld, 31st ISE Meeting, Venice, 1980.
10. P. Lorbeer and W. J. Lorenz, *Electrochim. Acta*, *25*: 375 (1980).
11. R. H. Hausler, *International Conference on Corrosion Inhibition*, National Association of Corrosion Engineers, Dallas, 1983, paper 19.
12. J. O'M. Bockris and D. A. J. Swinkels, *J. Electrochem. Soc.*, *111*: 736 (1964).
13. L. I. Antropov, in *First International Congress of Metallic Corrosion*, Butterworths, London, 1962, p. 147.
14. W. Lorenz, *Z. Phys. Chem.*, *219*: 421 (1962).
15. W. Lorenz, *Z. Phys. Chem.*, *224*: 145 (1963).
16. W. Lorenz, *Z. Phys. Chem.*, *244*: 65 (1970).
17. K. J. Vetter and J. W. Schulze, *Ber. Bunsenges Phys. Chem.*, *76*: 920, 927 (1972).
18. G. W. Schulze and K. J. Vetter, *J. Electroanal. Chem. Interfacial Electrochem.*, *44*: 63 (1973).
19. K. J. Vetter and J. W. Schulze, *J. Electroanal. Chem. Interfacial Electrochem.*, *53*: 67 (1974).
20. J. W. Schulze and K. D. Koppitz, *Electrochim. Acta*, *21*: 327, 337 (1976).
21. A. Frignani, G. Trabanelli, F. Zucchi, and M. Zucchini, *Proc. 5th Eur. Symp. Corros. Inhibitors*, Ann. Univ. Ferrara, N.S., Sez. V, Suppl. 7, 1185 (1980).
22. N. Hackerman and R. M. Hurd, in *First International Congress on Metallic Corrosion*, Butterworths, London, 1962, p. 166.
23. R. G. Pearson, *J. Am. Chem. Soc.*, *85*: 3533 (1963); *Science*, *151*: 172 (1966).
24. L. Horner, *Chem. Ztg.*, *100*: 247 (1976).
25. R. C. Ayers, Jr., and N. Hackerman, *J. Electrochem. Soc.*, *110*: 507 (1963).
26. P. F. Cox, R. L. Every, and O. L. Riggs, Jr., *Corrosion*, *20*: 299t (1964).
27. L. P. Hammett, *Chem. Rev.*, *17*(1): 125 (1935).
28. F. M. Donahue and K. Nobe, *J. Electrochem. Soc.*, *112*: 886 (1965).
29. F. M. Donahue, A. Akiyama, and K. Nobe, *J. Electrochem. Soc.*, *114*: 1006 (1967).
30. F. M. Donahue and K. Nobe, *J. Electrochem. Soc.*, *114*: 1012 (1967).
31. A. Akiyama and K. Nobe, *J. Electrochem. Soc.*, *117*: 999 (1970).
32. A. I. Altsybeeva, S. Z. Levin, and A. P. Dorokhov, *Proc. 3rd Eur. Symp. Corros. Inhibitors*, Ann. Univ. Ferrara, N.S., Sez. V, Suppl. 5, 501 (1971).

33. I. L. Rosenfeld, Yu. I. Kuznetsov, I. Ya. Kerbeleva, V. M. Brusnikina, B. V. Bochorov, and A. A. Lyashenko, *Prot. Met.*, *14*: 495 (1978).
34. V. Carassiti, F. Zucchi, and G. Trabanelli, *Proc. 3rd Eur. Symp. Corros. Inhibitors*, Ann. Univ. Ferrara, N.S., Sez. V, Suppl. 5, 525 (1971).
35. Z. Szklarska-Smialowska and M. Kaminsky, in *Proceedings of the Fifth International Congress on Metallic Corrosion*, National Association of Corrosion Engineers, Houston, 1974, p. 555.
36. B. M. Larkin and I. L. Rosenfeld, *Prot. Met.*, *12*: 235 (1976).
37. G. Trabanelli and F. Zucchi, *Rev. Coat. Corros.*, *1*(2): 97 (1972).
38. F. Zucchi, G. L. Zucchini, G. Trabanelli, and V. Carassiti, *Br. Corros. J.*, *4*: 267 (1969).
39. R. R. Annand, R. M. Hurd, and N. Hackerman, *J. Electrochem. Soc.*, *112*: 138 (1965).
40. R. R. Annand, R. M. Hurd, and N. Hackerman, *J. Electrochem. Soc.*, *112*: 144 (1965).
41. T. P. Hoar and R. P. Khera, *Proc. 1st Eur. Symp. Corros. Inhibitors*, Ann. Univ. Ferrara, N.S., Sez. V, Suppl. 3, 73 (1961).
42. N. Hackerman, D. D. Justice, and E. McCafferty, *Corrosion*, *31*: 240 (1975).
43. W. J. Lorenz and H. Fischer, *Proc. 3rd Int. Congr. Met. Corros.*, *2*: 99 (1969).
44. G. Trabanelli, F. Zucchi, G. L. Zucchini, and V. Carassiti, *Electrochim. Met.*, *2*: 463 (1967).
45. A. Frignani, G. Trabanelli, F. Zucchi, and M. Zucchini, *Proc. 4th Eur. Symp. Corros. Inhibitors*, Ann. Univ. Ferrara, N.S., Sez. V, Suppl. 6, 652 (1975).
46. E. J. Duwell, J. W. Todd, and H. C. Butzke, *Corros. Sci.*, *4*: 435 (1964).
47. I. N. Putilova, N. V. Rudenko, and A. N. Terentev, *Russ. J. Phys. Chem.*, *38*: 263 (1964).
48. F. Zucchi, G. L. Zucchini, and G. Trabanelli, *Proc. 3rd Eur. Symp. Corros. Inhibitors*, Ann. Univ. Ferrara, N.S., Sez. V, Suppl. 5, 415 (1970).
49. I. I. Podobaev, A. G. Voskresenkii, and G. F. Semikolenkov, *Prot. Met.*, *3*: 88 (1967).
50. G. W. Poling, *J. Electrochem. Soc.*, *114*: 1209 (1967).
51. I. N. Putilova, S. A. Balezin, and V. P. Barrannik, *Metallic Corrosion Inhibitors*, Pergamon, London, 1960.
52. W. J. Lorenz and F. Mansfeld, *Corros. Sci.*, *21*: 647 (1981).
53. A. A. Aksüt, W. J. Lorenz, and F. Mansfeld, *Corros. Sci.*, *22*: 611 (1982).
54. N. Sato, *Metall.*, *33*: 1039 (1979).

55. O. Hollander, G. E. Geiger, and W. C. Ehrhardt, *Corrosion/82*, National Association of Corrosion Engineers, Houston, 1982, paper 226.

56. P. F. Lynch, C. W. Brown, and R. Heidersbach, *Corrosion/82*, National Association of Corrosion Engineers, Houston, 1982, paper 21.

57. D. T. Larson, *Corros. Sci.*, *19*: 657 (1979).

58. F. Mansfeld, in *Advances in Corrosion Science and Technology* (M. G. Fontana and R. W. Staehle, eds.), vol. 6, Plenum, New York, 1976, p. 163.

59. A. Caprani, I. Epelboin, Ph. Morel, and H. Takenouti, *Proc. 4th Eur. Symp. Corros. Inhibitors*, Ann. Univ. Ferrara, N.S., Sez. V, Suppl. 6, 517 (1975).

60. I. Epelboin, K. Keddam, and H. Takenouti, *J. Appl. Electrochem.*, *2*: 71 (1972).

61. F. Mansfeld, S. Tsai, W. J. Lorenz, and H. Meckel, *Corrosion/81*, National Association of Corrosion Engineers, Houston, 1981, paper 250.

62. M. Duprat, F. Dabosi, F. Moran, and S. Rocher, in *Proceedings 8th International Congress on Metallic Corrosion*, Dechema, Mainz, 1981, vol. II, p. 1218.

63. M. Duprat, F. Dabosi, and F. Moran, *Corros. Sci.*, *23*: 1047 (1983).

64. A. Bonnel, F. Dabosi, C. Deslouis, M. Duprat, K. Keddam, and B. Tribollet, *J. Electrochem. Soc.*, *130*: 753 (1983).

65. F. Dabosi, C. Deslouis, M. Duprat, and K. Keddam, *J. Electrochem. Soc.*, *130*: 761 (1983).

66. L. I. Antropov, *Prot. Met.*, *2*: 235 (1966).

67. L. I. Antropov, *Corros. Sci.*, *7*: 607 (1967).

68. L. I. Antropov, I. S. Pogrebova, and G. I. Dremova, *Prot. Met.*, *8*: 105 (1972).

69. L. I. Antropov, I. S. Pogrebova, and G. I. Dremova, *Prot. Met.*, *7*: 1 (1971).

70. N. Hackerman and I. D. Sudbury, *Trans. Electrochem. Soc.*, *93*: 191 (1948).

71. T. P. Hoar and R. D. Holliday, *J. Appl. Chem.*, *3*: 502 (1953).

72. W. J. Lorenz, G. E. Eichkorn, G. Albert, and H. Fischer, *Electrochim. Acta*, *13*: 183 (1968).

73. A. V. Pevneva, V. V. Kuznetsov, E. A. Selezneva, and V. P. Zhivopistsev, *Prot. Met.*, *13*: 381 (1977).

74. Z. A. Foroulis, *Proc. 2nd Eur. Symp. Corros. Inhibitors*, Ann. Univ. Ferrara, N.S., Sez. V, Suppl. 4, 285 (1966).

75. I. L. Rosenfeld, Yu. I. Kuznetsov, and A. V. Belov, *Prot. Met.*, *13*: 375 (1977).

76. G. Davolio and E. Soragni, *Proc. 3rd Eur. Symp. Corros. Inhibitors*, Ann. Univ. Ferrara, N.S., Sez. V, Suppl. 5, 219 (1971).

77. I. N. Putilova, *Proc. 2nd Eur. Symp. Corros. Inhibitors*, Ann. Univ. Ferrara, N.S., Sez. V, Suppl. 4, 139 (1966).
78. N. Hackerman, R. M. Hurd, and R. R. Annand, *Corrosion*, *18*: 37t (1962).
79. L. Cavallaro, G. P. Bolognesi, and L. Felloni, *Werkst. Korros.*, *10*: 81 (1959).
80. G. Trabanelli, F. Zucchi, G. Gullini, and V. Carassiti, *Werkst. Korros.*, *20*: 1012 (1969).
81. Y. Saito and K. Nobe, *Electrochemical Society Meeting 77*, Atlanta paper 121.
82. J. O'M. Bockris, J. McBreen, and L. Nanis, *J. Electrochem. Soc.*, *112*: 1025 (1965).
83. A. S. Afanasev, R. A. Eremeeva, and S. G. Tyr, *Prot. Met.*, *13*: 382 (1977).
84. E. S. Ivanov, *Prot. Met.*, *13*: 384 (1977).
85. A. D. Mercer, in *Corrosion* (L. L.. Shreir, ed.), vol. 2, Newnes-Butterworths, London, 1976, p. 18.2.
86. H. Leidheiser, Jr., and I. Suzuki, *J. Electrochem. Soc.*, *128*: 242 (1981).
87. J. G. Thomas, *Proc. 5th Eur. Symp. Corros. Inhibitors*, Ann. Univ. Ferrara, N.S., Sez. V, Suppl. 7, 453 (1980).
88. U. R. Evans, in *The Corrosion and Oxidation of Metals*, Edward Arnold, London, 1960; 1st supplementary volume, 1968; 2nd supplementary volume, 1976.
89. D. M. Brasher and A. D. Mercer, *Br. Corros. J.*, *3*: 120, 130, 136, 144 (1968).
90. D. E. Davies and Q. J. M. Slaiman, *Corros. Sci.*, *11*: 671 (1971).
91. M. Fischer, *Z. Phys. Chem. (Leipzig)*, *258*: 897, 987 (1977).
92. M. Fischer, *Z. Phys. Chem. (Leipzig)*, *260*: 121 (1979).
93. J. E. O. Mayne and C. L. Page, *Br. Corros. J.*, *10*: 99 (1975).
94. C. L. Page and J. E. O. Mayne, *Corros. Sci.*, *12*: 679 (1972).
95. G. B. Hatch and P. H. Ralston, *Mater. Prot. Perform.*, *11*(1): 39 (1972).
96. A. Marshall, *Corrosion/81*, National Association of Corrosion Engineers, Houston, 1981, paper 192.
97. T. C. Breske, *Mater. Perform.*, *16*(2): 17 (1977).
98. P. Hersch, J. B. Hare, A. Robertson, and S. M. Sutherland, *J. Appl. Chem.*, *11*: 251, 265 (1961).
99. P. Wormwell, *Chem. Ind.*, 556 (1953).
100. K. F. Lorking and J. E. O. Mayne, *J. Appl. Chem.*, *11*: 170 (1961).
101. H. Böhni and H. H. Uhlig, *J. Electrochem. Soc.*, *116*: 906 (1969).
102. G. Trabanelli, F. Zucchi, G. Brunoro, and V. Carassiti, *Werkst. Korros.*, *24*: 602 (1973).
103. J. B. Cotton and R. B. Scholes, *Br. Corros. J.*, *2*: 1 (1967).

104. F. Mansfeld, T. Smith, and E. P. Parry, *Corrosion*, 27: 289 (1971).
105. F. Mansfeld and T. Smith, *Corrosion*, 29: 105 (1973).
106. R. Walker, *Corrosion*, 31: 97 (1975); 32: 339, 414 (1976).
107. T. Notoya and J. W. Poling, *Denki Kagaku*, 47(10): 592 (1979).
108. P. G. Fox, G. Lewis, and P. J. Boden, *Corros. Sci.*, 19: 457 (1979).
109. J. I. Bregman, in *Corrosion Inhibitors*, Macmillan, New York, 1963.
110. T. Notoya, *Corros. Eng. Jpn.*, 27: 661 (1978).
111. F. Zucchi, G. Brunoro, and G. Trabanelli, *Werkst. Korros.*, 28: 834 (1977).
112. F. Zucchi, G. Brunoro, and M. Zucchini, *Mater. Chem.*, 3: 91 (1978).
113. T. Notoya and J. W. Poling, *Corrosion*, 35: 193 (1979).
114. R. Cigna and G. Gusmano, *Br. Corros. J.*, 14: 223 (1979).
115. G. Brunoro, F. Zucchi, M. Maja, and P. Spinelli, *Proc. 5th Eur. Symp. Corros. Inhibitors*, Ann. Univ. Ferrara, N.S., Sez. V, Suppl. 7, 977 (1980).
116. R. Spinelli, M. Maja, G. Brunoro, and G. Trabanelli, *Werkst. Korros.*, 31: 918 (1980).
117. N. D. Hobbins and R. F. Roberts, *Surf. Technol.*, 9:235 (1979).
118. G. W. Poling, *Corros. Sci.*, 10: 359 (1970).
119. M. Ohsawa and W. Suëtaka, *Corros. Sci.*, 19: 709 (1979).
120. N. Morito and W. Suëtaka, *J. Jpn. Inst. Met.*, 37: 216 (1973).
121. G. Lewis, *Corrosion*, 38: 119 (1982).
122. A. Weisstuch, D. A. Carter, and C. C. Nathan, *Mater. Prot. Perform.*, 10(4): 11 (1971).
123. D. C. Zecher, *Mater. Perform.*, 15(4): 33 (1976).
124. F. Tirbonod and C. Fiaud, *Corros. Sci.*, 18: 139 (1978).
125. B. W. Samuels, K. Sotoudeh, and R. T. Foley, *Corrosion*, 37: 93 (1981).
126. R. L. Leroy, *Corrosion*, 34: 98 (1978).
127. C. Brunoro, G. Trabanelli, and F. Zucchi, *Proc. 4th Eur. Symp. Corros. Inhibitors*, Ann. Univ. Ferrara, N.S., Sez. V, Suppl. 6, 443 (1975).
128. G. Schmitt, *Br. Corrosion J.*, 19: 165 (1984).
129. A. H. Roebuck, in *Corrosion Inhibitors* (C. C. Nathan, ed.), National Association of Corrosion Engineers, Houston, 1973.
130. W. Machu, *Proc. 2nd Eur. Symp. Corros. Inhibitors*, Ann. Univ. Ferrara, N.S., Sez. V, Suppl. 4, 153 (1966).
131. B. A. Miksic and R. H. Miller, *Proc. 5th Eur. Symp. Corros. Inhibitors*, Ann. Univ. Ferrara, N.S., Sez. V, Suppl. 7, 217 (1980).

132. C. Fiaud and G. Maurin, *Proc. 5th Eur. Symp. Corros. Inhibitors*, Ann. Univ. Ferrara, N.S., Sez. V, Suppl. 7, 1223 (1980).
133. I. L. Rosenfeld, in *Atmospheric Corrosion of Metals* (E. C. Greco, ed.), National Association of Corrosion Engineers, Houston, 1972.
134. G. Trabanelli and F. Zucchi, in *Proceedings Corrosion Week '74* (L. Prockl, ed.), OMKDK-Technoinform, Budapest, 1974, p. 289.
135. I. L. Rosenfeld, V. P. Persiantseva, and P. B. Terentief, *Corrosion*, 20: 222t (1964).
136. H. R. Baker, *Ind. Eng. Chem.*, 46: 2592 (1954).
137. K. Schwabe, *Dechema Monogr.* 45: 273 (1962).
138. E. Fiegna, G. Gilli, G. Trabanelli, and G. L. Zucchini, *La Tribune du Cebedeau*, no. 300 (1968).
139. G. Trabanelli, A. Fiegna, and V. Carassiti, *Trib. Cebedeau*, no. 288 (1967).
140. B. Mazza, P. Pedeferri, G. Re, and D. Sinigaglia, *Proc. 4th Eur. Symp. Corros. Inhibitors*, Ann. Univ. Ferrara, N.S., Sez. V, Suppl. 6, 552 (1975).
141. F. Mansfeld, *Proc. 5th Eur. Symp. Corros. Inhibitors*, Ann. Univ. Ferrara, N.S., Sez. V, Suppl. 7, 191 (1980).
142. F. Mansfeld and T. Tsai, *Corros. Sci.*, 20: 853 (1980).
143. R. Cigna, G. Gusmano, M. Marabelli, and S. Massa, *Proc. 5th Eur. Symp. Corros. Inhibitors*, Ann. Univ. Ferrara, N.S., Sez. V, Suppl. 7, 41 (1980).
144. J. A. Collins and M. L. Monack, *Mater. Prot. Perform.*, 12(6): 11 (1973).
145. G. Trabanelli, in *Proceedings 7th International Congress on Metallic Corrosion*, ABRACO, Rio de Janeiro, 1978, p. 83.
146. R. W. Staehle, *Proc. 4th Eur. Symp. Corros. Inhibitors*, Ann. Univ. Ferrara, N.S., Sez. V, Suppl. 6, 709 (1975).
147. D. M. Brasher and A. D. Mercer, in *Proceedings 3rd International Congress on Metallic Corrosion*, MIR, Moscow, 1969, vol. 2, p. 21.
148. J. Venczel and G. Wranglen, *Corros. Sci.*, 7: 461 (1967).
149. G. Brunoro, F. Zucchi, and M. Zucchini, *Mater. Chem.*, 5: 135 (1980).
150. F. Zucchi, G. Brunoro, and M. Zucchini, *Mater. Chem.*, 3: 91 (1978).
151. L. C. Rowe, in *Corrosion Inhibitors* (C. C. Nathan, ed.), National Association of Corrosion Engineers, Houston, 1973, p. 173.
152. P. D. Thompson and M. B. Hayden, *Corrosion/77*, paper 63, National Association of Corrosion Engineers, 1977.
153. E. Beynon, *Corrosion/77*, paper 171, National Association of Corrosion Engineers, 1977.

154. J. L. Popplewell, *Corrosion/77*, paper 170, National Association of Corrosion Engineers, 1977.
155. I. L. Rosenfeld, in *Localized Corrosion*, National Association of Corrosion Engineers, Houston, 1974, p. 394.
156. M. Pourbaix, *Rapport Technique*, no. 230, Cebelcor, Bruxelles, 1975.
157. H. P. Godard, in *The Corrosion of Light Metals*, Wiley, New York, 1967, p. 45.
158. H. H. Strehblow, *Werkst. Korros.*, *27*: 792 (1976).
159. H. H. Strehblow and M. B. Ives, *Corros. Sci.*, *16*: 317 (1976).
160. K. J. Vetter and H. H. Strehblow, in *Localized Corrosion*, National Association of Corrosion Engineers, Houston, 1974, p. 240.
161. F. Hunkeler and H. Böhni, *Werkst. Korros.*, *32*: 129 (1981).
162. S. Szklarska-Smialowska, in *Localized Corrosion*, National Association of Corrosion Engineers, Houston, 1974, p.. 329.
163. H. H. Strehblow and B. Titze, *Corros. Sci.*, *17*: 461 (1977).
164. K. J. Vetter and H. H. Strehblow, *Ber. Bunsenges. Phys. Chem.*, *74*: 1024 (1970).
165. H. Böhni and H. H. Uhlig, *J. Electrochem. Soc.*, *116*: 906 (1969).
166. S. Matsuda and H. H. Uhlig, *J. Electrochem. Soc.*, *111*: 156 (1964).
167. M. Leckie and H. H. Uhlig, *J. Electrochem. Soc.*, *113*: 1262 (1966).
168. F. Zucchi, private communication.
169. J. Horvath, T. M. Salem, B. Abd El-Nabey, El-Sayed Khalil, L. Hackl, and A. Rauscher, *Proc. 4th Eur. Symp. Corros. Inhibitors*, Ann. Univ. Ferrara, N.S., Sez. V, Suppl. 6, 743 (1975).
170. V. P. Grigoriev, V. V. Ekilik, G. N. Ekilik, and T. B. Kudryashova, *Proc. 4th Eur. Symp. Corros. Inhibitors*, Ann. Univ. Ferrara, N.S., Sez. V, Suppl. 6, 369 (1975).
171. A. P. Kravaeva, I. P. Marshakov, and S. M. Mel'nik, *Prot. Met.*, *4*: 191 (1968).
172. F. Zucchi, G. Brunoro, and G. Trabanelli, *Metall.. Ital.*, *69*: 493 (1977).
173. M. G. Fontana and N. D. Greene, in *Corrosion Engineering*, McGraw-Hill, New York, 1978.
174. G. Pini and J. Weber, *Proc. 4th Eur. Symp. Corros. Inhibitors*, Ann. Univ. Ferrara, N.S., Sez. V, Suppl. 6, 110 (1975).
175. D. J. Godfrey, in *Corrosion* (L. L. Shreir, ed.), vol. 1, Newnes, London, 1977, chap. 8.8.
176. C. S. O'Dell, B. F. Brown, and R. T. Foley, *Corrosion*, *36*: 183 (1980).

177. G. Trabanelli, A. Frignani, F. Zucchi, and M. Zucchini, *Z. Phys. Chem.*, *264*(4): 813 (1983).
178. R. N. Parkins, *Proc. 4th Eur. Symp. Corros. Inhibitors*, Ann. Univ. Ferrara, N.S., Sez. V, Suppl. 6, 595 (1975).
179. R. N. Parkins, in *Comprehensive Treatise of Electrochemistry* (J. O'M. Bockris, B. E. Conway, E. Yeager, and R. E. White, eds.), vol. 4, Plenum, New York, 1981, p. 307.
180. C. S. O'Dell and B. F. Brown, in *Corrosion Control by Coatings* (H. Leidheiser, Jr., ed.), Science Press, Princeton, 1979, p. 339.
181. H. H. Lee and H. Uhlig, *J. Electrochem. Soc.*, *117*: 18 (1970).
182. L. Pinard, *Mem. Sci. Rev. Metall.*, *69*: 425 (1972).
183. F. Zucchi, A. Frignani, M. Zucchini, and G. Trabanelli, *Ann. Chim. Rome*, *68*: 15 (1978); *Corros. Sci.*, *20*: 791 (1980).
184. F. Zucchi, G. Trabanelli, A. Frignani, and M. Zucchini, *Corros. Sci.*, *18*: 87 (1978).
185. G. Trabanelli, A. Frignani, M. Zucchini, and F. Zucchi, in *Proceedings 9th International Congress on Metallic Corrosion*, National Research Council of Canada, 1984, vol. 2, p. 230.
186. H. Spähn, in *Corrosion Fatigue*, National Association of Corrosion Engineers, Houston, 1972, p. 40.

4

COATINGS

HENRY LEIDHEISER, JR.

Sinclair Laboratory No. 7, Lehigh University, Bethlehem, Pennsylvania

4.1 INTRODUCTION

Corrosion protection by coatings is a growing area of science and technology. This method of protection allows one to design a substrate
with the desired physical and mechanical properties and to utilize a
coating that is resistant to the environment to which the part is to be
exposed. One thus combines the best of two worlds. This chapter
will concentrate on protective coatings largely utilized at temperatures below 200°C. High-temperature protective coatings will not be
discussed since this subject will be treated in Chapter 9.

Coatings are broken down for convenience into four different types:
metallic coatings, polymeric coatings, conversion coatings, and cementitious coatings. Each of these coating types will be discussed separately since the principles governing the performance of each type are
different. The unusual inorganic coatings, such as silicon nitride,
used in the semiconductor device industry will not be discussed.

4.2 METALLIC COATINGS

Many methods are available for forming a coating of a corrosion-resistant metal on a corrosion-prone substrate metal. Brief descriptions
will be given of these different methods.

Cathodic sputtering: A partial vacuum is required. The part to
be coated is attached to the anode and an inert gas such as argon at
a low pressure is admitted to the system. A discharge is initiated and
the positively charged gas ions are attracted to the cathode. The col-
lision of a gas ion dislodges atoms from the cathode, which are in turn

attracted to the anode and coat the part. One of the major advantages of cathode sputtering is that nonconducting as well as conducting substrates can be coated. Major disadvantages include the heating of the substrate and low deposition rates.

Diffusion coating: This method requires a preliminary coating step followed by thermal treatment and diffusion of the coating metal into the substrate. A commercial material known as galvannealed steel is made by coating steel with zinc followed by heat treatment and the formation of an iron-zinc intermetallic coating by diffusion.

Electrophoretic deposition: Finely divided materials suspended in an electrolyte develop a charge as a result of asymmetry in the charge distribution caused by the selective adsorption of one of the constituent ions. Coatings may be formed by immersing the substrate metal in the electrolyte and applying a potential. If the particles have a negative charge they will be deposited on the anode, and if they have a positive charge they will be deposited on the cathode. Commercial applications of this method in the case of metals are limited.

Electroless plating: This method, also known as immersion plating or chemical plating, is based on the formation of metal coatings resulting from chemical reduction of metal ions from solution. The solution must contain a reducing agent, and the surface on which the deposit occurs must be catalytically active and remain catalytically active as deposition proceeds. Metals commonly plated by this technique include copper, silver, cobalt, and palladium. The silvering of mirrors falls in this classification. Typical reducing agents include hypophosphite, amine boranes, formaldehyde, borohydride, and hydrazine. Electroless nickel deposits formed with hypophosphite as a reducing agent contain phosphorus, and many of the properties are determined by this alloying constituent.

Electroplating: One of the more versatile methods for forming a metallic coating is by electrodeposition, i.e., making the metal to be coated the cathode in an electrolytic cell and applying a potential between the cathode on which the plating occurs and the anode, which may be the same metal or an inert material such as graphite. The method is applicable to all metals which can be electrolytically reduced from the ionic state to the metallic state when present in an electrolyte. Certain metals, such as aluminum, titanium, sodium, magnesium, and calcium, cannot be electrodeposited from aqueous solution because the competing cathodic reaction, $2H^+ + 2e^- = H_2$, is strongly thermodynamically favored and occurs in preference to the reduction of the metal ion. These metals can be electrodeposited from conducting organic solutions or molten salt solutions in which the H^+ ion concentration is negligible.

The mass of electrodeposited coatings can be accurately controlled because the amount deposited is a function of the number of coulombs passed. If the current efficiency for the metal deposition process is less than 100%, as when some hydrogen is also formed, it is necessary to know the current efficiency before the deposit mass can be calculated. Equal in importance to the total mass deposited is the distribution of the deposit. The ability of an electroplating bath to deposit uniform thicknesses at all sites on the cathode is characterized by the term throwing power. A bath with good throwing power has a greater tendency to deposit the metal in uniform thickness than one with poor throwing power. Quantitative measurements of throwing power may be obtained in a Hull cell, in which the cathode is oriented nonparallel to the anode, or in a Haring-Blum cell, in which two cathodes are located at different distances from the anode.

The morphology of the electrodeposit may be controlled by the addition of other materials to the plating bath. Bright electrodeposits of nickel may be obtained, for example, by the addition of organic materials such as sulfonates. Nickel plating baths that have the ability to fill in minor scratches and depressions (leveling) are obtained by the use of small amounts of coumarin or another organic compound in the bath.

Many alloys may be electrodeposited. Some of the more important include copper-zinc, copper-tin, lead-tin, cobalt-tin, nickel-cobalt, nickel-iron, and nickel-tin. The copper-zinc alloys are used to coat steel wire used in tire cord; lead-tin alloys are known as terneplate and have many corrosion-resistant applications; nickel-tin alloys have been proposed as a substitute for gold in protecting electronic components from corrosion.

Explosion bonding: As its name implies, this method involves the development of a bond between two metals by the exertion of a strong force that compresses the two metals sufficiently to develop a strong interfacial interaction.

Flame spraying: A fine metal powder or wire is passed through a flame whose temperature is sufficient to melt the metal and maintain it in the molten condition until it strikes the part to be coated. The method is used with aluminum and with zinc. The density of flame-sprayed aluminum is lower than that of pure aluminum because of voids in the coating.

Fusion bonding: Low-melting materials such as tin, lead, zinc, and aluminum may be applied as a coating by cementing the metal as a powder to the surface and then heating the part to a temperature above the melting point of the coating metal.

Gas plating: Some metal compounds can be decomposed by heat to form the metal. Outstanding examples are metal carbonyls, metal halides, and metal methyl compounds. Nickel deposits may be obtained by thermal decomposition of nickel carbonyl, a compound with high volatility and high toxicity. Old production processes for titanium and zirconium were based on the formation of iodides, which were transported in the gas phase to a hot wire, where the iodide was decomposed to form the metal.

Hot dipping: Large tonnages of steel are coated with zinc by immersion of the metal as a continuous sheet into a molten bath of zinc to form galvanized steel. The establishment of a good bond at the zinc-steel interface requires the presence of a small amount of aluminum in the bath. The thickness of the zinc coating is controlled by rigid control of the temperature of the galvanizing bath, the speed of transit through the bath, the temperature of the steel sheet before it enters the bath, and the use of air or gas jets, which exert a wiping action on the molten zinc as the sheet emerges from the bath. Tinplate was formerly manufactured by hot dipping, but practically all commercial tinplate is now made by electrodeposition because of the ability to control the thickness of the tin.

Metal cladding: Clad and mechanically formed coatings are composites of two or more metals that have been joined together in the solid state without the use of intermediate binders. Good bonding requires that the mating surfaces be free of oxides, grease, moisture, and other contaminants before the cladding step. The most common method of manufacture involves roll bonding, in which there is a 50—75% reduction in thickness. Bonding may be carried out with or without prior heating of the metals.

Plasma spraying: This method is analogous to flame spraying except that the energy is applied by forms of heating other than a flame.

Vacuum and vapor deposition: This technique receives its greatest use in the formation of metallic coatings on nonconductive substrates. Rhodium coatings on mirrors and aluminum coatings on plastics are the more common vacuum-deposited coatings.

The selection of a metal coating and the choice of an application method are made largely in terms of the end use. Among the properties that must be taken into consideration are:

Corrosion resistance
Crystal size
Density
Ductility
Electrical properties
Hardness
Magnetic properties
Optical properties
Preferred orientation of crystallites
Strength
Stress (compressive or tensile in coating)

Surface morphology
Thermal properties
  Expansion coefficient
  Melting point
  Thermal conductivity
Wear

Each of these properties assumes special importance under certain circumstances. For example, hot-dipped zinc coatings exhibit different
degrees of preferred orientation of the zinc grains depending on the
manufacturing parameters. Products with a high degree of (0001)
preferred orientation show better paint adherence when the painted
galvanized steel is deformed during industrial forming operations. Internal stresses in nickel electrodeposits may result in poorer corrosion-
resistant properties. Many palladium electrodeposits have fine cracks,
and the number and spacing of the cracks are functions of the conditions used in the plating operation. Flame-sprayed aluminum deposits
exhibit different degrees of porosity depending on the rate of deposition and other parameters.

The quality of a metal deposit on a metallic or nonmetallic substrate
is strongly a function of the quality of the interfacial bond between
the two materials. The quality of the bond, in turn, is a function of
the controlled chemistry of the mating surfaces. Undesirable constituents at the interface include oxides, organic materials such as
greases and oils, and inorganic foreign matter.

Important principles in the use of metallic coatings for corrosion
protection are demonstrated by tinplate, by nickel-chromium electrodeposits used on the external trim of automobiles, and by gold deposits on copper. These systems will now be discussed.

4.2.1 Tinplate

The ability to support a growing world population is due in no small
way to our ability to store foodstuffs for transport to remote locations and for use at times considerably later than harvest. Cecil
and Woodroof [1] studied the storage life of commercial packaged
food products over a period of many years and concluded that "corrosion of containers, ... assuming use of the best commercial container materials, would still limit storage life of most items at temperatures above freezing to shorter periods than those estimated solely by
projection of changes in product quality." Among the methods for
long-term storage of foods of many types is the use of tinned steel
containers. Such containers consist of sheet steel protected with a
very thin coating of tin. The inherent strength and cheapness of

iron are thus combined with the corrosion resistance of the expensive material tin used in small quantities.

The effectiveness of the tinned container in many environments is a consequence of the fact that tin sacrificially protects iron and yet itself corrodes at a very low rate under mild pH and anaerobic conditions. In fruit acids, such as citric acid, iron is several hundredths of a volt cathodic to a tin electrode in the absence of air. This potential difference becomes somewhat greater in the presence of a low Sn(II) concentration. In the absence of iron, tin corrodes at a low rate in this medium even at the boiling point. The plain tinned container is not satisfactory for any corrosive product in which tin is cathodic. In such a case, the large expanse of tin cathode would promote intense corrosion on the small areas of steel exposed at pores or breaks in the tin coating and localized pitting would occur.

An interesting aspect of the iron-tin system is the strong inhibiting effect of tin ions on the dissolution of iron in moderately corrosive media [2—5] under anaerobic conditions. In boiling 0.2 M citric acid, for example, iron corrodes at a rate of 15 mg/cm^2 hr. This rate is decreased to 0.06 mg/cm^2 hr when the Sn(II) concentration is 0.001 M. The importance of Sn(II) in the iron-tin galvanic couple in boiling 0.2 M citric acid has been shown in laboratory experiments by Buck and Leidheiser [6]. Iron and tin samples were coupled in chambers that were connected by a fritted disk to restrict diffusion between the chambers yet allow charges to be transmitted readily. The results of these experiments are summarized in Table 1. Simple coupling between the metals in the absence of Sn(II) in contact with the iron leads to a small decrease in the iron corrosion rate (15 to 10 mg/cm^2 hr) and an increase in the tin corrosion rate (0.02 to 0.3 mg/cm^2 hr). The presence of tin ions is necessary in order to reduce the iron corrosion rate significantly. In the tin can, the stagnant conditions existing in the can on the shelf result in the presence of tin ions in the vicinity of iron exposed at defects in the tin coating. The polarization diagram which describes these effects is given in Fig. 1. The corrosion rate of tin in the absence of coupling to iron is at point L; with coupling to iron the corrosion rate moves to point K because of the activity of iron in the hydrogen evolution reaction. The corrosion rate of iron is at point J in the absence of Sn^{2+} ions in the electrolyte. The mechanism of inhibition is controversial, but there is evidence, albeit not conclusive, that FeSn$_2$ is formed when iron is immersed in a solution containing Sn^{2+} and the potential is made sufficiently negative by electrical contact with elemental tin [7]. In drawing the diagram it is assumed that the exchange current density for the hydrogen evolution reaction is reduced by tin ions, but to a lesser extent than the anodic reaction.

Commercial tinplate used in the manufacture of containers is prepared by electrolytic deposition of tin followed by thermal treatment

TABLE 1 Corrosion Rates of Tin and Iron in Boiling 0.2 M Citric Acid
Under Various Experimental Conditions

Experimental conditions	Rate of corrosion (mg/cm^2 hr)	
	Iron	Tin
Absence of foreign metal cations; not coupled	14.7	0.018
Presence of 10^{-3} M cations of other metal; not coupled	0.062	0.018
Both metals in same vessel, bathed by cations formed during corrosion, no metallic cations present initially; coupled externally	0.83	0.30
Metals in separate vessels connected by a porous disk, metals protected from exposure to cations of other metal; coupled externally	10.0	0.27

above the melting point of tin. During this thermal processing the intermetallic compound $FeSn_2$ forms at the interface between the iron
and the tin. The quality of the $FeSn_2$ formed determines in a major
way the corrosion behavior of tinplate, particularly when the amount
of free tin is small. The best performing tinplate is that in which the
$FeSn_2$ uniformly covers the steel so that the area of iron exposed is
very small in case the tin should dissolve. Good coverage requires
good and uniform nucleation of $FeSn_2$. Many nuclei form when electrodeposition of tin is carried out from the alkaline stannate bath. In
fact, $FeSn_2$ is detected on the steel surface after plating and before
thermal treatment [8].

The growth of $FeSn_2$ at the iron-tin interface on thermal treatment
follows a parabolic growth law, $w^2 = kt$. This parabolic growth rate
is observed at all temperatures in the range 175--316°C and is independent of tin coating weight, different steel bases, and the type of
acid treatment applied before electrolytic tinning. An abrupt change
in the rate constant occurs at the melting point of tin, but the activation energy (29.3 kcal/mole) is the same above and below the melting
point of tin.

$FeSn_2$, in all but the strongest oxidizing environments, is chemically inert compared to either iron or tin. The compound is stable

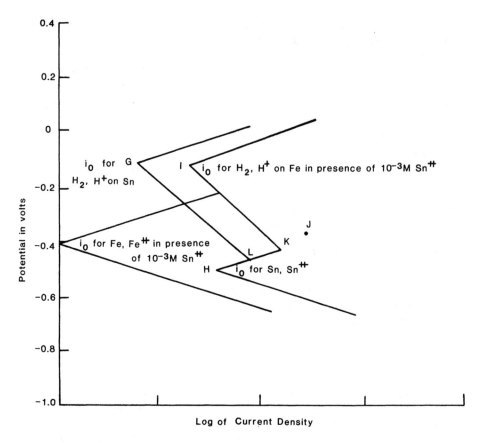

FIGURE 1 Schematic polarization diagram of iron and tin in a fruit acid containing 10^{-3} M Sn^{2+}.

up to 496°C, at which temperature it decomposes to FeSn and Sn. It crystallizes with a tetragonal lattice with a = 6.53 Å and c = 5.32 Å. The corrosion potential of the alloy in deaerated citric acid solution is of the order of -0.45 V relative to the Saturated Calomel Electrode. This potential is more noble than that of either tin or iron in the same medium, and the alloy might be expected to serve as the cathode in a galvanic couple with either iron or tin. It is doubtful that FeSn$_2$ plays any role in the electrochemistry of tinplate, however, because of its extreme inertness. It is much more likely that its function is purely as a barrier between the active iron and tin members of the couple. It serves to reduce greatly the surface area of iron that is sacrificially protected.

4.2.2 Chromium Electrodeposits for Outdoor Exposure

The chromium-plated bumpers and chromium trim on automobiles remain bright and shiny for many years, and we take the corrosion resistance and stain resistance as a matter of course. It has not always been so. Immediately after World War II it was not unusual for the chromium-plated bumpers of the most expensive cars to show severe signs of rust within a few months of winter exposure in the northern part of the United States. The shortage of strategic metals and the consequent efforts to extend the supply by economizing on the amount of metal used were part of the reason for the problem, but the more basic reason was lack of sufficient knowledge of the corrosion process to control the attack by the atmosphere. When these facts were recognized, an aggressive industrial program was mounted to obtain a better understanding of the corrosion process and ways to control it.

As with many problems in corrosion, significant progress was hampered by the lack of an accelerated corrosion test that would properly reflect service experience. Under the auspices of the American Electroplaters' Society a survey was made of the regions in the country where the corrosion of chrome trim was most severe, and it was found that a common constituent in the soil on city streets was copper. The standard salt spray test was then modified to include the addition of copper ions to the salt spray solution. This test, known as the copper-accelerated acetic acid salt-spray (CASS) test, is standard ASTM test B 368, which specifies a spray of 5% NaCl solution acidified to pH 3.1 to 3.3 with acetic acid to which is added 1 g/gal $CuCl_2 \cdot 2H_2O$. The test chamber is maintained at 120°F. Another accelerated test, known as the Corrodkote test, involves the application of a synthetic soil to the chromium-plated part while maintaining a relative humidity of 80–90% at 100°F for up to 20 hr. The composition of the synthetic soil is: cupric nitrate, 0.035 g; ferric chloride, 0.165 g; ammonium chloride, 1.0 g; kaolin, 30.0 g; and water, 50 ml. The CASS and Corrodkote tests produce corrosion in 16–20 hr comparable to that in outdoor exposure for 1 year in the Detroit area. The function of the copper ions and the ferric ions in these two tests is to provide a ready cathodic reaction ($Cu^{2+} + 2e^- = Cu$) and ($Fe^{3+} + e^- = Fe^{2+}$) for the corrosion reaction.

A third test, known as the electrolytic corrosion (EC) test, involves making the plated part the anode for 1 min in a solution containing nitric acid, sodium nitrate, and sodium chloride. The impressed potential is maintained at a constant value until the current density achieves a predetermined value. The current density is then maintained constant by reducing the potential as required. Indicators are applied to the surface or are used in the solution to obtain a measure of the extent of substrate corrosion. Two minutes of electrolysis is claimed to produce corrosion comparable to that experienced outdoors in Detroit in 1 year.

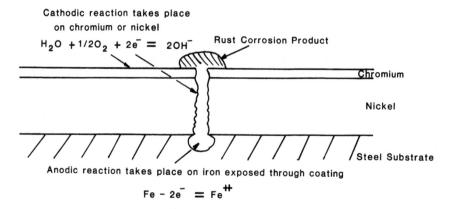

FIGURE 2 Steel corrodes at breaks in a nickel/chromium coating during exposure to the atmosphere.

Chromium-plated parts on automobiles are made on steel substrates
with an intermediate layer of nickel or, in some cases, layered deposits of copper and nickel. The nickel provides the corrosion protection
to the steel substrate and the thin chromium deposit provides the
bright appearance and stain-free surface. In such a system the nickel
must cover the iron completely because the iron will be the anode and
nickel the cathode. At breaks or pores in the coating the situation
will be as pictured in Fig. 2. This figure illustrates the reason for
the corrosion of chrome trim experienced after World War II.
    The corrosion problem was intensified by the fact that concurrently,
because of cost pressures, the automobile industry formed the nickel
electrodeposits by using addition agents in the plating bath that resulted in a bright deposit. Thus the labor costs associated with the
former buffing operation were avoided. The chromium plating baths
yielded a bright deposit when the deposition was carried out on a
bright substrate and when the deposit was sufficiently thin. It was
recognized in the industry that sulfur included in a nickel deposit
made the nickel more active from a corrosion standpoint. This is demonstrated in Fig. 3, where the polarization curves for sulfur-containing nickel and sulfur-free nickel deposits are compared. Note that the
critical current density for passivity is approximately two orders of
magnitude greater for the sulfur-containing nickel and that the current density is greater at all potentials from -0.25 to +0.30 V. This is
a rather discouraging fact from a corrosion standpoint. However, it
occurred to investigators that this apparent disadvantage of bright
nickel could be put to very good use. Figure 4 shows how this has
been done. The nickel coating between the steel substrate and the

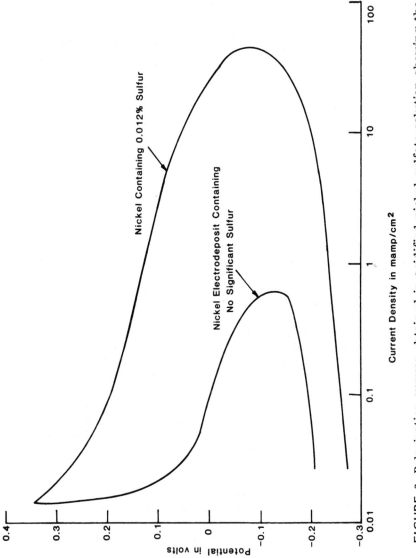

FIGURE 3 Polarization curves obtained in acidified nickel sulfate solution showing the much greater activity of nickel electrodeposits containing sulfur. (Data taken from Ref. [43].)

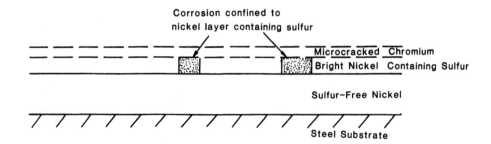

FIGURE 4 Schematic diagram showing how a duplex nickel electrode-posit is used to prevent corrosion of a steel substrate.

chromium exterior is made up of a so-called duplex nickel deposit. The inner layer is plated with sulfur-free nickel and the outer layer is plated in the presence of an addition agent that results in sulfur being incorporated in the deposit. When corrosion occurs through openings in the chromium coating, the attack is limited to the bright nickel layer containing sulfur. The corrosion spreads laterally be-tween the chromium and sulfur-free nickel deposits because the outer members of this sandwich (chromium and sulfur-free nickel) are cath-odic to the sulfur-containing nickel. A likely disadvantage of this method of corrosion control is the undermining of the chromium and the possibility that the brittle chromium deposits will flake off the surface. In order to avoid this eventuality, methods have been de-veloped for plating chromium in such a way that minute cracks or por-ous areas are formed that do not detract from the bright appearance of the chromium; such deposits are known as microcracked or micro-porous chromium. These microcracks or micropores form very uni-formly over the exterior of the plated material and serve to distribute the corrosion process over the entire surface. The net effect of mi-crocracked and microporous chromium and duplex nickel has been to extend very greatly the lifetime of chromium-plated steel on outdoor exposure. Some manufacturers have gone a step further and use triplex nickel, but the principles described schematically in Fig. 4 still apply.

4.2.3 Gold Coatings on Copper

Gold electrodeposits are commonly used in electronic applications to protect copper connectors and other copper components from corro-sion. Because of the great cost of gold, it is desirable to obtain the corrosion protection with the minimum thickness of gold. As the thickness of an electrodeposit is decreased, there is a tendency for the deposit to provide inadequate coverage of the substrate. Thus it is important to have a rapid method for appraising the coverage of copper by a gold electrodeposit. Such a method was developed by Morrissey [9], using corrosion principles as a guide. He observed that gold serves as the cathode and copper serves as the anode in aerated 0.1 M NH_4Cl solution and that, at a high cathode/anode surface area fraction, the corrosion potential is linearly related to the area fraction of copper as shown in Fig. 5.

This relationship is readily developed from the considerations of Stern [10], who showed that for binary galvanic couples controlled by activation polarization, the corrosion potential is given by

$$\phi_{corr} = -\frac{E_A \beta_C}{\beta_A + \beta_C} - \frac{\beta_A \beta_C}{\beta_A + \beta_C} \log A_A i_{0A} + \frac{\beta_A \beta_C}{\beta_A + \beta_C} \log A_C i_{0C}$$

where E_A is the corrosion potential of pure anode material, A_A and A_C are the area fractions of the anode and cathode, i_{0A} and i_{0C} are the exchange current densities for the anodic and cathodic processes, and β_A and β_C are the Tafel slopes of the logarithmic polarization curves for the anodic and cathodic processes. In the case of a specific system in which the cathodic surface area is very large, approaching 1, the values of E_A, β_C, β_A, i_{0C}, and i_{0A} are constants and the equation reduces to

$$\phi_{corr} = K_1 - K_2 \log A_A$$

The corrosion potential under these conditions is thus a linear function of the logarithm of the anode area fraction exposed.

Morrissey confirmed that the logarithm of the amount of copper corrosion of inadequately gold-plated specimens was linearly related to the corrosion potential. These observations along with the theoretical treatment provided a basis for determining the area fraction of copper exposed through the gold deposit by simply measuring the corrosion potential of the gold-plated copper in 0.1 M NH_4Cl. The applications of this simple method for determining the degree of coverage of the copper include selection of the optimum plating conditions and determination of the minimum thickness of gold for adequate coverage. Morrissey and Weisberg [11] were also able to show that gold deposits with a (111) preferred orientation exhibit the maximum substrate coverage for thin gold deposits.

4.3 POLYMERIC COATINGS

Protective polymeric coatings fall broadly into three different classes: lacquers, varnishes, and paints. Varnish is a term applied to coatings

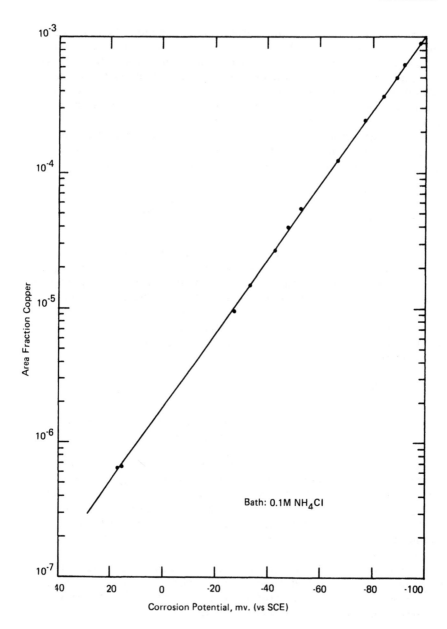

FIGURE 5 Data showing that the fractional exposed area of copper in copper/gold system is linearly related to the corrosion potential at low exposed copper areas. (Data taken from Ref. [9].)

which are solutions of either a resin alone in a solvent (spirit varnishes) or combinations of an oil and a resin in a solvent (oleoresinous varnishes). The term lacquer is generally limited to a composition whose basic film former is nitrocellulose, cellulose acetatebutyrate, ethyl cellulose, acrylic resin, or another resin that dries by solvent evaporation. The term paint is applied to more complex formulations of a liquid mixture that dry or harden to form a protective coating. Typical formulation constituents include a liquid vehicle, which may be water or an organic solvent, pigments to give the coating color, fillers to maintain coating thickness at low cost, wetting agents to promote penetration of the paint into scratches and pores in the substrate, antifoaming agents, antimildew agents, catalysts to cause polymerization of some coatings, corrosion inhibitors, and constituents which control the rheological properties. Other constituents may also be included to obtain a special property in the coating.

A polymeric coating protects a metal substrate from corroding by two mechanisms: (a) serving as a barrier for the reactants, water, oxygen, and ions, and (b) serving as a reservoir for corrosion inhibitors that assist the surface in resisting attack. Coatings which contain large quantities of metallic zinc also provide corrosion protection by galvanic action. The barrier properties of the coating are improved by increased thickness, by the presence of pigments and fillers that increase the diffusion path for water and oxygen, and by the ability to resist degradation. The common degradation mechanisms of the coating include abrasion and impact, cracking or crazing at low or high temperature, bond breakage within the polymer matrix because of hydrolysis reactions, oxidation or ultraviolet light, and freeze-thaw cycling. Such degradation allows access of reactants to the coating/ substrate interface without the necessity for diffusion through the polymer matrix.

Corrosion of a metal beneath a polymeric coating is an electrochemical process that follows the same principles as corrosion of an uncoated metal. What makes the process different from crevice corrosion, for example, is the fact that the reactants often reach the metal through a solid. Also, the small volumes of liquid that are involved in the early stages of corrosion result in extreme values of pH and ion concentrations. The total corrosion process may be thought of in terms of the following components:

Transport through the coating of water, oxygen, and ions
Development of an aqueous phase at the coating/metal interface
Activation of the metal surface for the anodic and cathodic reactions
Deterioration of the coating/metal interfacial bond

4.3.1 Types of Corrosion Beneath Organic Coatings on Metals

Six specific types of corrosion beneath organic coatings will be discussed: blistering, early rusting, flash rusting, anodic undermining,
filiform corrosion, and cathodic delamination. Each of these will be
treated separately. Another type of coating deterioration known as
loss of adhesion when wet, or simply wet adhesion, which may or may
not be related to corrosion, will also be discussed.

Blistering

An up-to-date review of blistering has been given by Funke [12].
Blistering is one of the first signs of breakdown in the protective nature of the coating. The blisters are local regions where the coating
has lost adherence from the substrate and where water may accumulate and corrosion may begin. Five mechanisms, operative under different circumstances, are used to explain blister formation which occurs prior to the corrosion process.

Blistering by Volume Expansion Due to Swelling

All organic coatings absorb water and those used in corrosion protection are usually in the range of 0.1 to 3% water absorption upon
exposure to liquid water or an aqueous electrolyte. Water absorption leads to swelling of the coating, and when this occurs locally
for any reason, blisters may form and water may collect at the interface.

Blistering Due to Gas Inclusion or Gas Formation

Air bubbles or volatile components of the coating may become incorporated in the film during film formation and leave a void. Such
blisters are not necessarily confined to the interface, but when they
are, they can serve as a corrosion precursor site.

Electroosmotic Blistering

Water may move through a membrane or capillary system under the
influence of a potential gradient. Potential gradients, such as may
exist with a galvanic couple, can lead to a blister.

Osmotic Blistering

The driving force for osmotic blistering is the presence at the coat-
ing/substrate interface of a soluble salt. As water penetrates the
coating to the interface, a concentrated solution is developed with
sufficient osmotic force to drive water from the coating surface to the
interface, and a blister is formed. The osmotic mechanism is probably the most common mechanism by which blisters form.

An outstanding example of osmotic blistering was cited by an un-
identified discussion participant at the Corrosion 81 meeting in

Toronto. A ship was painted in Denmark and made a voyage immediately thereafter across the Atlantic and into the Great Lakes. When
it reached port, a blister pattern in the form of a handprint was observed above the water line. Apparently, the paint was applied over
a handprint. No blistering occurred during exposure to seawater because of the high salt content of the water, but when the ship was
exposed to fresh water the osmotic forces became significant and
blistering occurred.

Blistering Due to Phase Separation During Film Formation

A special type of osmotic blistering can occur when the formulation
includes two solvents, the more slowly evaporating one of which is
hydrophilic in nature. When the hydrophilic solvent is in low concentration, the phase separation process occurs at a later stage in film
formation and may occur at the coating/substrate interface. Water
diffuses into the hydrophilic solvent, or into the void left by the hydrophilic solvent, and blisters are initiated. Glycol ethers or esters,
which have low volatility, are prone to cause such blister formation.

In all the above cases, the blister provides a locale for collection
of water at the coating/substrate interface. Oxygen penetrates
through the coating, leaching of ionic materials from the interface or
from the coating occurs, and all the constituents are available for
electrochemical corrosion. Oxygen is necessary for the cathodic reaction:

$$H_2O + \frac{1}{2}O_2 + 2e^- = 2OH^-$$

but it is also consumed in the conversion of Fe(II) to Fe(III). The
ferric corrosion products tend to concentrate on the inside dome of
the blister and at the periphery of the blister, where the oxygen
concentration is highest. The cathodic region is at the periphery of
the blister and the anodic region is in the center of the blister, where
the oxygen concentration is lowest.

Early Rusting

This term is applied to a measles-like rusting that occurs after the
coating has dried to the touch. It only occurs after the coated metal
is exposed to high-moisture conditions. A typical condition under
which it is observed has been cited by Grourke [13]. A steel tank
was abrasively cleaned and was then painted with an acrylic latex
late in the afternoon during the summer. Rust spots were most prom-
inent on the bottom of the horizontally mounted tank and were observed
up to the liquid level within the tank. The top half of the tank exhibited no rust spots.

The three conditions which lead to early rusting are (a) a thin latex coating (less than 40 µm), (b) a cool substrate temperature, and
(c) high-moisture conditions. Early rusting can be duplicated in the
laboratory under these conditions. The severity of the problem tends
to increase as the activity of the steel is increased. For example,
early rusting is more severe on panels given a white abrasive blast
than on those given less adequate cleaning with a power tool.

Early rusting occurs with latex coatings because of the mechanism
by which they lose water. Film formation occurs through coalescence
of the latex particles. Particle-particle contact occurs because of water evaporation, particle-particle deformation then occurs as a consequence of surface tension and capillary forces, and finally diffusion
of polymer chains occurs among latex particles and the film hardens.
Early rusting occurs under conditions that slow down the drying and
allow water-soluble iron salts to be leached through the paint film.
If moist conditions do not exist during the latex drying process,
early rusting does not occur.

In summary, early rusting is a consequence of moist conditions occurring before the latex coating has dried sufficiently. Water ingress
and egress occur readily before particle coalescence has been com-
pleted, and movement of soluble iron salts through the film, followed
by water evaporation, leads to the rust staining. Success in preventing early rusting has been achieved through the use of soluble inhibitors in the formulation.

Flash Rusting

Brownish rust stains may appear on blast-cleaned steel shortly
after priming with a water-based primer. This phenomenon is known
as flash rusting. Work done by the Paint Research Association has
shown that this defect may be avoided by removing the contaminants
remaining after blast cleaning or by a chemical treatment before application of the primer.

The steel or ceramic grit on the surface apparently leaves crevices
and/or galvanic cells are set up between the steel grit and the steel
base sufficient to activate the corrosion process as soon as the surface is wetted by the water-based paint. The staining is a result of
the soluble corrosion products penetrating the coating and being oxidized to the ferric form within or on the surface of the coating.

The possible adverse effect of steel grit blasting on the performance of paints is a subject worthy of investigation.

Anodic Undermining

Figure 6 shows six planes along which delamination may occur to
separate the organic coating from the metal. Anodic undermining

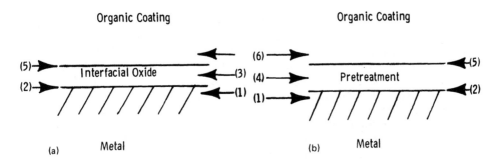

FIGURE 6 Six planes along which delamination may occur in an organic coating/metal substrate system. (a) Without pretreatment; (b) with pretreatment.

represents that class of corrosion reactions underneath an organic coating in which the major separation process is the anodic corrosion reaction under the coating. An outstanding example is the dissolution of the thin tin coating between the organic coating and the steel substrate in a food container. In such circumstances the cathodic reaction may involve a component in the foodstuff, or a defect in the tin coating may expose iron, which then serves as the cathode. The tin is selectively dissolved and the coating separates from the metal and loses its protective character.

Aluminum is particularly susceptible to anodic undermining. Koehler [14] cites an example from laboratory studies in which pairs of organic-coated aluminum panels were sealed to opposite ends of a cylindrical cell filled with 0.05 M citrate solution at pH 3.5 containing 0.5% NaCl. The two panels were connected to a power source and a current of 0.09 μamp/cm^2 was passed for 1 week. Underfilm corrosion was observed on the anode and no underfilm damage occurred on the cathode.

Anodic undermining of organic coatings on steel may occur under circumstances where the steel is made anodic by means of an applied potential. In the absence of an applied potential, coatings on steel fail largely by cathodic delamination.

Anodic undermining has not been studied as extensively as cathodic delamination because there do not appear to be any mysteries. Galvanic effects and principles which apply to crevice corrosion provide a suitable explanation for observed cases of anodic undermining.

Filiform Corrosion

Filiform corrosion is a type of attack in which the corrosion manifests itself as threadlike filaments. It represents a specialized form of anodic undermining. It generally occurs in humid environments and is most common under organic coatings on steel, aluminum, magnesium, and zinc (galvanized steel). A good treatment of this subject has been presented by Ruggeri and Beck [15]. In some cases, filiform corrosion will develop on uncoated steels on which small amounts of contaminating salts have been accidentally deposited. It has also been observed on thin electrodeposits of tin, silver, and gold and under conversion coatings such as phosphate. The most annoying types of filiform corrosion are those which occur under paint films that are designed to retain an aesthetic appearance on metals exposed to the atmosphere and those which occur under the lacquers that protect the interior of food containers.

The threads which form in filiform corrosion exhibit a wide variety of appearances from nodular shapes such as those on aluminum to the very fine, sharply defined threads observed under clear lacquers on steel. The widths of the filaments are of the order of 0.05 to 0.5 mm, and under laboratory conditions the filaments grow at the rate of 0.01 to 1 mm/day. The rate of growth of the filaments is approximately constant over long periods of time. Filiform corrosion requires a relatively high humidity, generally over 55% at room temperature, but insufficient work has been done to determine an exact lower limit. It can be encouraged to develop by scratching through the coating to the metallic substrate and then maintaining the panel at a relative humidity of 70 to 85%. It is developed in the laboratory in some cases simply by putting powders of a salt on the surface of the coating or by putting salt crystals on the metal substrate and applying the coating over the contaminated surface. Preston and Sanyal [16], for example, obtained filiform corrosion on steels at 99% relative humidity by contaminating the steel surface with the following powders before applying the coating: sodium chloride, calcium sulfate, ammonium sulfate, sodium nitrate, zinc chromate, flue dust, cinders, iron oxide, and carborundum.

Hoch [17] made a detailed study of the character of filiform corrosion on steel, magnesium, and aluminum substrates. In the case of iron, the very leading edge of the filament had a pH of approximately 1, whereas the liquid immediately behind the leading edge had a pH of 3 to 4, just what would be expected on the basis of hydrolysis of Fe^{2+} ions. Magnesium and aluminum also exhibited very low pH values at the leading edge and higher pH values in the liquid adjoining the leading edge. Insoluble corrosion products formed in all three cases a short distance back from the leading edge. In the cases of aluminum and magnesium, hydrogen bubbles were observed at the leading edge, indicating that the liquid in the leading edge was low in oxygen and the hydrogen evolution reaction took precedence.

The following mechanism appears to account satisfactorily for the filiform corrosion of aluminum coated with an organic lacquer. First, a highly localized defect forms in the coating. This defect may arise at the edge of a scratch through the coating, at an inclusion in the coating, at a defect in the metal surface, or as a result of the presence of a local high concentration of electrolyte which causes penetration of the coating. In the presence of high relative humidity, water penetrates through the coating. In the presence of electrolyte, which has either penetrated the coating or was inadvertently occluded beneath the coating, a tiny liquid aggregate forms because of the high affinity of ions such as Na^+ and Cl^- for water. Once sufficient molecules are present to have a liquidlike identity, additional water diffusing through the coating is retained because of the low vapor pressure of concentrated electrolyte solutions. Minor corrosion of the substrate occurs, yielding additional dissolved ions and promoting further retention of diffusing water species. As the liquid increases in dimension and corrosion occurs, local conditions cause an imbalance in the oxygen supply at some point in the microscopically circular corrosion area. The oxygen-deficient area becomes the anode and the periphery becomes the cathode. The circular droplet then assumes an elliptical shape and the conditions for filamentary growth are present. Once the filament has been nucleated, an oxygen concentration cell is developed and the propagation of the filament proceeds because of a highly effective anode at the head and a cathodic area present in the areas surrounding the head. Immediately at the leading edge of the growing head, aluminum is dissolved to yield a highly concentrated Al^{3+} ion solution. Hydrolysis occurs with the following reactions liberating H^+ and thus generating the very low pH:

$$Al(H_2O)_6^{3+} = Al(H_2O)_5OH^{2+} + H^+$$

$$Al(H_2O)_5OH^{2+} = Al(H_2O)_4(OH)_2^+ + H^+$$

$$Al(H_2O)_4(OH)_2^+ = Al(H_2O)_3(OH)_3 + H^+$$

The last reaction, which occurs at some distance from the leading edge, results in the precipitation of hydrated aluminum oxide and the pH is reduced to the range of 3 to 4 because of dilution from incoming water.

The oxygen deficiency at the very leading edge along with the low pH also permits the competing cathodic reaction, $2H^+ + 2e^- = H_2$, to occur to a limited extent, and a small amount of hydrogen gas is generated.

Koehler [18] emphasized the importance of the anion in filiform corrosion. He noted that the filaments formed on steel in the presence of sulfate contaminants beneath the coating were less numerous and were much finer than those observed with chloride contaminants. He also noted that the head of the filament contained the anion of the contaminating salt, but not the cation. The anions apparently migrated to provide charge compensation for the ferrous ions formed in the active region at the head of the filament.

The fascinating question about filiform corrosion is why the corrosion occurs in the form of filaments as opposed to circular spots. No complete answer is possible at the present time, but it does appear that the limited availability of oxygen, by diffusion through the coating, and the limited availability of water, by diffusion through the coating under high relative humidity conditions, are the determining factors. At very high relative humidities or on exposure to liquid water, filiform corrosion passes over to more general corrosion and the filamentary character is lost.

Much attention has been paid in paint laboratories to reducing filiform corrosion. Phosphate conversion coatings, followed by chromate rinses and distilled water rinses, provide some protection, but do not completely eliminate the filiform corrosion of iron. The properties of the coating also have an effect on the extent and character of filiform corrosion. Coatings that are highly permeable to water and to oxygen are especially susceptible to filiform attack. Coatings that are very brittle and are ruptured by pressures generated by the corrosion process lose the entrapped moisture, and pitting attack often results. In the case of magnesium, filiform corrosion occasionally converts to a virulent pitting attack.

Cathodic Delamination

Many coated steel products are subject to scratches or dents with consequent exposure of the steel to the environment. If the coated materials are continuously immersed in an electrolyte, as, for example, ships, underground pipelines, and the interior of vessels holding an aqueous solution, it is possible to protect the exposed areas by an applied cathodic potential. One of the undesirable consequences of cathodic protection is that the coating adjoining the defect may separate from the substrate metal. This loss of adhesion is known as cathodic delamination. This type of delamination may also occur in the absence of an applied potential. The separation of the anodic and cathodic corrosion half-reactions under the coating provides regions which are subject to the same driving force as when the cathodic potential is applied externally. An excellent pictorial example of the separation of the anodic and cathodic reactions is shown on the cover of the April 1983 issue of *Materials Performance*.

It is generally believed that the major driving force for cathodic delamination in corrosion processes in the presence of air is the cathodic reaction, $H_2O + 1/2O_2 + 2e^- = 2OH^-$. When an applied potential is used, the important reaction may be $2H^+ + 2e^- = H_2$, if the driving force is sufficient. Figure 7 shows a typical cathodic polarization curve for steel in 0.5 M NaCl saturated with air. The regions of

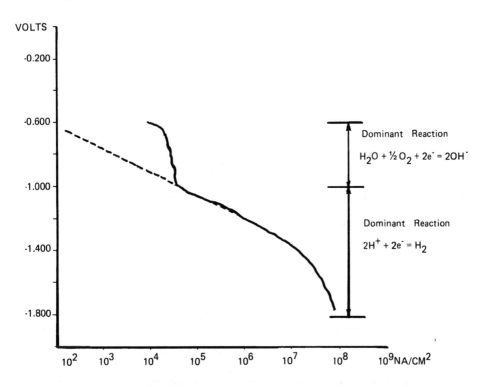

FIGURE 7 Cathodic polarization curve for iron in 0.5 M NaCl solution, pH 6.5, open to the air.

dominance of the two cathodic reactions are noted on the figure, and the cathodic polarization curve for hydrogen evolution in the absence of oxygen is shown by the dashed line. It is apparent from this figure that at a potential of -0.8 V (vs. SCE), the dominant reaction is the oxygen reduction reaction. Polarization at -0.8 V of polymer-coated steel containing a defect in the absence of oxygen leads to no significant delamination from the defect, whereas in the presence of air there is significant delamination.

Studies indicate that the pH beneath the organic coating where the cathodic reaction occurs is highly alkaline, as the cathodic equations indicate. Ritter and Krüger [19] reported that the pH at the delaminating edge is as great as 14 as measured by pH-sensitive electrodes inserted through the metal substrate from the back side. Other studies which integrate the pH over a larger volume of liquid beneath the coating yield pH values of 10 to 12. Cathodic polarization curves on steel in 0.5 M NaCl at pH values of 6.5, 10, and 12.5

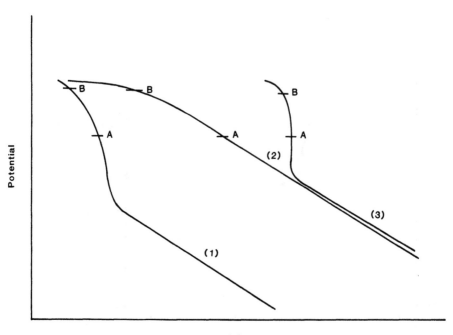

Log of Current Density

FIGURE 8 Schematic representation of the tendency toward cathodic
delamination in metals exhibiting different types of curves for the
oxygen reduction reaction. See text for explanation.

are approximately the same, suggesting that the cathodic behavior beneath the coating may be rationalized in terms of the cathodic polarization curve that is applicable at the exposed defect.

Figure 8 represents the extremes of three types of polarization
curves that are observed on metals whose behavior during cathodic
treatment has been studied. Point A on each curve represents the
potential at the defect and point B represents the assumed potential
at the delaminating front. Curve (1) has the shape of the polarization curve of aluminum in 0.5 M NaCl. The surface is not active for
the oxygen reduction reaction and the rate of delamination is low, as
indicated by the location of point B. Curve (2) is a hypothetical
curve somewhat comparable to the polarization curve for tin in 0.5
M NaCl. The curve has a steep slope so that the current density
falls off greatly with increase in potential. Thus the current density at the delaminating front is low. Curve (3) is typical of iron
and copper in 0.5 M NaCl. The oxygen reduction reaction is catalyzed over a wide potential range and the current density remains

the same over this range. Cathodic delamination occurs at a relatively
high rate because the current density at point B is high relative to
comparable points on curves (1) and (2).

The cathodic reaction which occurs at the delaminating front generates hydroxyl ions, which appear to be the major destructive influence on the organic coating/substrate bond. The value of the pH
at the delaminating front is determined by the following factors: the
rate at which the reaction occurs, the shape of the delaminating front,
the rate of diffusion of hydroxyl ions away from the delaminating front,
and buffering reactions which may involve the interfacial oxide or the
polymer.

All the evidence presently available indicates that the cathodic delamination process occurs because of the high pH generated by the
cathodic reaction. The real question is what is the consequence of
a high pH on the interface. The experimental evidence suggests that
the strong alkaline environment may attack the oxide at the interface
or may attack the polymer. Attack of the oxide was seen by Ritter
and Kruger [19] using ellipsometric techniques in the case of polybutadiene coatings on steel, and surface analysis techniques in the
hands of Dickie and colleagues [20] gave clear evidence that carboxylated species are present at the interface as a result of hydroxyl
ion attack of the polymer. Dickie has also recently shown that coatings more resistant to alkaline attack exhibited better performance
when scribed and submitted to salt spray. Koehler [21] believes
that the disbonding process is a consequence of displacement of the
coating by a high-pH aqueous film between the coating and the substrate.

It is proposed that the major mechanism for the delamination process is solubilization of the thin oxide coating at the interface between
the organic coating and the metal. Dissolution of the oxide, as the
major delaminating mechanism, was proposed previously by Gonzalez
et al. [22] to account for effects observed with food can lacquers on
steel when immersed in solutions containing strong complexing agents
for Fe^{2+} ions. In this process the oxide itself participates in the cathodic reaction according to

$$\gamma\text{-}Fe_2O_3 + 6H^+ + 2e^- = 2Fe^{2+} + 3H_2O$$

and the complexing agent serves to drive the reaction to the right by
complexation of the ferrous ions. In the cathodic delamination process
being proposed here, the dissolution of the oxide breaks the bond between the coating and the substrate metal and the high pH leads to localized attack of the polymer at the interface. The presence of oxidized organic species on the metal surface after delamination has occurred may be the result of a posteriori adsorption of oxidized species
or of islands of organic left on the surface. The X-ray photoelectron

spectroscopy (XPS) technique used in Ref. [18] illuminates a large area, and spatial resolution is lacking to determine whether the organic material is present over the entire surface or is islandlike in nature.

Since the important delaminating process is a consequence of the hydroxyl ions generated by the cathodic reaction, $H_2O + 1/2O_2 + 2e^- = 2OH^-$, the delamination may be prevented by any of the following:

Preventing reactant water from reaching the reaction site
Preventing reactant oxygen from reaching the reaction site
Preventing electrons from reaching the reaction site
Preventing cation counterions from reaching the reaction site
Reducing the catalytic activity of the surface for the cathodic reaction

Water and oxygen reach the reaction site largely by diffusion through the coating, so the rate of reaction may be reduced by reducing the permeability of the coating for these constituents. It appears unlikely on the basis of present knowledge that it will be possible to eliminate completely the diffusion of these constituents through the coating or through defects in the coating that are present when the coating is prepared or that occur during normal service. The electrons, however, reach the reaction site through the metal phase and through any interfacial oxide or other film that exists at the metal/coating interface. Any type of metal surface film that is a poor electronic conductor may be able to limit the access of electrons to the reaction site. The low rate of cathodic delamination from aluminum surfaces, compared to that from zinc and steel, is probably a consequence of the poor electronic conductivity of the aluminum oxide at the interface.

The cathodic reaction generates anions, and cation counterions must be available to balance the charge locally. Hydrogen ions do not perform this function, or else the pH would not rise as dramatically as it does. The major transport medium for the cations may be through the liquid layer that forms at the coating/substrate interface or through the coating itself.

The cathodic reaction is a catalyzed one and the chemical character of the interfacial region determines whether the reaction will occur. It has been shown that cobalt ions have the ability to poison the zinc oxide surface on zinc for the oxygen reduction reaction [23], and it has also been shown that this poisoning leads to a lower rate of cathodic delamination. Some commercial pretreatments include the function of poisoning the surface for the cathodic reaction.

4.3.2 Loss of Adhesion of a Coating When Wet

This phenomenon shows up in a number of ways. The best example
is exhibited by the "hot-water test" used in characterizing pipeline
coatings. A candidate coating on a steel substrate is immersed in water at 80 to 100°C for 10 to 30 days and the adherence of the coating
is determined at the end of this time by the use of a strong and sharp
knife. A similar test approximately 1 hr in duration is used with certain coil coated products. Poor wet adhesion is also exhibited by coatings that dry too rapidly and trap organic solvents at or near the interfacial region. It is also apparent in the removal of epoxy powder
coatings with methyl ethyl ketone. After the coated metal is soaked
in methyl ethyl ketone for one or more days, the coating may readily
be stripped from the substrate if the time after removal from the organic liquid is correctly selected. In too short a time, the coating is
gummy and cannot be separated from the substrate. In too long a
time, adherence to the substrate is regained. When the correct time
is chosen, the coating may be removed in much the same way as a
weakly adhering adhesive tape. At the correct time there is apparently a layer of liquid which is intermediate between the coating and
the substrate, and the coating has a physical state intermediate between a soft clay and a rigid coating.

The major problem in studying wet adhesion is the difficulty in
quantitatively defining the degree of adherence. The cross-cut test
is used with some degree of success with coatings less than about 100
μm thick but is not suitable for very thick coatings. Impact tests are
useful with coatings in the same thickness range as the cross-cut test.
Unfortunately, there is no satisfactory test for coatings 100 μm or
greater in thickness. A vertical tear-off test, commonly used as a
coating adherence test, is not applicable to thick, inflexible coatings,
but it can be used to determine wet adhesion of thin coatings in some
cases. Funke [24] proposed a test which as yet has not received wide
acceptance. He suggested that the water absorption of free films and
of the same material on a metal be determined at 90% relative humidity.
Loss of adhesion on exposure to high humidity is indicated by a crossover in the two absorption curves. The crossover time indicates when
the coated metal begins to lose adhesion because of the presence of
water at the metal/coating interface. Coatings which exhibit good wet
adhesion do not show a crossover.

The major unsolved problem related to wet adhesion is the mechanism by which water affects the adhesion. This problem is related
to the very basic question concerning factors that control adhesion.
The way in which adhesion between two materials is viewed has gone
through a number of fashions. Some years ago adhesion was viewed

as an interaction between polar groups; later it became the rule to
treat adhesion in terms of dispersion forces. More recently, adhesion of organic coatings has been viewed as an acid/base interaction.
This approach has much merit because the acid/base character of a
surface can be quantitatively assessed and the adsorption of organic
polymers on metals and metal oxides can be studied to determine the
appropriateness of this viewpoint. The effect of water on the degree
of polymer adhesion shows promise of providing the basis for understanding wet adhesion.

4.3.3 Nature of the Interfacial Region Between an Organic Coating and a Metal

Much research has been carried out with the objective of characterizing the metal surface before the application of the coating. Examples include the work of Schwab and Drisko [25] on the effect of
surface profile, the work of Mansfeld et al. [26] on surface chemistry, the work of Zurilla and Hospadaruk [27] on characterizing the
oxygen reduction capacity of phosphated surfaces, and the work of
Iezzi and Leidheiser [28] on the effect of parameters which control
the ability of a steel surface to accept uniform phosphating.

The chemical nature of the intact organic coating/metal substrate
interfacial region has been little studied, largely because of the difficulty in devising experimental techniques to make such a study.
Optical techniques such as ellipsometry are useful with thin transparent coatings, but they cannot be used with opaque coatings or
those which contain pigments or fillers. Ellipsometric techniques
can provide information on the thickness of the oxide film at the
interface and, with good fortune, the optical parameters may be
used to determine the composition of the oxide. Ritter and Kruger
[19] found that an oxide exists at the interface and that the optical
properties of the interface change when cathodic delamination or corrosion occurs.

Emission Mössbauer spectroscopy is a technique which may be utilized to identify the chemical nature of the emitting atom. Chemical
compounds yield characteristic spectra which can be used as finger-
prints to identify the compound. Leidheiser, Simmons, and Kellerman [29] applied Mössbauer techniques to show chemical changes beneath a coating, and Leidheiser, Musić, and Simmons [30] were able
to show changes in the nature of the oxide film on cobalt when a polybutadiene coating was applied. Baking the coating at 200°C resulted
in conversion of a portion of the Co^{3+} to Co^{2+}. The emission Mössbauer technique has the severe limitation that it is only applicable
with relative ease to cobalt and tin, and the complete interpretation
of the spectrum is often difficult and/or ambiguous.

Major advances in organic coatings protective against corrosion are dependent on the development of new techniques for studying in a nondestructive way the chemical nature of the interfacial region between the metal and the organic coating. The interfacial region is where the action is, particularly in the case of cathodic delamination. The oxide at the interface is the catalytic surface for the oxygen reduction reaction; it is the medium through which the electrons are supplied for the oxygen reduction reaction; and it provides the bonding which results in the adherence between the coating and the metal. Commercial systems which provide the maximum resistance against corrosion include an inorganic coating between the organic coating and the metal substrate. This inorganic coating, often called a pretreatment or a conversion coating, replaces the normal metal oxide and provides to the organic coating a substrate with different chemical properties—a poorer catalyst for the oxygen reduction reaction, a less conductive interfacial region, and, in some cases, a rougher interface that improves organic coating/substrate adherence and resistance to deterioration under service conditions. The more common inorganic coatings include phosphates, chromates, and mixed metal oxides. With the exception of the research of Machu, little has been published on the science associated with interfacial inorganic coatings. It is a fertile area for research.

4.3.4 Properties of the Coating

Polymers form the matrix of organic coatings. They are the retainers for pigments, fillers, corrosion inhibitors, and other additives present for specific purposes. The polymer selected is based on both end-use requirements and the ability to apply the coating in the desired manner. Emulsion polymers, or latexes, are suspended in an aqueous medium, and they form a coating by loss of water by evaporation and coalescence of the individual particles into a continuous film. Some materials, such as butadiene, are applied to the substrate in a solvent and the polymerization process occurs as the solvent is removed. Cross-linking by oxidation also occurs at elevated temperature in the case of butadiene. Other coatings based on condensation polymers are polymerized in situ. A good example is the epoxy-polyamine coating, in which the two constituents are mixed just prior to application and the polymerization process occurs over a period of time. Other polymers are dissolved in solvents and the polymer forms the coating as the solvent is evaporated. Production line painting often involves the use of heat or other radiation to cause the film-forming process to occur more rapidly.

The properties of polymeric coatings depend not only on the size, shape, and chain structure of the individual units, but also on the

spatial shape of the polymer molecules. Linking of many carbon atoms
and freedom of rotation about carbon-carbon bonds permit the molecule to assume a variety of spatial shapes such as spirals, coils, and
tangles. This wide latitude in shape also leads to a variety of ways
in which individual molecules are oriented with respect to their neigh-
bors. Three classes of arrangements are recognized:

1. Segments of the molecule are randomly distributed regardless of
    whether they belong to the same molecular chain or another chain.
    Such a material is termed amorphous or glassy and the properties
    are uniform in all directions.
2. Segments of the molecule possess a degree of lateral order through
    the folding of individual chains. The volume element over which
    this occurs may be considered a single crystal. The individual
    crystals may be randomly oriented or they may be aligned in the
    same direction. In the latter case, the coating may have physical
    properties that differ in different directions.
3. Segments of the molecule may show lateral order through the paral-
    lel arrangement of extended chains. As in 2, these parallel ar-
    rangements may be unoriented with respect to neighboring vol-
    umes or there may be a degree of spatial orientation. Materials
    of this type are obtained when a polymer melt is solidified under
    shear or stress.

An unusual case of corrosion in which the rate of corrosion appears
to be related to segmental motion of portions of the polymer chain is
shown in some interesting research by Yializis et al. [31]. These
workers observed that the rate of corrosion of aluminum-coated polypropylene capacitors either in the dry state or when immersed in a
dielectric fluid was a function of the frequency of an applied ac potential. A sharp maximum in corrosion rate occurred at 3.5×10^3 Hz.
The following explanation for this phenomenon is offered. On exposure to the atmosphere and during corona discharge prior to metalli-
zation, polypropylene dissolves significant quantities of oxygen and
water. Appreciable amounts of chloride ion remain in the polypropylene from the manufacturing operation. When the metallized capacitor is exposed to an ac voltage, segmental motions occur in the polypropylene. Over a limited frequency range, the motions of segments
of the polymer are such as to allow diffusion of water and oxygen to
occur along special pathways in the polymer. Since the aluminum coating is essentially opaque to the passage of water, chloride ions, and
oxygen, sufficient reactants accumulate at the aluminum/polymer interface to allow the following reactions to occur.

Cathodic: $H_2O + \frac{1}{2}O_2 + 2e^- = 2OH^-$

Anodic: $Al - 3e^- = Al^{3+}$

The chloride ion provides sufficient conductivity in the aqueous phase
to allow the electrochemical reactions to occur and prevents the formation of a passive film of aluminum oxide at the aluminum/polypropylene
interface.

An important characteristic of polymers and organic coatings is the
glass transition temperature, T_g. It is that temperature at which a
discontinuity occurs in a physical property as a function of temperature when the polymer exists in the amorphous condition. Typical
physical properties which show discontinuities at T_g include the coefficient of expansion and the specific heat. It is interpreted as that
temperature above which the polymer has sufficient thermal energy
for isomeric rotational motion or for significant torsional oscillation
to occur about most of the bonds in the main chain which are capable
of such motion. Values of T_g are obtained by many different experimental techniques, the more common of which include dilatometry, dielectric measurements, spectroscopy, calorimetry, and refractive index measurements. Standish and Leidheiser [32] outlined a simple
technique for the determination of T_g of coatings on a metal, using
dielectric measurements as a function of temperature. The value of
T_g is important because physical properties such as water and oxygen permeability and ductility differ above and below T_g.

The mechanical properties of coatings which are not highly loaded
with pigments or fillers depend on molecular weight, crystallinity, and
the three-dimensional arrangement of the branches. An increase in
molecular weight makes a polymer harder and stronger. The higher
the degree of crystallinity, the stronger the polymer. Chain polymers containing two different groups, R and R', have different mechanical properties depending on the arrangement of the branches. Cur-
ing or cross-linking causes a polymer to become harder, more brittle,
and less soluble.

Polymer coatings are exposed to the environment and thus are subject to degradation by environmental constituents. The main agencies
by which degradation occurs are thermal, mechanical, radiant, and
chemical. Polymers may also be degraded by living organisms such
as mildew. Deterioration takes the form of discoloration, cracking
and crazing, loss of adherence to the substrate, or change in a physical property such as resistivity or mechanical strength. The mode
of degradation may involve depolymerization, generally caused by
heating, splitting out of constituents in the polymer, chain scission,
cross-linking, oxidation, and hydrolysis. Polymers are subject to
cracking on the application of a tensile force, particularly when exposed to certain liquid environments. This phenomenon is known as
environmental stress cracking or stress corrosion cracking and there
are many analogies to similar phenomena observed with metals.

4.3.5 Zinc-Rich and Zinc-Pigmented Coatings

The term zinc-rich applies to coatings which contain up to about 95% metallic zinc in the dry film. They are electrically conductive and protect the metal substrate electrochemically in much the same way as zinc protects steel in galvanized steel. The vehicle is usually a silicate in the case of the so-called inorganic zincs, and various organic polymers are used to obtain so-called organic zincs.

Zinc-pigmented paints are usually a mixture of metallic zinc (80%) and zinc oxide (20%), the latter of which is added to reduce the rate of settling of zinc both in the container and after application to a surface. Galvanic protection does not occur in the case of the zinc metal-zinc oxide paints, whose films do not exhibit good electrical conductivity.

The galvanic action of the zinc when present at concentrations of approximately 95% protects the steel at holidays, cut edges, and scribe marks, and a similar galvanic action probably applies during the first stages of corrosion beneath the coating. The major action of the zinc, however, appears to be sealing of the paint film by precipitation of zinc compounds so that it has improved resistance to penetration by active environmental species. There is also evidence that the zinc pigment and the zinc oxide prevent deterioration of the inorganic and organic binders and assist in maintaining flexibility and desirable mechanical properties of the coating.

4.3.6 Corrosion Inhibitors

It is thought that corrosion inhibitors in an organic coating function in the same way as those added to a liquid environment. Corrosion inhibitors used in coatings include oxidizing agents such as chromate, inorganic salts that function in the same manner as benzoates, metallic cations, of which lead is the most widely used, and organic compounds. All organic coatings are permeable to water, and water contents for many coatings at 100% relative humidity are in the range 0.1 to 3%. Thus, when the coating is wet, some fraction of the inhibitor is solubilized and can be transported to the metal surface. The coating simply serves as a reservoir for the inhibitor.

Six technical requirements for an ideal corrosion inhibitor to be used in organic coatings are:

1. The inhibitor must be effective at pH values in the range 4 to 10 and ideally in the range 2 to 12.
2. The inhibitor should react with the metal surface so that a product is formed which has much lower solubility than the unreacted inhibitor.
3. The inhibitor should have low but sufficient solubility.

4. The inhibitor should form a film at the coating/substrate interface that does not reduce the adhesion of the coating.
5. The inhibitor must be effective as both an anodic and a cathodic inhibitor.
6. The inhibitor should be effective against the two important cathodic reactions, $H_2O + 1/2O_2 + 2e^- = 2OH^-$ and $2H^+ + 2e^- = H_2$.

A critical problem facing the coatings industry at the present time is the need to replace chromates and lead compounds as corrosion inhibitors. Both of these classes of materials have been identified as hazardous and the pressures to remove them from formulations are growing. Adequate proven substitutes have not yet been universally accepted, so there is increased interest in developing accelerated tests that can be used to select corrosion inhibitors.

4.3.7 Methods for Monitoring the Corrosion of Polymer-Coated Metals

Electrical methods for studying the protective properties of coatings are numerous and many have produced important results. Electrical measurements that provide data which are useful in predicting the lifetime of a coating include measurements of coating conductivity, of impedance as a function of frequency, of equivalent ac resistance at constant frequency, and of the ratio of capacitive to resistive components at constant frequency. The ac properties of a coating have also been used to estimate the amount of water taken up by the coating. Scanning techniques have proved useful in characterizing the electrical homogeneity of coatings. The rate of diffusion of sodium chloride through coatings has been determined by using measured values of the dc resistance and the membrane potential of the film. Radiotracer techniques, however, are much preferred for measurements of this type. Dielectric techniques are also useful in determining the glass transition temperature of coatings, in determining the effects of coating composition and structure, and in the quality control of coating components.

Impedance methods are gaining wide acceptance as a means for appraising the properties of organic coatings [33]. Plots of log impedance vs. log frequency yield straight lines when the coating is behaving as a pure capacitor, i.e., there is no significant ionic transport through the coating. The resistance of the coating can be obtained from the same plot at the low frequency values. Information about water uptake and equivalent dc conductivity of the coating can be obtained by measurements at a single frequency [34]. It is an empirical observation that corrosion beneath the coating becomes appreciable when the equivalent dc resistance of the coating is less than

10^7 ohm/cm^2 measured area [35]. Highly protective coatings exhibit an equivalent dc resistance of 10^9 to 10^{11} ohm/cm^2 measured area.

Corrosion potential measurements and their applicability to coated metals have been summarized by Wolstenholme [36]. As a generalization, it can be concluded that movement of the corrosion potential in the noble direction is indicative of an increasing cathodic/anodic surface area ratio and is indicative that oxygen and water are penetrating the coating and arriving at the metal/coating interface. Movement of the corrosion potential in the active direction is indicative that the anodic/cathodic surface area ratio is increasing and that the overall corrosion rate is becoming significant. Increasingly positive potentials with time suggest that alkaline conditions caused by the oxygen reduction reaction are developing locally at the metal/ coating interface and that delamination is of concern. Increasingly active potentials indicate rusting (in the case of steel) beneath the coating and represent a signal that the coating lifetime is limited.

4.3.8 Transport Properties of Coating for Water, Oxygen, and Ions

The corrosion reaction under the coating generally involves the cathodic reaction, $H_2O + 1/2O_2 + 2e^- = 2OH^-$. Thus, oxygen and water are required to be present at the interface along with ionic constituents that provide sufficient conductivity in the interfacial region to permit separation of the anodic and cathodic regions. The following comments will provide some information on the transport properties of organic coatings for water, oxygen, and ions.

Most of the information developed on transport properties in organic materials has been obtained on films after removal from the metal substrate. Removal is accomplished by using a substrate pretreatment that reduces the adhesion or by using an intervening layer between the metal and the coating that can be solubilized so that the coating can be freed from the metal. Free films are attractive for experimental study because the boundary conditions on each side of the film can be controlled and a variety of experimental techniques can be applied with convenience. Measurements made with the coating still adherent to the substrate represent the real world, but there are restrictions on the techniques which can be applied. It must be recognized that measurements made on free films are not extrapolatable with 100% certainty to conditions which exist in an adherent film.

The transport properties are often characterized in terms of the diffusion coefficient D, the permeability coefficient P, or the flux J through the film. The diffusion coefficient is defined as the constant in Fick's first law of diffusion:

$$dJ/dt = -D(dc/dx)$$

TABLE 2 Diffusion Data for Water Through Organic Films[a]

Polymer	Temperature (°C)	P ($\times 10^9$) [cm^3 (STP) cm] (cm^2/sec cm Hg)	D ($\times 10^9$) (cm^2/sec)
Epoxy	25	10—44	2—8
	40	—	5
Phenolic	25	166	0.2—10
Polyethylene (low density)	25	9	230
Polyisobutylene	30	7—22	—
Polymethyl methacrylate	50	250	130
Polystyrene	25	97	—
Polyvinyl acetate	40	600	150
Polyvinyl chloride	30	15	16
Vinylidene chloride/acrylo-nitrile copolymer	25	1.7	0.32

[a]Data taken from Ref. [44].

where dJ/dt is the flux per unit time and dc/dx is the change in concentration with distance in the x direction. The permeability is given as the product D × S, where S is the solubility coefficient of the diffusing species in the film. The flux is the actual mass transported across the film per unit time.

All organic polymers and organic coatings are permeable to water; they differ only in degree of permeability. Typical values for the permeability and the diffusion coefficient for water in a number of different polymers are given in Table 2. Note that the diffusion coefficient, for example, differs in this group by as much as three orders of magnitude. The equilibrium concentration of water in a coating exposed to 100% relative humidity is generally in the range 0.5 to 3%, but values as high as 15% have been determined for hydrophilic polymers such as cellulose acetate. The flux of water through a free film 100 μm thick is typically of the order of 1—10 mg/cm² day for many classes of coatings. If the same value applies to a coating on a metal, it can easily be seen that the flux of water through the coating is not rate-determining in the corrosion process.

TABLE 3 Flux of Oxygen Through Representative
Free Films of Paint, 100 μm Thick

Paint	J $(mg/cm^2 \text{ day})$
Alkyd (15% PVC Fe_2O_3)	0.0069
Alkyd (35% PVC Fe_2O_3)	0.0081
Alkyd-melamine	0.001
Chlorinated rubber (35% PVC Fe_2O_3)	0.017
Cellulose acetate	0.026 (95% RH)
Cellulose nitrate	0.115 (95% RH)
Epoxy melamine	0.008
Epoxy coal tar	0.0041
Epoxy-polyamide (35% PVC Fe_2O_3)	0.0064
Vinyl chloride/vinyl acetate copolymer	0.004 (95% RH)

The rate of oxygen migration through a coating is generally very much lower than that of water. Data for free films of paint, 100 μm thick, are given for representative systems in Table 3. It has been shown in some instances that the flow of oxygen increases with the water content of the film, probably because the water acts to swell the polymer.

Parks [37] measured the rate of diffusion of Na^+ and Cl^- ions through a pigmented alkyd paint on steel using radiotracers dissolved in 0.5 M NaCl and in companion experiments measured the rate of diffusion of water into the same coating using capacitance measurements. The steady-state values for ^{22}Na, ^{36}Cl, and water diffusion yielded diffusion coefficients of 1.9×10^{-11}, 0.6×10^{-11}, and 0.87×10^{-11} cm^2/sec, respectively. Water entered the coating at a much higher rate during the first 3 hr when the samples were first immersed in the electrolyte, and it appeared that the ionic diffusion occurred to a significant extent only after water had penetrated the coating. The rate of migration of sodium through the coating was increased approximately one order of magnitude when the metal substrate was polarized cathodically to -0.8 V vs. SCE. Values of D were of the order of 10^{-11} cm^2/sec and were six orders of magnitude less than values measured in electrolyte solutions.

4.3.9 Development of an Aqueous Phase at Coating/Metal Interface

All organic coatings absorb water and water will pass through them at a rate which differs with the coating system. Many coatings stand aggressive environments for many years without serious deterioration because aqueous-phase water does not form at the metal/coating interface. It is only when an aqueous phase capable of supporting the electrochemical corrosion reaction is present at the interface that the corrosion reaction can occur. As stated earlier in the discussion of blisters, there are a number of mechanisms whereby aqueous phase water can condense at the interface, the most common of which probably involves the presence of a soluble ionic species which retains water as a result of osmotic forces.

An important unanswered question is how much water constitutes aqueous phase water. Although the number of water molecules is unknown, an answer can be given in terms of the consequences. The number necessary is equivalent to a volume that will support an electrochemical reaction. A 1 M solution of sodium chloride is 55 M in water. It is thus hypothesized from the standpoint of corrosion that an aqueous phase exists when 50 or more molecules of water are in physical contact or in a cluster. There are, of course, other approaches to the question. One of these is based on an answer to the question of how many water molecules constitute a stable cluster with a significant lifetime.

The detection of aqueous phase water at the coating/substrate interface is a challenging problem. The total amount of water in a coating is readily determined by radiotracer measurements with tritium, gravimetric measurements, or capacitance measurements, but none of these methods discriminates between aqueous phase water and mobile gaslike water. A technique which shows some promise is based on dielectric measurements at frequencies in the range 10^8 to 10^{10} Hz. Aqueous phase water exhibits an absorption at approximately 10^9 Hz [38]. Other less difficult techniques for detecting aqueous phase water beneath an opaque coating are badly needed.

All organic coatings absorb water, as discussed above, and the amount absorbed may often reach a value of the order of 3%. Water absorption may lead to swelling of the coating, and when this occurs locally for any reason, blisters may form and water may collect at the interface. Air bubbles or volatile components of the coating may become incorporated in the film during film formation and leave a void. Such blisters are not necessarily confined to the interface, but when

they are, they can serve as a corrosion precursor site. Water may
move through a membrane or capillary system under the influence of
a potential gradient by electroosmosis. Potential gradients, such as
may exist in a galvanic couple or with an applied potential, have the
capability of leading to a blister.

4.4 CONVERSION COATINGS

The term conversion coating is used to describe coatings in which the
substrate metal provides ions which become part of the protective
coating. Conversion coatings on steel such as phosphate coatings,
chromate coatings, and mixed oxide coatings are generally used beneath a paint, whereas anodized coatings on aluminum are generally
used with no further protection. This section will cover two types of
conversion coatings, zinc phosphate coatings and anodic oxide coatings on aluminum.

4.4.1 Phosphate Coatings

Commercial phosphating processes are proprietary in nature and
changes in formulation are constantly being made in response to new
demands on the process and to environmental considerations. We will
confine our remarks to the zinc phosphating of steel. A typical phosphating solution contains Zn^{2+} ions, phosphoric acid, other metal ions.
titanium compounds in colloidal form, accelerators such as a nitrite or a
nitrate, and viscosity control agents. The steel surface must be free
of greases, oils, and other carbonaceous material before immersion in
the phosphating solution or before spray application. The presence
of oxide at the surface does not appear to be deleterious to the quality of the phosphate layer formed. Baths operated above 50°C have
pH values of approximately 2 and those operated below 50° have pH
values of approximately 3.

   The development of a zinc phosphate coating is basically a result
of a corrosion process. The anodic reaction is $Fe - 2e^- = Fe^{2+}$ and
the cathodic reaction is $2H^+ + 2e^- = H_2$. The function of accelerators is to react with the hydrogen formed, remove the hydrogen
from the surface, and thus allow the cathodic reaction, which is
rate-limiting initially, to proceed at a higher rate. Very little gaseous hydrogen is evolved when accelerators are present. The cathodic reaction, by its consumption of H^+ ions, causes the pH to increase at the local cathodes and insoluble $Zn_3(PO_4)_2 \cdot 4H_2O$ precipitates on the surface as a consequence of the reaction

$$3Zn(H_2PO_4)_2 + 4OH^- = Zn_3(PO_4)_2 \cdot 4H_2O \text{ (hopeite)}$$

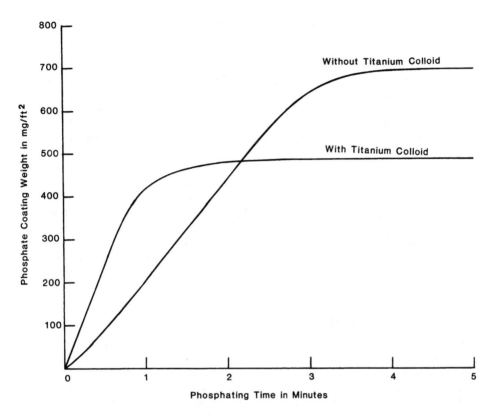

FIGURE 9 Effect of a titanium colloid in increasing the rate of nucleation of a phosphate coating on steel. (Data taken from Ref. [39].)

The titanium colloid, adsorbed on the steel surface, assists in nuclea-
ting the insoluble zinc phosphate (see Fig. 9). Some of the ferrous
ions formed in the corrosion process also precipitate out on the cathodic regions in the form of $Zn_2Fe(PO_4)_2 \cdot 4H_2O$ (phosphophyllite) or
$Fe_3(PO_4)_2 \cdot 8H_2O$ (vivianite). As the surface becomes covered with
phosphate the area available for the anodic reaction is decreased and
the current density on the cathodic regions becomes insufficient to
maintain the high pH needed to precipitate the phosphate. The reaction is self-stifling and the reaction rate falls off with time as shown
in Fig. 9. An equation due to Machu [40] describes the rate at any
moment in terms of the fraction of the anodic surface still available as
the surface becomes coated with the phosphate,

$$K = (2.303/t) \log(F_A^0/F_A)$$

where F_A^0 is the initial anodic surface area and F_A is the anodic surface area at time t.

During phosphating the potential of the steel moves in the noble direction, indicating that the surface area on which the anodic reaction is occurring is being reduced in area fraction at a higher rate than the surface area on which the cathodic reaction is occurring. It is to be expected, however, that a considerable amount of free iron surface will remain exposed at the end of the phosphating procedure. Typically 0.1 to 1% of the surface is not covered by phosphate, and this part of the surface is reduced in activity by passivating treatment with a chromate solution in a second step.

A zinc phosphate coating typically consists of a dense agglomeration of thin crystals lying both in the plane of the metal surface and at various angles to the surface. The number of crystals per square centimeter is of the order of 10^6 and the total thickness of the phosphate layer is 3–5 μm. The phosphate coating is very rough with much open volume. The roughness and open volume provide excellent anchor points for paint and good adhesion between the phosphate layer and paint is achieved.

4.4.2 Anodizing of Aluminum and Its Alloys

The ability to alter the surface properties of aluminum by an electrochemical process known as anodizing has provided the metal finisher with a means for obtaining this industrially important metal in a wide variety of surface conditions. A secondary process known as sealing has vastly increased the usefulness of anodizing and extended the range of applicability to the preparation of finishes of various colors and aesthetically pleasing appearance.

Anodizing, as its name implies, is the electrochemical treatment of a metal in an electrolyte when the metal serves as anode. Many electrolytes are used and have been described in the patent literature, but the majority of processes is based on four solutions: sulfuric acid, chromic acid, boric acid, and oxalic acid and other organic acids.

The most widely used solution for anodization of aluminum is based on sulfuric acid, and it has had wide acceptance since being patented in 1937. A large range of operating conditions can be utilized depending on the specific requirements for the finished part. Hard, protective coatings are obtained which serve as a good base for dyeing, but the coating must be sealed for maximum corrosion resistance. It is not recommended for work containing joints which can retain the electrolyte upon removal from the bath and serve later as sites for corrosion. The anodic coating is devoid of coloring and is clear and transparent on pure aluminum. Alloys containing manganese or silicon and the heterogeneous aluminum-magnesium alloys yield coatings which

range from gray to brown and may be patchy in some cases. The
great adsorptive power of the coating makes it an excellent base for
dyes, especially if the coating is subsequently sealed in nickel or cobalt acetate solution. Coatings formed in sulfuric acid are especially
suitable for the protection of electropolished surfaces, where retention of reflectivity is important.

Coatings produced in chromic acid are unattractive compared to
coatings produced in sulfuric acid. They are generally opaque, gray,
and iridescent, but the quality is dependent on the concentration and
purity of the electrolyte. Colorless and transparent coatings can be
obtained in the presence of 0.03% sulfate. The coatings are generally
thin and of low porosity and hence difficult to dye. In concentrated
solutions and at elevated temperatures, black coatings can be obtained; addition of titanium, zirconium, and thallium compounds to
the electrolyte results in the formation of attractive opaque surfaces.
Chromic acid anodizing plays an important and indispensable part in
aircraft finishing. It is the only anodizing process which can be
safely employed on structures containing blind holes, crevices, or
difficult-to-rinse areas. A specialized application of the chromic acid
electrolyte is in the detection of cracks on items such as propeller
blades. The acid seeps into the crack and allows the crack to be
observed after anodization. Fatigue strength is generally increased
by chromic acid anodizing, whereas sulfuric acid electrolytes may
produce decreases in fatigue strength.

Boric acid electrolytes are used almost exclusively for the preparation of thin dielectric films on electrolytic condensers. The film
is iridescent and oxides in the range, 2500—7500Å, are generally obtained. The coating is essentially nonporous.

Oxalic and other organic acids are applicable electrolytes both for
protective and for decorative purposes. The unsealed coatings are
generally yellow in color. They are harder and more abrasion-re-
sistant than conventional sulfuric acid coatings, but they are not
superior to the specially hard coatings obtained under certain conditions in solutions based on sulfuric acid.

The anodic films produced in chromic acid and boric acid electrolytes are nearly pure Al_2O_3, but those formed in sulfuric acid contain considerable amounts of sulfate ion and small amounts of water.
The actual sulfate concentration varies considerably depending on
the conditions of formation, but 15% sulfate is common. The sulfate
concentration is less at higher formation temperatures and lower
current densities. The presence of sulfate has been confirmed by
radioactive tracer studies, X-ray fluorescence measurements, electron probe microanalysis, infrared spectroscopy, and chemical analysis. The sulfate ion is very tightly bound since only a small fraction is removed by extensive leaching. Anodic films formed in oxalic
acid and in phosphoric acid also contain included anions, but the

concentration is much less than in films formed in sulfuric acid. Very little of the anion is removed during sealing in boiling water.

The water content of unsealed films is rather low. Normally prepared films lose on the order of 2% water on heating at 200°C and negligible additional amounts above 200°. The water in the unsealed films is probably physically adsorbed on the interior of the pores or retained in pores by capillary action. Sealed films typically lose 2% in weight after heating at 100°C, with continuing weight loss with increase in temperature up to 600°C. The total weight loss of a film formed at 25°C and sealed for 60 min in water at the boiling point was 13%.

When alloys of aluminum are anodized, the amount of alloying element that enters the anodic film is as follows: chromium—high; copper—low; iron—generally low; magnesium—appreciable; manganese—low; silicon—high; and zinc—considerable.

As a rule, anodic films formed in sulfuric, chromic, boric, or oxalic acid are sufficiently noncrystalline that they do not give X-ray or electron diffraction patterns characteristic of crystalline materials. In some cases, however, weak diffraction patterns are obtained. Kormany [41] examined several hundred anodic films and concluded that three possible solid forms are obtained: (a) amorphous aluminum oxide only, (b) amorphous aluminum oxide plus crystalline $Al_2O_3 \cdot 3H_2O$, and (c) amorphous aluminum oxide plus crystalline $Al_2O_3 \cdot H_2O$.

The anodized film consists of two major components, the nonporous barrier layer adjoining the metal and a porous layer extending from the barrier layer to the outer surface of the film. In the case of thick films especially there may be an identifiable transition region between the barrier layer and the porous layer. Boric acid electrolytes yield barrier-type films only, whereas both barrier and porous layers are obtained in sulfuric, chromic, and oxalic acids. Electron microscope studies of the barrier layer indicate that it consists of a hexagonal distribution of cells that continue up into the porous layer. The central portion of each cell is amorphous, whereas the outer portion has a partially crystalline nature. In the case of aluminum anodized in boric acid at 20°C and with an applied voltage of 500 V, the number of cells was $1.4 \times 10^8/cm^2$.

4.5 CEMENTITIOUS COATINGS

Three main corrosion-resistant cements are being used today: alkali silicates, portland cement, and calcium aluminates. The alkali silicates form a hard coating by a polymerization reaction involving repeating units of the structure $-\underset{\underset{OH}{|}}{\overset{\overset{OH}{|}}{Si}}-O-$, whereas the calcium aluminates and portland cements consume water and form hard hydrated products.

Cementitious coatings are applied by casting, troweling, or spraying. The spraying process, known under the trade term Shotcrete or Gunite, is especially applicable to systems with unusual geometries or with many sharp bends or corners. It has the advantage that there are no seams, which are often the weak points from the standpoint of corrosion protection.

Cementitious coatings provide corrosion resistance to substrates such as steel by maintaining the pH at the metal/coating interface above 4, a pH range where steel corrodes at a low rate. Surface preparation prior to application of the cement is based on the same principles that govern the application of polymeric coatings. The surface must be free of mill scale, oil, grease, and other chemical contaminants. The surface should be roughened by sandblasting and the coating should be applied immediately after surface preparation. In situations where adhesion between the substrate and the coating is poor or where thermal expansion characteristics are incompatible, an intervening bonding coating is used.

The suitability of four types of cement coatings for use in preventing attack by electrolytes and organic compounds is summarized in Table 4 from data provided by Hall [42].

TABLE 4 Chemical Resistance of Four Cementitious Coatings [42][a]

Generic cement type	Silicate/ silica	Calcium aluminate	Modified silicate	Portland
pH range	0-7	4.5-10	0-9	7-12
H_2O resistance	E	E	E	E
H_2SO_4	E	X	E*	X
HCl	E	X	G	X
H_3PO_4	G*	P	G*	X
HNO_3	E	X	E	X
Organic acids	E	F	E	P
Solvents	E	G	E	C
NH_4OH	F	F	E	G
NaOH	X	F	F*	F
$Ca(OH)_2$	X	F	F*	G
Amines	X	F	G*	G

[a] E, excellent; G, good; F, fair; P, poor; X, not recommended; *, consult manufacturer.

REFERENCES

1. S. R. Cecil and J. G. Woodroof, *Food Technol.*, *17*: 639 (1963).
2. R. M. Lueck and H. T. Blair, *Trans. Electrochem. Soc.*, *54*:
    257 (1928).
3. E. F. Kohman and N. H. Sanborn, *Ind. Eng. Chem.*, *20*: 1373
    (1928).
4. T. P. Hoar, *Trans. Faraday Soc.*, *30*: 472 (1934).
5. W. R. Buck III and H. Leidheiser, Jr., *Z. Elektrochem.*, *62*:
    690 (1958).
6. W. R. Buck III and H. Leidheiser, Jr., *J. Electrochem. Soc.*,
    108: 203 (1961).
7. W. R. Buck III, A. N. J. Heyn, and H. Leidheiser, Jr., *J.
    Electrochem. Soc.*, *111*: 386 (1964).
8. W. R. Buck III and H. Leidheiser, Jr., *J. Electrochem. Soc.*,
    112: 243 (1965).
9. R. J. Morrissey, *J. Electrochem. Soc.*, *117*: 742 (1970).
10. M. Stern, *Corrosion*, *14*: 329t (1958).
11. R. J. Morrissey and A. M. Weisberg, *Trans. Inst. Met. Finish.*,
    53: 9 (1975).
12. W. Funke, *Prog. Org. Coat.*, *9*: 29 (1981).
13. M. J. Grourke, *J. Coat. Technol.*, *49*(632): 69 (1977).
14. E. L. Koehler, in *Localized Corrosion* (R. W. Staehle, B. F.
    Brown, J. Kruger, and A. Agrawal, eds.), National Association
    of Corrosion Engineers, Houston, 1974, p. 117.
15. R. T. Ruggeri and T. R. Beck, *Corrosion*, *39*: 452 (1983).
16. R. St. J. Preston and B. Sanyal, *J. Appl. Chem.*, *6*: 26 (1956).
17. G. M. Hoch, in *Localized Corrosion* (R. W. Staehle, B. F. Brown,
    J. Kruger, and A. Agrawal, eds.), National Association of Cor-
    rosion Engineers, Houston, 1974, p. 134.
18. E. L. Koehler, *Corrosion*, *33*: 209 (1977).
19. J. J. Ritter and J. Kruger, *Surf. Sci.*, *96*: 364 (1980).
20. J. S. Hammond, J. W. Holubka, and R. A. Dickie, *J. Coat.
    Technol.*, *51*(655): 45 (1979).
21. E. L. Koehler, *Corrosion*, *40*: 5 (1984).
22. O. D. Gonzalez, P. H. Josephic, and R. A. Oriani, *J. Electro-
    chem. Soc.*, *121*: 29 (1974).
23. H. Leidheiser, Jr. and I. Suzuki, *J. Electrochem. Soc.*, *128*:
    242 (1981).
24. W. Funke, in *Corrosion Control by Coatings* (H. Leidheiser, Jr.,
    ed.), National Association of Corrosion Engineers, Houston,
    1981, p. 70.
25. L. K. Schwab and R. W. Drisko, in *Corrosion Control by Organic
    Coatings* (H. Leidheiser, Jr., ed.), National Association of Cor-
    rosion Engineers, Houston, 1981, p. 222.

26. F. Mansfeld, J. B. Lumsden, S. L. Jeanjaquet, and S. Tsai, in *Corrosion Control by Organic Coatings* (H. Leidheiser, Jr., ed.), National Association of Corrosion Engineers, Houston, 1981, p. 227.

27. R. W. Zurilla and V. Hospadaruk, *SAE Trans.*, *87*: 762 (1978).

28. R. A. Iezzi and H. Leidheiser, Jr., *Corrosion*, *37*: 28 (1981).

29. H. Leidheiser, Jr., G. W. Simmons, and E. Kellerman, *J. Electrochem. Soc.*, *120*: 1516 (1973).

30. H. Leidheiser, Jr., S. Musić, and G. W. Simmons, *Nature*, *297*: 667 (1982).

31. A. Yializis, S. W. Cichanowski, and D. G. Shaw, paper presented at IEEE Meeting, Boston, May 1980.

32. J. V. Standish and H. Leidheiser, Jr., *J. Coat. Technol.*, *53*(678): 53 (1981).

33. F. Mansfeld, M. W. Kendig and S. Tsai, *Corrosion*, *38*: 478 (1982).

34. R. E. Touhsaent and H. Leidheiser, Jr., *Corrosion*, *28*: 435 (1972).

35. H. Leidheiser, Jr., *Prog. Org. Coat.*, *7*: 79 (1979).

36. J. Wolstenholme, *Corros. Sci.*, *13*: 521 (1973).

37. J. Parks, thesis, Lehigh Univ., January 1985.

38. D. Eadline, thesis, Lehigh Univ., October 1984.

39. J. V. Laukonis, in *Interface Corrosion for Polymer Coatings* (P. Weiss and G. D. Cheever, eds.), Elsevier, Amsterdam, 1968, p. 182.

40. W. Machu, *Interface Corrosion for Polymer Coatings* (P. Weiss and G. D. Cheever, eds.), Elsevier, Amsterdam, 1968, p. 128.

41. I. T. Kormany, *Tavkozlesi Kut. Int. Kozlemen*, *9*: 113 (1964); known through *Chem. Abstr.*, *64*: 1619 (1964).

42. G. R. Hall, in *Encyclopedia of Materials Science and Engineering*, Pergamon, London, in press.

43. G. A. Di Bari and J. V. Petrocelli, *J. Electrochem. Soc.*, *112*: 99 (1965).

44. A. J. Kinloch, in *Adhesion 3* (K. W. Allen, ed.), Applied Science, London, 1979, pp. 1–12.

5

ATMOSPHERIC CORROSION

VLADIMIR KUCERA and EINAR MATTSSON

Swedish Corrosion Institute, Stockholm, Sweden

5.1 INTRODUCTION

Atmospheric corrosion of metals has been studied all over the world for many decades, in several countries as the first subject chosen when starting up corrosion research. The reason is, of course, the extensive use of metals for outdoor structures such as buildings, bridges, pylons, fences, cars, and ships. Most of the studies have been field tests at sites in different types of climate, and much useful information has been obtained from them. Advanced research tools have also been employed for laboratory investigations of atmospheric corrosion and its mechanisms. Still, this complex phenomenon has not yet been fully clarified. In this chapter the main features of atmospheric corrosion will be surveyed for the commonly used metals (steel, zinc, aluminum, and copper). Efforts will be made to correlate the fundamental corrosion properties of these metals with test results and with application aspects, using a previous treatise as starting point [1].

5.2 CONDITIONS FOR ATMOSPHERIC CORROSION

Atmospheric corrosion proceeds in a relatively complicated system consisting of metal, corrosion products, surface electrolyte, and atmosphere. It is generally an electrochemical process which takes place in corrosion cells. The cells can operate only when an electrolyte is present on the metal surface. Atmospheric corrosion may thus, at least from the practical point of view, be considered as a

discontinuous process. It has been described by Barton and Bartonova [2] by the equation

$$K = \sum_{1}^{n} \tau_n v_k(n)$$

where

K = accumulated corrosion effect
τ_n = time of wetness, i.e., the period with an electrolyte layer on
        the surface
v_k = average corrosion rate during the individual periods of wet-
        ting

The total corrosion effect over a period is thus determined by the
total time of wetness and the composition of the electrolyte, which
usually, together with the temperature, determines the corrosion rate.
The parameters affecting the time of wetness and the composition of
the electrolyte film will now be briefly surveyed.

5.2.1 Time of Wetness

There is no unambiguous definition of time of wetness. The principal
meaning of the term is the length of time during which the metal surface is covered by a film of water that renders significant atmospheric
corrosion possible.

The time of wetness varies with the climatic conditions at the site.
It depends on the relative humidity of the atmosphere, the duration
and frequency of rain, fog, and dew, the temperature of the air and
the metal surface, as well as the wind speed and hours of sunshine.
According to Tomashov [3] and Rozenfeld [4], the total time of wetness, τ_{wet}^{tot}, may be divided into the periods when the metal is mois-
tened due to adsorption of water vapor on the surface, τ_{ads}, and the
periods when the surface is covered by a phase layer of water, τ_{pha},
due to rain, wet snow, fog, or dew:

$$\tau_{wet}^{tot} = \tau_{ads} + \tau_{pha}$$

There is no sharp boundary between these two categories of time of
wetness, and it may be difficult to distinguish between them experimentally.

Adsorption Layers

The amount of water adsorbed on a metal surface depends on the relative humidity of the atmosphere and on the chemical and physical

properties of the corrosion products. The metal surface may be wetted if hygroscopic salts, deposited or formed by corrosion, absorb water from the atmosphere. Such absorption occurs above a certain relative humidity, called the critical relative humidity. Bukowiecki [5] showed by laboratory experiments that its value depends on the metal and the surface contaminants. Usually the corrosion sharply increases when the relative humidity rises above the value at which the salt starts to absorb water and dissolve. This critical relative humidity corresponds to the vapor pressure above a saturated solution of the salt present (Fig. 1). Capillary condensation may also contribute to the formation of adsorption layers of electrolyte on the surface [3,6], although its practical importance in the corrosion process has not yet been established.

The amount of water on the corroding surface is of great importance for the corrosion rate. Several laboratory investigations [7–10] have shown that the corrosion rate above the critical value sharply increases with increasing relative humidity (Fig. 2).

Barton et al. [11] roughly estimated the amount of water on the metal surface as follows:

Conditions	Amount of water (g/m^2)
Critical relative humidity	0.01
100% relative humidity	1
Covered by dew	10
Wet from rain	100

Phase Layers

Phase layers may arise from precipitation of rain, fog, or wet or melting snow, or from dew formed by condensation on cold metallic surfaces.

Dew

Dew formation occurs when the temperature of the metal surface is below the dew point of the atmosphere. This may occur outdoors during the night, when the surface temperature may decrease by radiant heat transfer between the metal structure and the sky. Another reason for dew formation may be the conditions in the early morning, when the temperature of the air increases faster than the temperature of the metal, especially if the mass, and thus the heat capacity, of the metal is great. Dew may also form when metal products are brought into warm storage after cold transport.

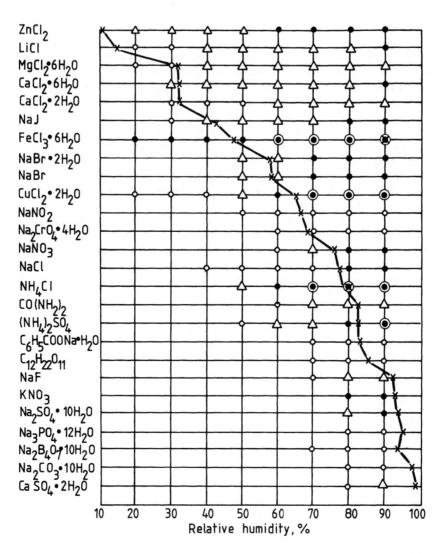

FIGURE 1 Corrosivity of different salts toward steel at exposure to an excess of the salt in 7-day laboratory exposures at different relative humidities, 20°C [5]. × = Relative humidity above a saturated solution of the salt. Corrosion loss = O, 0--1 mg (no corrosion); △, 2-5 mg (very slight corrosion); ●, 6-20 mg; ⊙, 21-100 mg; ⊗, >100 mg.

Dew is considered a very important cause of corrosion, especially under sheltered conditions. The amount of water on a surface covered by dew is about 10 g/m², which is considerably more than that

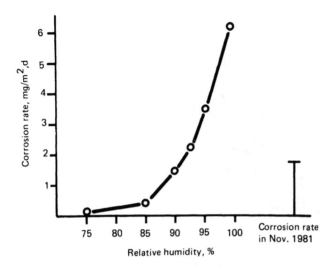

FIGURE 2 Corrosion rate of steel, preexposed for 1 month in an urban atmosphere, when exposed in the laboratory at different relative humidities [10].

on surfaces covered by adsorption layers. Periods of dew are considered very corrosive, as the washing effect is usually slight or negligible. Laboratory experiments performed by Knotkova et al. [12] and Ericsson and Sydberger [13], who sprayed specimens periodically with distilled water, showed that wetting increased the corrosion rates very substantially up to some orders of magnitude (Fig. 3).

One factor contributing to the high corrosivity of dew may be the large content of atmospheric contaminants in dew. pH values lower than 3 have been reported in heavy polluted industrial atmospheres [7]. This applies especially to inert substrates such as painted surfaces, on which the dry deposition between periods of rain can lead to accumulation of corrosive, fairly acid pollutants. Concentrations of 0.35 g chlorides/liter and 0.20 g sulfates/liter have been reported by Strekalov [14]. These were about 100 times higher than those occurring in rainwater at the same location. The extreme corrosivity of the electrolyte formed by dew condensation may be seen in the effect of runoff of condensed water from painted roofs. Such runoff may cause perforation of guttering of galvanized steel plate within 3 years.

Rain

Rain creates even thicker layers of electrolyte on the surface than dew. The thickness of the water layer retained on the surface has

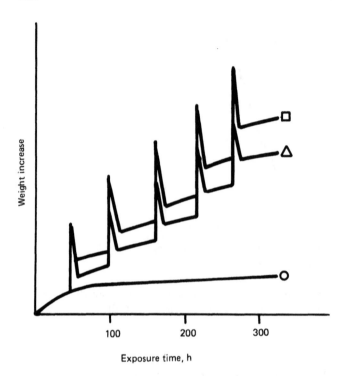

FIGURE 3 Schematic appearance of weight increase curves for steel samples at 90% relative humidity with and without periodic wetting [13]. ○ = No wetting; △ = periodic immersion; □ = periodic spraying.

been estimated to be about 100 g/m^2. Precipitation in the form of rain affects corrosion by giving rise to a phase layer of moisture on the material surface and by adding corrosion stimulators in the form of, e.g., H$^+$ and SO$_4^{2-}$. On the other hand, rain also washes away pollutants deposited on the surface during the preceding dry period. Whereas the first two processes promote corrosion, the third—at least in the case of steel—decreases corrosion. The significance of the two latter processes is dependent on the ratio between the dry and wet deposition of pollutants. As shown in Table 1, the corrosion on the skyward side of steel plates in the strongly polluted atmosphere of Kvarntorp is substantially lower than that on the downward side. In a strongly polluted atmosphere, where dry deposition is considerably greater than wet deposition of sulfur pollutants, the washing effect of rain predominates. At the two other less polluted sites the situation is reversed, which indicates that the corrosive action of rain there is more important [15].

TABLE 1 Corrosion on Skyward and Groundward Sides of Steel Plates at Three Test Sites in Sweden [15]

Test site	Corrosion rate (g/m² year)		Skyward (% of total)
	Skyward	Groundward	
Kiruna	73	59	55
Bohus Malmön	145	107	58
Kvarntorp	279	466	37

The pH value of precipitation seems to be of significance for metals whose corrosion resistance may be ascribed to a protective layer of basic carbonates or sulfates, as on zinc or copper. If the pH of the rainwater falls to values close to 4 or even lower, this may lead to accelerated dissolution of the protective coatings. It may be mentioned that pH values as low as 3.7 and 2.9 on a monthly and daily basis, respectively, have been found in Norway [16].

Especially high acidity and high concentrations of sulfate and nitrate can be found in fog droplets in areas with high air pollution. In California the pH of the fog water has been found to be in the range 2.2 to 4.0 [17]. The processes controlling the fog water chemistry appear to be condensation of water vapor on and its evaporation from preexisting aerosol and scavenging of gas-phase nitric acid.

Measurement of Time of Wetness

For practical purposes the time of wetness is usually determined on the basis of meteorological measurements of temperature and relative humidity. The period when the relative humidity $\geq 80\%$ at temperatures $>0°C$ is often used for estimation of the actual time of wetness. The time of wetness determined by this method may not necessarily be the same as the "actual" time of wetness, because wetness is influenced by the type of metal, pollution of the atmosphere, presence of corrosion products, and degree of coverage against rain. The expression for the time of wetness mentioned above, although not based on a detailed theoretical model, usually shows a good correlation with corrosion data from field tests under outdoor conditions. This implies that this parameter corresponds to the kinetically decisive time periods during which corrosion proceeds. Under sheltered and indoor conditions, however, other criteria seem to be valid, which have not yet been fully clarified.

The time of wetness may, however, also be measured directly with electrochemical cells. The cell consists of thin metal electrodes (0.5—1 mm) separated from each other by a thin insulator (0.1—0.2 mm). Originally, galvanic cells of the Fe-Cu type were used by Tomashov and co-workers [18] and Pt-Zn cells were used by Sereda [19]. When the surface becomes wetted a current starts to flow, and the time of wetness is defined as the time when the cell current or the electromotive force exceeds a certain value. Later, electrolytic cells with electrodes of steel, zinc, copper, aluminum, or passive metals (e.g., stainless steel or gold) were used [20—22]. Not even the direct electrochemical measurement of time of wetness will give an unambiguous value, as the result depends, among other factors, on the type of cell, its dimensions, the presence of corrosion products, and the definition of the threshold value of current or voltage which is considered the lower limit for the time of wetness. Most of the electrochemical techniques indicate mainly the time of wetness caused by phase layers of electrolyte, and they usually give lower values than calculations from meteorological data. Values between 1000 and 2700 hr/year are often reported from sites in the temperate climatic zone [23—25].

5.2.2 Composition of Surface Electrolyte

The electrolyte film on the surface will contain various species deposited from the atmosphere or originating from the corroding metal. For the thermodynamic as well as the kinetic conditions of the corrosion process the composition of the electrolyte is often of decisive importance. In the following a brief survey will be presented of the origin, transformation reactions, deposition mechanisms, and levels of the main atmospheric constituents and pollutants.

Oxygen

Oxygen is readily absorbed from the air, so that at least the outer region of the thin water film on the metal surface may be considered saturated with oxygen.

SO_x

The main part of anthropogenic SO_x pollution is caused by combustion of fossil fuels, i.e., oil and coal in industrialized regions, which cover less than 5% of the earth's surface. Averaged over the surface of the globe, anthropogenic and natural emissions of SO_x into the atmosphere are of about the same magnitude, about 100 million tons of S per year. Emissions of SO_2 in Europe sharply increased after 1950 due to the rise of oil consumption and amounted to about 25 million tons of sulfur by 1970 (Fig. 4) [26,27].

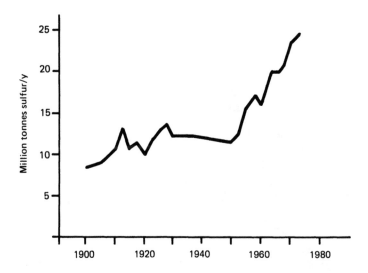

FIGURE 4 Anthropogenic emissions of sulfur dioxide (expressed as sulfur) in Europe during the 20th century [26].

Most of the sulfur derived from the burning of fossil fuels is emitted in gaseous form as SO_2. Both the chemical composition and the physical state of the pollutants change during their transport in the atmosphere. The sulfur dioxide is oxidized on moist particles or in droplets of water to sulfuric acid:

$$SO_2 + H_2O + \frac{1}{2}O_2 \rightarrow H_2SO_4$$

The lifetime of SO_2 in the atmosphere is usually 1/2 to 2 days, which corresponds to a mean transport distance of a few hundred kilometers. The sulfuric acid can be partly neutralized, particularly with ammonia derived from the biological decomposition of organic matter. This results in the formation of particles containing $(NH_4)_2SO_4$ and different forms of acid ammonium sulfate such as NH_4HSO_4 and $(NH_4)_3H(SO_4)_2$ [28].

The main processes of deposition of sulfur compounds are [29]:

Adsorption of gas (SO_2) on material surfaces⎱
Impaction of particles (sulfates)⎰ —dry deposition
Removal of gas and aerosols by precipitation—wet deposition

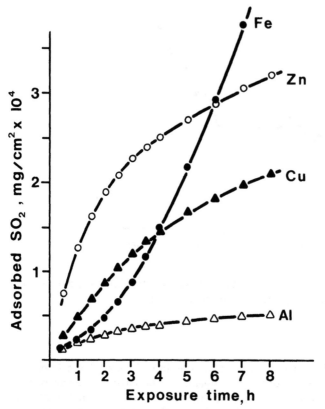

FIGURE 5 Deposition rate of SO_2 on polished samples of carbon steel, zinc, copper, and aluminum as a function of time for exposure in 0.1 ppm SO_2 at 90% relative humidity [30].

Dry deposition consists mainly of the adsorption of sulfur dioxide. The deposition is proportional to the concentration in the atmosphere. The deposition rate differs on different materials, as shown in Fig. 5. Whereas SO_2 is absorbed virtually quantitatively on rusty steel surfaces at high relative humidities, the deposition on copper, and particularly on aluminum, is much less [30].

The rate of dry deposition of particulate sulfur compounds seems to be about one order of magnitude less than that of sulfur dioxide. Sulfate is deposited primarily by wet deposition and has a lifetime of 3–5 days. This corresponds to a mean transport distance of the order of a thousand kilometers [27]. In urban and industrial areas which are close to the emission sources the sulfate concentrations are thus usually substantially lower than the SO_2 concentrations. Dry deposition of sulfates is therefore of minor importance compared to SO_2 in areas with large emissions. This is in accordance with findings from the literature. The corrosive effect of sulfates has been demonstrated in several laboratory investigations [31–34]. In field tests, however, the effect has not been found.

From the practical point of view, atmospheric corrosion is confined to areas with a large population, many structures, and severe pollution. Thus the atmospheric corrosion caused by S pollutants is usually restricted to an area close to the source, as illustrated by Fig. 6 [15]. In such areas, dry deposition of SO_2 is likely to be the most important sulfur deposition process for the atmospheric corrosion of structural metals.

The deposition rate depends, as already pointed out, on the concentration in the air. The concentration may, of course, vary considerably, and thus it is difficult to give reliable ranges. Several authors [7,35,36] have compiled data on major constituents; the values from the United States given by Rice et al. [36] may serve as an example (Table 2). The deposition rates for SO_2 in various types of atmosphere are of the following order of magnitude [37,38]:

Type of atmosphere	Deposition rate (mg SO_2/m^2 day)
Rural	<10
Urban	10–100
Industrial	up to 200

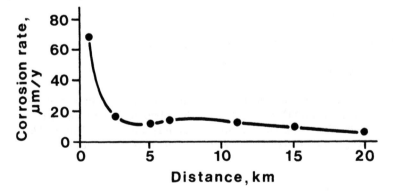

FIGURE 6 Corrosion rate of carbon steel as a function of distance from the emission source [15].

TABLE 2 Typical Indoor and Outdoor Ranges of Inorganic Pollutants in the United States [36]

Pollutant	SO_2	NO_2	H_2S	O_3	HCl	Cl_2	NH_3
Outdoor range, $\mu g/m^3$	3–185	20–160	1–36	10–90	0.3–5	<5% of HCl levels except where local source exists	6–12
Indoor range, $\mu g/m^3$	1–40	3–60	0.2–1	7–65	0.08–0.3	0.004–0.015	10–150

NO_x

Emissions of NO_x originate mainly from different combustion processes, road traffic and energy production being the main sources. Another source of NO_x which may be of importance in microclimates is electric discharge phenomena. European NO_x emission has increased from rather low values 100 years ago to the present value of about 6 million tons of nitrogen per year. Since 1940 the atmospheric emission of NO_x relative to SO_2 (in equivalents) has increased markedly, which stresses the increasing importance of NO_x emissions [26,27].

In combustion processes most of the nitrogen oxides are emitted as NO. In the atmosphere oxidation to NO_2 takes place successively according to

$$2NO + O_2 \rightarrow 2NO_2$$

and at greater distances from the source by the influence of ozone

$$NO + O_3 \rightarrow NO_2 + O_2$$

Nitrogen dioxide may be considered the main nitrogen pollutant near emission sources. The NO_2/NO ratio in the atmosphere varies with time and distance from the source and is usually between 10 and 100. In addition, NO_x may be oxidized to HNO_3 according to the total reaction

$$2NO + H_2O + \frac{3}{2}O_2 \rightarrow 2HNO_3$$

This reaction, however, has a very low rate. Therefore in the vicinity of the emission source the contents of HNO_3 and nitrates are very low [39,40].

The mechanisms of deposition of nitrogen compounds have not yet been clarified in all details. At long distances from the emission source wet deposition of nitrates seems to predominate. Near the source wet deposition is probably substantially lower than dry deposition, as NO and NO_2 have a low solubility in water and HNO_3, which is highly soluble in water, has not yet formed [41,42]. In two field corrosion tests [43,44] the deposition of NO_X was measured. The amounts of deposited NO_X and NO_3^- were found to be 10 to 100 times lower than the amount of SO_2.

Chlorides

Chlorides are deposited mainly in the marine atmosphere as droplets or as crystals formed by evaporation of spray carried by the wind from the sea. Other sources of chloride emission are coal burning and municipal incinerators. Most coals have a chloride content of 0.09–0.15%. In high-chlorine coals, values of 0.7% are found. In the burning of coal most of the chlorine is emitted as gaseous HCl.

In marine environments chloride deposition usually decreases strongly with increasing distance from the shore, as the droplets and crystals settle by gravitation or may be filtered off when the wind passes through vegetation. Deposition rates in marine areas are reported to be in the range 5 to 1500 mg $NaCl/m^2$ day, using the wet candle method [38,45,46].

In urban and industrial areas the deposition rates of chloride are usually less than 10% of that of SO_2, as illustrated in Table 3.

TABLE 3 Deposition Rates (mg/m^2 day) of Chloride and SO_2 at Different Places

Place	Period	Cl^-	SO_2	Cl^-/SO_2 (%)
Essen	Sept. 1978–Sept. 1979	3.9	93.4	4.1
Duisburg	Sept. 1978–Sept. 1979	12.9	126.0	10.2
Prague-Letnany	June 1978–May 1979	4.6	73.0	6.3
Stockholm-Vanadis	July 1975–May 1979	1.9	26.0	7.3

CO$_2$

Carbon dioxide occurs in the atmosphere in a concentration of 0.03 to
0.05% by volume, varying slightly with the time of day and the sea-
son of the year due to its cycle in nature [47]. At equilibrium, the
percentages mentioned correspond to a concentration of the order of
10^{-5} mole/liter in the water film, if its pH value is 6 or lower.

Concentrations of Different Species

The concentrations of the various species in the electrolyte on the
surface vary greatly with respect to parameters such as: deposition
rates, corrosion rate, intervals between rain washing, presence of
rain shelter, and drying conditions.

To determine what concentration of the corroding metal may occur
in the electrolyte film, the supply of corrosion products has been calculated on the assumption of a corrosion rate of 1 μm/year and the
amounts of water on the surface mentioned earlier. The following results were obtained:

	Amount of water (g/m^2)	Supply of corrosion products (mole/liter day)
Critical relative humidity	0.01	30
100% relative humidity	1	0.3
Covered by dew	10	0.03
Wet from rain	100	0.003

One may conclude that the concentration in the electrolyte film will be
low during a rainy period, while a highly concentrated solution may
form after a long period without rain washing.

The pH value of the water film is also difficult to specify. A moisture film in contact with an atmosphere highly polluted with SO$_x$ may
initially have a pH value as low as 2. Due to acid rain or fog the moisture film may also have a low pH value. Because of reaction with the
metal and the corrosion products the pH value will usually increase.
When a steady state has been reached the pH is generally of the order of 5–6 [48].

5.2.3 Temperature

The influence of temperature on atmospheric corrosion is complex and
has not been the subject of many systematic investigations [10,11,35,

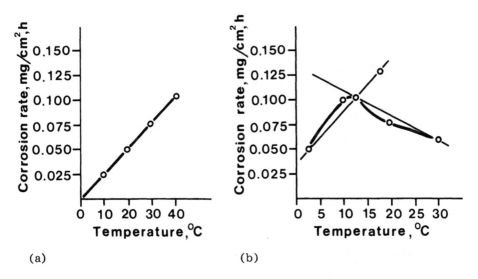

FIGURE 7 Influence of air temperature on the atmospheric corrosion rate of carbon steel [50]: (a) during rain and (b) at drying-out of a film of water at a constant relative humidity of 75%.

49–51]. It seems, however, that the influence of temperature on atmospheric corrosion is greater on carbon steel than on, e.g., zinc and copper [11,52]. On the one hand, an increase in temperature stimulates the corrosion attack by enhancing the rate of electrochemical and chemical reactions as well as diffusion processes. Thus, under constant-humidity conditions a temperature increase promotes corrosion (Fig. 7). On the other hand, an increase in temperature leads to more rapid evaporation of surface moisture films created, e.g., by dew or rain. Consequently, the time of wetness is shortened and the corrosion rate decreases (Fig. 7) [50]. The solubility of oxygen and corrosive gases in the electrolyte layer also decreases with increasing temperature. At temperatures below 0°C the electrolyte film may freeze. This leads to a very pronounced decrease in the corrosion rate, which may be illustrated by the low corrosion in subarctic and arctic regions [53]. While the corrosion rate of carbon steel during 4 years of exposure was 11–13 μm/year in rural areas in southern Scandinavia, it was only 4 μm/year at the subarctic site Gällivare in northern Sweden. Thus, even if the temperature may influence the corrosion rate strongly under certain conditions, it seems generally not to be of decisive importance during long-term exposure in the temperate climatic zone.

5.3 THERMODYNAMIC AND KINETIC CONCEPTS

The results surveyed in the preceding section will be used for thermodynamic and kinetic considerations to explain the mechanism of atmospheric corrosion of the common structural metals. In the sequel
the basic concepts will be presented briefly.

   The thermodynamic possibilities for atmospheric corrosion reactions
on a metal surface and the formation of different corrosion products
will be surveyed in so-called potential-pH diagrams. This type of
diagram was developed by Pourbaix [54], who worked out an atlas of
diagrams for various metals in contact with pure H_2O. As an example, the diagram for the system $Cu-H_2O$ is shown in Fig. 8. The

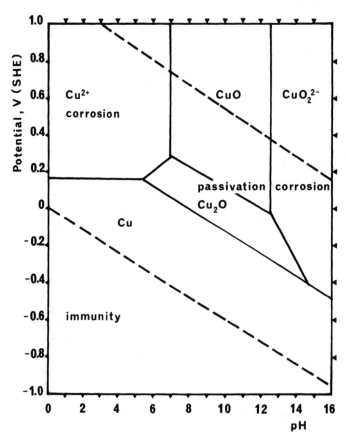

FIGURE 8 Potential-pH diagram, $Cu-H_2O$; 25°C, 10^{-6} M Cu [54].

potential-pH diagram shows the stability domains for various species with respect to the redox potential and the pH value of the corrosive agent. The stability domain for water is enclosed between the two dashed lines. The positions of the borderlines between solid species and species in solution depend on the concentrations (or rather activities) of the species in solution.

In atmospheric corrosion the metal ion concentration in the water film on the metal surface is fluctuating and is generally rather high. Therefore, in the diagrams used here it has been set at 10^{-1} mole/liter. When constructing diagrams representative of atmospheric corrosion, one must also take into account species other than those in the metal-water system dealt with in the Pourbaix atlas. Thus one must consider H_2CO_3, SO_4^{2-}, Cl^-, and NO_3^-, which may be dissolved in the moisture film and under certain conditions may form solid phases or complexes with the corroding metal. The stability domains for these species can be calculated from stability constants available in the literature, assuming appropriate concentration values for the species involved; the constants used here have generally been taken from a selection made by Hogfeldt and Sillén [55]. In the potential-pH diagram the potential and the pH value are proportional to the *logarithm* of the concentrations (activities). Therefore, even if the concentrations in the moisture film fluctuate and deviate considerably from the values assumed, the stability domains calculated may still give approximate information about which species are stable on the surface. It should be pointed out, however, that the potential-pH diagram gives information only about the thermodynamic stability and possibilities for reactions, not about the kinetics.

Regarding the kinetics, the corrosion rate depends on the amount and composition of the electrolyte on the surface. The instantaneous corrosion rate varies strongly with time, its maximum value being several powers of 10 greater than its minimum value (Fig. 9 and Table 4). Nevertheless, the cumulative attack over a long period does not depend so strongly on the time of wetness, for when the surface becomes wet, the large amount of corrosive surface contaminants accumulated during a long dry period will generally cause a higher corrosion rate than the smaller amount accumulated during a shorter dry period. In fact, the cumulative damage, averaged over the surface and a period (t) covering one or several years, generally follows a continuous curve (Fig. 10). This damage will be called the penetraton depth (p). Regarding the corrosion rate, two different concepts may be distinguished in addition to the instantaneous corrosion rate mentioned: the mean corrosion rate during a certain exposure time (t_1), i.e., p_1/t_1, and the differential corrosion rate at a certain exposure time (t_1), i.e., $(dp/dt)_{t=t_1}$. Generally the differential corrosion rate approaches a constant value (the steady-state corrosion rate) after some years of exposure.

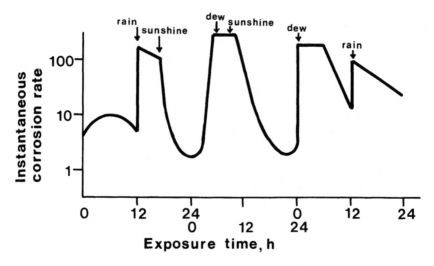

FIGURE 9 Instantaneous corrosion rate during a few days of exposure [13].

5.4 CARBON STEEL

Carbon steel is the most widely used structural material, and its corrosion behavior under atmospheric conditions has been the subject of a great number of investigations [56,57]. Among the topics which have been studied, the composition and structure of corrosion products and the influence of different environmental parameters on the corrosion rate may be mentioned especially. Several authors have used these investigations as a basis for explanation of the corrosion mechanism.

5.4.1 Fundamentals

The thermodynamic reaction possibilities for the rusting of steel can be surveyed in the potential-pH diagram for the system $Fe-H_2O$ (Fig. 11). As can be seen, the metal is not stable in aqueous solutions. In the domain labeled FeOOH more or less effective protection by a passivating coating of FeOOH can be expected, whereas corrosion is likely to occur under conditions corresponding to the domain labeled Fe^{2+}, where soluble Fe^{2+} ions are stable.

Ferrous ions are the primary corrosion products formed by anodic dissolution. Ferrous ions are converted by precipitation and oxidation reactions into insoluble ferric oxide hydroxide, FeOOH, which constitutes the thermodynamically stable end product in atmospheric corrosion of steel [58].

TABLE 4 Corrosion Rate of Carbon Steel at Different Relative Humidities or at Wetting

Relative humidity/ wetting	Atmosphere/contamination	Corrosion rate $(g/m^2\ hr)$	Reference
87%	Laboratory, clean surface	0.002	
96%		0.009	[56]
100%		0.014	
90%	Laboratory, 1 ppm SO_2	0.01	
Dew		0.2–0.4	[13]
70–80%	Outdoor, industrial at-	0.0035	
80–90%	mosphere	0.035	[57]
90–100%		0.085	
Rain		0.7	
Dew	Laboratory, SO_4^{2-} con-taminated rust	1–2	[12]
Continuous wetting	Laboratory, 25°C	1.2	[10]

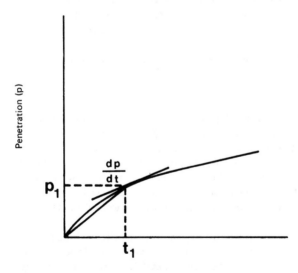

Exposure time (t)

FIGURE 10 Penetration depth vs. exposure time curve. p, Cumulative penetration depth averaged over the surface; p_1/t_1, mean corrosion rate during the period t_1; $(dp/dt)_{t=t_1}$, differential corrosion rate at exposure time t_1.

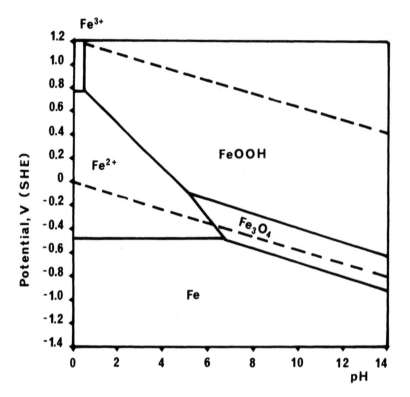

FIGURE 11 Potential-pH diagram, Fe-H_2O; 25°C, 10^{-1} M Fe [1].

Composition of Corrosion Products

The composition of the rust layer depends on the conditions in the surface electrolyte and thus differs in different types of atmosphere.
$Fe(OH)_2$ may form in neutral to basic solutions. In atmospheres polluted by SO_2, however, the surface electrolyte is usually weakly acidic
and $Fe(OH)_2$ does not precipitate. Some authors consider formation of
the so-called green rust with the proposed formula $2Fe(OH)_3 \cdot 4Fe(OH)_2$.
$FeSO_4 \cdot xH_2O$ as an intermediate product in the oxidation of $Fe(OH)_2$ to
FeOOH [59].

FeOOH can occur in four different crystalline modifications: α-
FeOOH (goethite), β-FeOOH (akageneite), γ-FeOOH (lepidocrocite),
and δ-FeOOH. Several authors consider γ-FeOOH to be the primary
crystalline corrosion product [60--62]. In weakly acidic solutions γ-
FeOOH is transformed into α-FeOOH, the process being dependent on
the sulfate concentration and the temperature [60]. α-FeOOH seems
to be the most stable modification of the ferric oxide hydroxides. The
solubility of α-FeOOH is approximately 10^5 times lower than that of γ-
FeOOH [63]. The relative amounts of α- and γ-FeOOH depend on the

type of atmosphere and on the length of exposure [60,62,64—66]. In freshly formed rust in SO_2-polluted atmospheres γ-FeOOH is usually slightly dominant. On prolonged exposure the ratio of γ to α decreases [67].

In marine atmospheres, where the surface electrolyte contains chlorides, β-FeOOH is found. β-FeOOH has been shown to contain up to 5% by weight of chloride ions in marine locations [64,68].

δ-FeOOH has not been reported in rust created under atmospheric conditions on carbon steel.

Magnetite, Fe_3O_4, may form by oxidation of $Fe(OH)_2$ or intermediate ferrous-ferric species such as green rust [68]. It may also be formed by reduction of FeOOH in the presence of a limited oxygen supply [66] according to

$$8FeOOH + Fe \rightarrow 3Fe_3O_4 + 4H_2O$$

Magnetite is usually detected in the inner part of the rust adhering to the steel surface in specimens which have been subject to prolonged exposure, where depletion of oxygen may occur [66,68,69]. An exception is rust formed in a marine atmosphere, where Fe_3O_4 has been found to be a main constituent.

The rust layer formed on unalloyed steel generally consists of two regions [70]:

1. An inner region, next to the steel/rust interface, often consisting primarily of dense amorphous FeOOH with some crystalline Fe_3O_4
2. An outer region consisting of loose crystalline α-FeOOH and γ-FeOOH.

The composition of the two layers may, of course, differ with the length of exposure and the exposure conditions, i.e., the type of atmosphere.

5.4.2 Corrosion Mechanism

Atmospheric corrosion of carbon steel has in the past often been considered a process of general corrosion. It was believed to proceed in cells with microscopic anodes and cathodes fluctuating at the surface in a statistically disordered way. Investigations made during the past two decades, however, have shown that the corrosion process, which is electrochemical in nature, takes place in cells of macroscopic dimensions with very distinct anodic and cathodic areas.

Initiation

In a dry, clean atmosphere the steel surface becomes covered by a 20—50 Å thick oxide film which practically prevents further oxidation. This oxide film consists of an inner layer of Fe_3O_4 and an outer layer of polycrystalline Fe_2O_3. In atmospheres containing small amounts of water vapor γ-FeOOH may also form [71].

The initiation of corrosion on a clean metal surface in noncontaminated atmospheres is a very slow process even in atmospheres saturated with water vapor. In this case initiation may occur at surface inclusions such as MnS, which dissolve when the surface becomes wet [72]. A more important factor for the initiation of corrosion is the presence of solid particles on the surface [31]. Settled airborne dust may promote corrosion by adsorbing SO_2 and water vapor from the atmosphere [3,73]. Of special importance are particles of hygroscopic salts, such as chlorides or sulfates, which form a corrosive electrolyte on the surface [5]. Carbonaceous particles can also start the corrosion process as they may form cathodes in microcells with the steel surface [3]. In addition, residues of mill scale may form cathodes on the steel surface, thus initiating the corrosion process when the surface becomes wet. It should, however, be stressed that in SO_2-polluted atmospheres at high relative humidities rusting is rapidly initated on polished steel samples in the absence of any particles [30].

Propagation

During the initiation period anodic spots surrounded by cathodic areas are created. In the presence of an electrolyte film on the metal surface, conditions are created for propagation of the corrosion process. The process is stimulated by SO_2, which is adsorbed and oxidized in the rust layer to SO_4^{2-}. In the corrosion cells sulfate accumulates at the anodes and thus creates so-called sulfate nests in the rust, which were first described by Schwarz [74]. In the initial stage the surface is covered by a great number of small sulfate nests. With increasing exposure period the nests grow larger and their number per unit area decreases [74-77]. The size and distribution of the nests depend on, among other factors the type of atmosphere and the degree of sheltering. After 4 months of outdoor exposure the average diameter of the nests was about 0.5 mm; on further prolonged exposure it increased to about 1 mm [74].

In atmospheres polluted with chlorides the corrosion of carbon steel proceeds in local cells which resemble the sulfate nests mentioned above [78,79]. They may arise around chloride particles deposited on the surface, where the concentrated chloride solution locally destroys the passivating film of FeOOH. In the anodic areas so formed the chlorides are concentrated by migration, while the surrounding surface covered by rust acts as a cathode.

When the surface becomes wetted by rain, dew, or moisture adsorption, the sulfate nests in combination with the surrounding area form corrosion cells (Fig. 12). The electrolyte is mostly very concentrated and has a low water activity. Anodes are located inside the sulfate nests, where the pH value and the redox potential become low. The conditions here correspond to a position in the Fe^{2+}

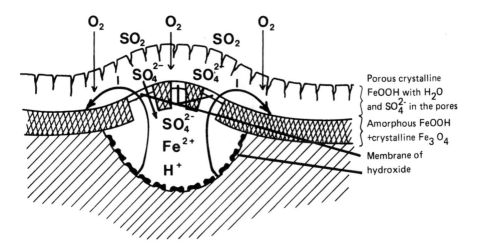

FIGURE 12 Sketch of corrosion cell at sulfate nest on steel [1].

domain in the potential-pH diagram and local attack will take place in the steel surface. The surrounding area acts as a cathode. This happens even if the surface is covered with oxide containing crystalline magnetite (Fe_3O_4), because magnetite is a good electronic conductor. The corrosion reaction may be described in terms of an electrochemical cell of the type

$$Fe/Fe^{2+}(aq)//OH^-/O_2(aq)"Fe_3O_4"$$

Besides magnetite, hydroxides containing both divalent and trivalent ions, i.e., green rust, may serve as cathodes as they possess appreciable electronic conductivity.

The following equations may in principle describe the reactions taking place in the corrosion cells.

At the cathode The main cathodic reaction is considered to be reduction of oxygen dissolved in the electrolyte film:

$$1/2 O_2 + H_2O + 2e^- \rightarrow 2OH^-$$

This process causes a local increase in pH at the cathodes and promotes precipitation of corrosion products at some distance from the anodes.

As soon as ferric corrosion products have been formed another cathodic process may take place:

$$Fe^{3+} + e^- \rightarrow Fe^{2+}$$

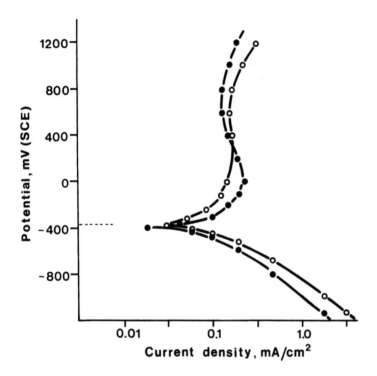

FIGURE 13 Potentiodynamic polarization curves of weathering steel after 5 months of outdoor exposure. Measurement performed in aerated ($\circ$) and deaerated ($\bullet$) 0.1 M Na_2SO_4 at 20°C [76].

The cathodic step then consists of reduction of ferric rust to magnetite according to

$$8FeOOH + Fe^{2+}(aq) + 2e^- \rightarrow 3Fe_3O_4 + 4H_2O$$

This process takes place during wet periods and has been verified through cathodic polarization measurements on rusty steel specimens in sulfate solutions [76]. As can be seen in Fig. 13, the cathodic polarization curves are only slightly affected by aeration, which indicates that under these circumstances reduction of ferric rust was the dominating cathodic reaction.

Rozenfeld [4] presented a theory stating that the acceleration of atmospheric corrosion of steel by SO_2 is due to its cathodic reduction to dithionite or sulfide. As will be discussed later, however, this process is not likely to occur even in polluted outdoor atmospheres, as the SO_2 content in the electrolyte film is not high enough.

<u>At the anode</u> The basic anode reaction

$$Fe \rightarrow Fe^{2+} + 2e^-$$

has different mechanisms in neutral and in acid solutions. At low concentrations of sulfates the mechanism of iron dissolution may be described as follows according to Heusler [80] and Bockris et al. [81]:

$$Fe + H_2O \rightarrow Fe(OH)^- \text{ ads} + H^+$$

$$Fe(OH)^- \text{ ads} \rightarrow Fe(OH) \text{ ads} + e^-$$

$$Fe(OH) \text{ ads} \rightarrow Fe(OH)^+ + e^-$$

$$Fe(OH)^+ \rightarrow Fe^{2+} + OH^-$$

If insoluble corrosion products are formed at anodic sites the corrosion rate may be reduced substantially due to *passivation* [2]. This may occur in rural atmospheres, where the content of sulfates and chlorides in the rust layer is very low.

In sulfate-containing solutions, however, the anodic dissolution proceeds according to a mechanism proposed by Florianovitch and Kolotyrkin [82]:

$$Fe + H_2O \rightarrow Fe(OH) \text{ ads} + H^+ + e^-$$

$$Fe(OH) \text{ ads} + H_2O \rightarrow /Fe(OH)_2/ \text{ ads} + H^+ + e^-$$

$$/Fe(OH)_2/ \text{ ads} + SO_4^{2-} \rightarrow FeSO_4 + 2OH^-$$

$$FeSO_4 \rightarrow Fe^{2+} + SO_4^{2-}$$

The pH-regulating effect of $FeSO_4$ will result in maintaining a relatively low pH at the anodic sites and thus preventing precipitation of iron hydroxides directly on the metal surface. This creates favorable conditions for corrosion in the active state, the sulfate accelerating the anodic dissolution of iron. Tanner [83] identified crystalline iron(II) sulfate at the steel/rust interface as tetrahydrate $FeSO_4 \cdot 4H_2O$. Thus a reservoir of soluble sulfates exists within the sulfate nests, contributing to their high stability.

In the rust layer

$$2Fe^{2+} + 3H_2O + 1/2\ O_2 \rightarrow 2FeOOH + 4H^+$$

This so-called oxidative hydrolysis plays an important role in most of the proposed mechanisms of atmospheric corrosion. According to Evans and Taylor [84,85], the magnetite produced by cathodic reduction is reoxidized by oxygen in the presence of water

$$Fe_3O_4 + 1/4\ O_2 + \frac{3}{2}H_2O \rightarrow 3FeOOH$$

The sulfate nest becomes enclosed within a semipermeable membrane of hydroxide formed through oxidative hydrolysis of the iron ions. The electric current in the corrosion cell causes migration of SO_4^{2-} ions into the nest. This will stabilize the existence of the nest.

Schikorr [10] proposed a theory of atmospheric corrosion of steel based on the "acid regeneration cycle." Sulfuric acid formed by oxidation of SO_2 absorbed in the rust layer attacks the steel according to the overall reaction

$$4H_2SO_4 + 4Fe + 2O_2 \rightarrow 4FeSO_4 + 4H_2O$$

Sulfuric acid is then re-formed by oxidative hydrolysis

$$2FeSO_4 + 1/2\,O_2 + 3H_2O \rightarrow 2FeOOH + 2H_2SO_4$$

Even if Schikorr's theory does not explain the detailed mechanism of the corrosion process, oxidative hydrolysis seems to be very important in the process of atmospheric corrosion of steel [2,69]. It should, however, be mentioned that according to Evans and Taylor [84] the oxidative hydrolysis of $FeSO_4$ is very slow and this reaction should affect the corrosion only during the initiation stage.

Role of SO_2 in the Corrosion Process

Since the classical work of Vernon [31], the decisive importance of SO_2 in the process of atmospheric corrosion of steel has been recognized. Under atmospheric conditions SO_2 may be adsorbed at the steel surface. The adsorption rate on rusty and especially on polished specimens depends on the relative humidity. Investigations by Sydberger and Vannerberg [30] showed that the adsorption rate increases with relative humidity (Fig. 14). At high humidities a limiting value is reached which corresponds to the situation where every SO_2 molecule reaching the surface is adsorbed. Johansson and Vannerberg [86,87] made an extensive analysis of the thermodynamic conditions for reactions of SO_2 in moisture layers on metal surfaces.

At low and medium SO_2 concentrations (<1–10 ppm) Under atmospheric conditions there is a strong driving force for the reaction

$$SO_2(aq) + H_2O + 1/2\,O_2 \rightarrow H_2SO_4$$

as all sulfur(IV) species in aqueous solutions are thermodynamically unstable. The rate of oxidation of SO_2 by oxygen in water is rather low. The reaction is, however, catalyzed by, e.g., $Fe^{2+}(aq)$ and $Mn^{2+}(aq)$ as well as by oxides and hydroxides of iron which are present at the steel surface. The reaction is strongly pH-dependent, the reaction rate decreasing with pH (Fig. 15) [87,88]. At low SO_2 concentrations the pH value will be higher than at high SO_2 levels, thus

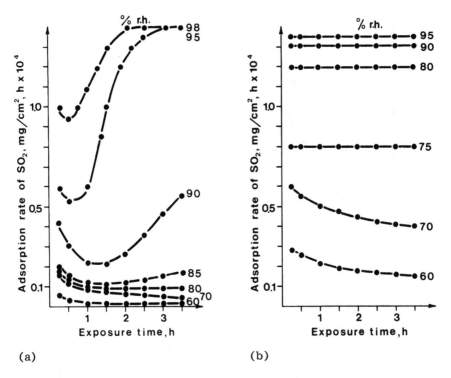

(a)   (b)

FIGURE 14 Adsorption rate of SO_2 on polished (a) and rusty (b) specimens of carbon steel as a function of time and relative humidity in an atmosphere with 10 ppm SO_2 at 20°C [30].

promoting the oxidation reaction. Apart from oxygen, Fe^{3+}(aq) and solid ferric oxide hydroxide may also oxidize SO_2 to sulfate.

At high SO_2 concentrations As already mentioned, Rozenfeld [4] found in laboratory experiments that SO_2 in high concentrations has a marked depolarizing effect on cathodic reactions at polished metal surfaces. A high SO_2 content also creates a low pH in the surface film and consequently a low rate of sulfate production. Thus the surface film will contain appreciable amounts of tetravalent sulfur. Under these conditions, at SO_2 contents greater than approximately 10 ppm, reduction of SO_2 to dithionite may take place according to the following reaction:

$$2SO_2(aq) + 2e^- \rightarrow S_2O_4^{2-}$$

The equilibrium potential for the reaction at pH 6 is about -370 mV [86]. Measurements of corrosion potentials performed by Pourbaix [89] on carbon steel exposed alternatively to wet and dry conditions

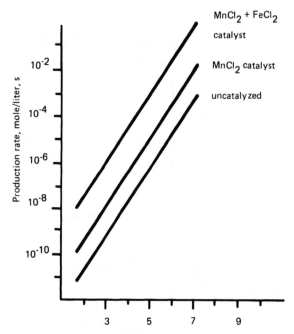

FIGURE 15 Oxidation rate of four-valent sulfur species in aqueous solutions as a function of pH. The solution is in equilibrium with air containing 1 ppm SO_2. The catalyst concentration is 10^{-5} M [86].

showed values between about -500 mV at the beginning of the exposure and -150 to +150 mV after prolonged exposure (Fig. 16). The necessary conditions for the reduction of SO_2 to dithionite may thus be fulfilled in severely SO_2-polluted atmospheres, where the protective rust layer may be partly destroyed, leading to a drop in the corrosion potential.

The reactive dithionite is then either oxidized or further reduced by the metal to sulfide

$$S_2O_4^{2-} + 8H^+ + 10e^- \rightarrow 2S^{2-} + 4H_2O$$

The situation may be illustrated by a potential-pH diagram, which shows that, at high SO_2 concentrations, FeS is the stable corrosion product in the pH range 3–6 (Fig. 17) [86,87]. Sulfide-containing corrosion products have, in fact, been reported in laboratory experiments at SO_2 > 10 ppm in several investigations [87,90,91]. At the SO_2 concentrations found in polluted outdoor atmospheres, however, which are usually in the range 0.01–0.2 ppm, reduction of SO_2 does not take place. This

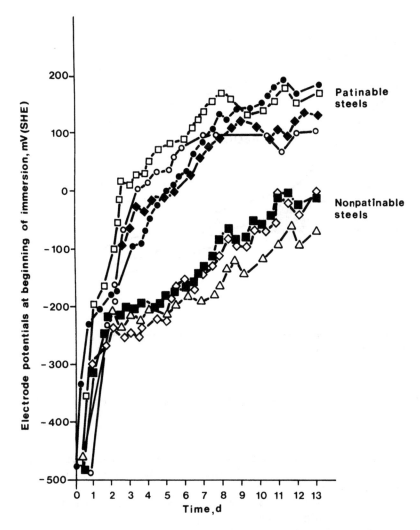

FIGURE 16 Corrosion potential as a function of time during alternating immersion and drying of electrodes of unalloyed carbon steel and weathering steels [89].

applies especially to rusty surfaces, which strongly catalyze the oxidation of SO_2.

In conclusion, sulfates influence the corrosion of steel in the following ways:

They give rise to sulfate nests, which create the basis for the electrochemical process in atmospheric corrosion of steel.

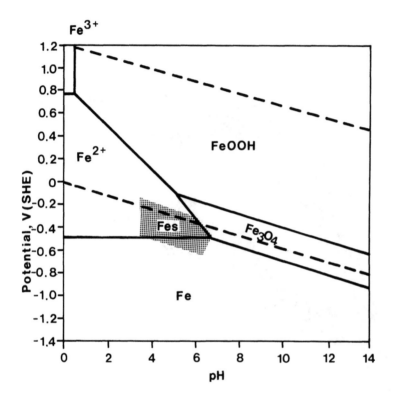

FIGURE 17 Potential-pH diagram, Fe-S-H$_2$O; 25°C, 1 M Fe, 10^{-2} M S [87].

   They increase the conductivity in the thin surface electrolyte films,
      thus increasing the current in the electrochemical cells, i.e.,
      the corrosion rate.
   They accelerate the anodic dissolution of iron in the active state in
      the neutral to slightly acid pH range.
   They influence the protective ability of the rust. If the rust formed
      becomes infected with sulfate nests, as may happen when the
      steel is first exposed during the winter season, the rust be-
      comes less protective for some time. If, on the other hand, the
      first rust is formed during the summer season, it will generally
      be less infected by sulfate and as a consequence will be more
      protective.

5.4.3 Corrosion as a Function of Time

During an initial period the corrosion rate is usually high because the
rust formed is porous and has poor protective properties. After the

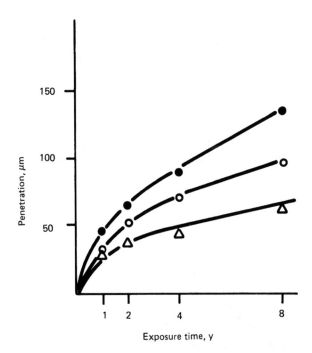

FIGURE 18 Penetration depth vs. exposure time for carbon steel in Mülheim/Ruhr (industrial atmosphere, ●), in Cuxhaven (marine atmosphere, ○), and in Olpe (rural atmosphere, △) [92].

initial period, the length of which is usually 1–5 years and depends on the corrosivity of the atmosphere, the protective properties improve and the corrosion rate decreases.

The penetration depth increases with the time of exposure, as shown in Fig. 18 [92]. The curves there follow a power law

$$p = kt^n$$

where p is the pentration depth, k is a constant, t is the time of exposure, and n is a constant.

Such a time dependence has been reported by many authors [93]. It suggests that transport of reactants through a growing protective rust layer determines the rate of the corrosion process. If the rust formation is diffusion-controlled, the diffusion coefficient is constant, and all the rust formed remains on the surface, n will have the value 0.5. In cases where the diffusion coefficient decreases, e.g., due to decreasing porosity in the rust, the n value will be lower than 0.5. Removal of rust from the surface, due to dissolution, flaking, or erosion, however, will lead to higher n values. In exposure tests of steel the n value has been found to vary between 0.4 and 0.8. In the long run a stationary state will be approached with constant thickness and porosity of the rust layer. Then the corrosion rate will become constant and the relationship between penetration and time linear.

The power law can also be written:

$$\log p = \log k + n \log t$$

This equation is represented by a straight line in a bilogarithmic diagram (Fig. 19) [92]. Pourbaix [93] suggested that the penetration

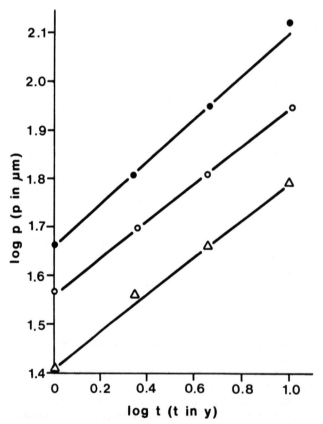

FIGURE 19 Bilogarithmic diagram of penetration depth (p) and exposure time (t) for carbon steel; the same results as in Fig. 18. Exposure sites: Mülheim/Ruhr (industrial atmosphere, ●), Cuxhaven (marine atmosphere, ○), and Olpe (rural atmosphere, △) [92].

depth for up to 20 to 30 years can be estimated by extrapolation of lines determined for an initial 4-year period. According to Burgmann and Grimme [92] and Bohnenkamp et al. [94] such an extrapolation may lead to underestimation of the penetration because of the increase in the n value mentioned.

5.4.4 Corrosion as a Function of Environmental Parameters

The rate of corrosion of carbon steel depends on interaction of several environmental parameters. Climatic parameters involving the temperature-humidity complex, different forms of precipitation, wind velocity, and pollutants in solid, gaseous, or liquid form may exert an influence on the corrosion process. Numerous field exposure programs have been performed to study the relation between the corrosion rate and the environmental parameters. An appropriate mathematical treatment of extensive long-term exposure programs shows that the most important parameters are the time of wetness and pollution of the atmosphere, mainly SO_2 and in marine locations also chlorides [35,95−100, 147]. This may be seen in Table 5, which gives some examples of equations describing the corrosion rate of carbon steel as function of environmental parameters.

These findings have opened the way for an engineering approach to atmospheric corrosion which gives guiding values and estimates of real values sufficient in practice for design and maintenance purposes. This knowledge is now being used for elaboration of standards and codes of practice expressing the corrosion aggressivity of atmospheres.

A rapid estimation of the corrosion rate can be made, for example, by using the nomogram in Fig. 20, which gives the corrosion rate as a function of the mean SO_2 pollution and the time of wetness expressed as hours of relative humidity $\geq 80\%$ at temperatures $\geq 0°C$ [78,101].

As can be seen from Figs. 18 and 19, as well as Fig. 20, the penetration is dependent on the type of atmosphere; the greater the SO_2 pollution, the higher the corrosion rate. Differences in meteorological parameters in different places within the temperate climatic zone are usually small. Accordingly, for technical purposes a further simplification is possible, and the corrosion rate within a defined climatic zone may be expressed as a function of SO_2 pollution alone. The corrosion rate intervals in different types of atmospheres given in Table 6 can be used as estimates [102].

Chlorides also play an important role in accelerating the corrosion rate of carbon steel. Their main importance is restricted to marine atmospheres, even though they may also be found in industrialized areas, where they may originate from, e.g., burning of coal.

Several investigations have shown the relation between chloride deposition and the corrosion rate of steel [46,97,103]. The investigation

TABLE 5 Corrosion of Carbon Steel as Function of Environmental Parameters

Reference	Equations	Parameters
Guttman and Sereda [95]	$K = 0.16 t_w^{0.7}(SO_2 + 1.78)$	SO_2 = SO_2 conc., ppm t_w = electrochemically monitored time of wetness, days K = corrosion loss, g per 3 by 5 in. panel
Haynie and Upham [96]	$K = 325\sqrt{t}\,\exp[0.00275SO_2 - (163.2/RH)]$	SO_2 = SO_2 conc., $\mu g/m^3$ t = exposure time, years RH = average relative humidity, % K = penetration, μm
Atteraas et al. [53, 97]	$K_1 = 5.28SO_2 + 176.6$ $K_4 = 18.5SO_2 + 292.5$	SO_2 = SO_2 conc., $\mu g/m^3$ K_1 = corrosion loss, g/m^2 per 1 year exposure K_4 = corrosion loss, g/m^2 per 4 years exposure
Hakkarainen and Ylasaari [98]	$K = 1.17 t_w^{0.66}(SO_2 + 0.048)$	t_w = time of wetness, i.e., total time, RH = >85%, hr SO_2 = SO_2 deposition, mg/dm^2 day K = penetration, μm
Barton et al. [100]	$K = 0.0152 t_w^{0.428}SO_2^{0.570}$	SO_2 = SO_2 deposition, mg/m^2 day t_w = time of wetness, i.e., total time, RH = >80% and t > 0°C, hr/day K = corrosion rate, g/m^2 day
Knotkova et al. [143]	$K_1 = 4.0SO_2 + 58$ $K_4 = 8.0SO_2 + 160$	SO_2 = SO_2 deposition, mg/m^2 day K_1 = corrosion loss, g/m^2 per 1 year exposure K_4 = corrosion loss, g/m^2 per 4 years exposure

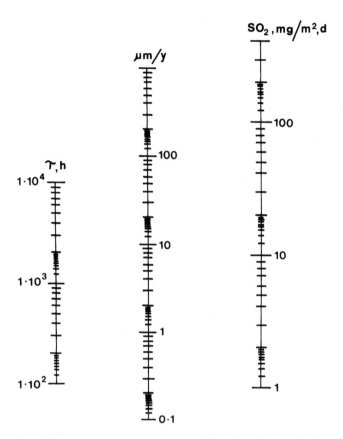

FIGURE 20 Nomogram expressing the steady-state corrosion rate of carbon steel (μm/year) as a function of SO_2 deposition (mg/m^2 day) and time of wetness (hr/year) [78].

by Ambler and Bain [103] in Nigeria may serve as an example (Fig. 21). The corrosion rate as well as the chloride deposition are usually high at the seashore but decrease sharply with distance from the sea.

The influence of nitrogen oxides on atmospheric corrosion has so far been investigated only on a limited scale. Systematic investigations in climate chambers have not revealed that 0.05 or 0.5 ppm of NO_2 has any significant effect on the corrosion of weathering steels or zinc [104]. Also, field tests of carbon steel in Japan failed to show an appreciable affect of NO_x on the corrosion rate [43]. This study also showed that the deposition rate of NO_x in comparison with SO_2 is very much smaller, of the order of one-hundredth. This may

TABLE 6   Corrosion Rates for Carbon Steel in Different Types
of Atmospheres [102]

Type of atmosphere	SO_2 deposition rate (mg $SO_2/$ m^2 day)	Steady-state corrosion rate (μm/year)	Examples	
			Place[a]	Steady-state corrosion rate (μm/year)
Rural	< 20	5−10	Flahult (S)	8
			Zvenigorod (USSR)	7
			Kasperske Hory (CS)	9
			Picherand (F)	6
Urban	20−110	10−30	Stockholm (S)	30
			Moscow (USSR)	20
			Prague (CS)	28
			St. Denis (F)	18
Industrial	110−(approx. 200)	30−60	Most (CS)	39
			Kopisty (CS)	52
			Auby (F)	50
			Duisburg (D)	70

[a]S, Sweden, CS, Czechoslovakia; F, France; D, West Germany.

be the main reason why NO_x seems to have a very limited influence on
corrosion of steel under outdoor conditions.

5.4.5 Application Aspects

Carbon steel is not able to develop a protective coating by itself.
Therefore unalloyed steel for use in outdoor atmospheres is generally
given additional surface protection, a coating of either antirust paint,
zinc, or aluminum.

   To ensure satisfactory life of an antirust paint coating one must
clean the steel surface carefully to remove rust and other contaminants before painting. Otherwise, corrosion cells may build up under the paint coating around sulfate nests or hygroscopic chloride

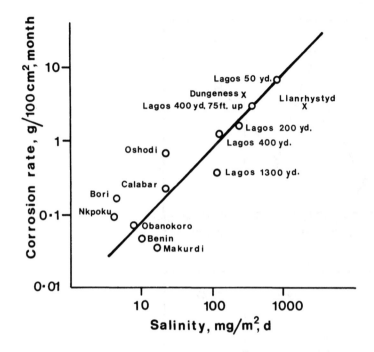

FIGURE 21 Corrosion of carbon steel as a function of salinity measured by wet candles on exposure in Nigeria [103].

inclusions while water and oxygen permeate through the paint, with the result that the paint coating delaminates and flakes off.

At present, cleaning before painting is generally carried out by blasting, wire brushing, or abrading [105]. It has been found, however, that not even thorough dry blasting will completely remove all SO_4^{2-} and Cl^- contamination from a rusty steel surface which has been exposed to a polluted atmosphere [106]. As shown by Gronvall et al. [107] and by Calabrese and Allen [108], salt residues may be entrapped by plastic deformation of surface irregularities into folds and pockets. The entrapped salt residues will initiate activity in corrosion cells, which may cause breakdown of a subsequently applied paint coating. In this respect it is important to note that rust formed at locations with different degrees of pollution contains different amounts of sulfates; values between 1 and 5% SO_4^{2-} are mentioned. However, seasonal variations in SO_4^{2-} have also been reported [108]. Rust formed during winter has been found to contain considerably more sulfate than summer rust. It has also been proved that paint applied on rusty surfaces in winter has a much shorter lifetime than paint on summer rust (Fig. 22). One can remove the water-soluble residues by washing the

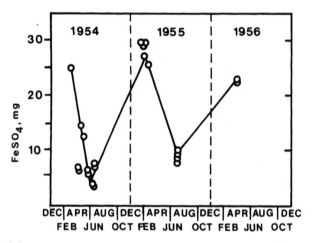

(a)

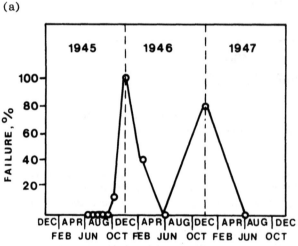

(b)

FIGURE 22 Seasonal variations of FeSO₄ in rust and failure frequency of paint coatings applied to rusty steel surfaces [109].

blasted surface with water. This complicated procedure is comparatively costly, so there is interest in replacing dry blasting by wet blasting, which also directly removes the entrapped salt residues if they are water-soluble.

Knowledge of the influence of environmental parameters on the corrosion rate of steel may also be used for the development of new reliable accelerated test methods. An ideal method should accelerate the corrosion rate compared with corrosion under conditions in practice without changing the mechanism of the corrosion reaction. The need for such methods may be illustrated by the widely used Kesternich test, where the two SO_2 levels prescribed correspond to 660 and 6600 ppm SO_2. There is an obvious risk that such high levels will lead to a corrosion mechanism other than that which occurs in outdoor atmospheres. This is confirmed by the fact that the corrosion rate of steel at 6600 ppm SO_2 in the Kesternich test is lower than the rate at 660 ppm SO_2 [110]. The need for more representative accelerated test methods has been recognized by several authors. Barton and co-workers [111–113] proposed testing in 100 ppm SO_2 at permanent condensation. However, results of Johansson and Vannerberg [87] suggest that one should not exceed 10 ppm SO_2 to avoid changes in the corrosion mechanism compared with that in outdoor conditions. In this connection it must be stressed that the decisive pollution parameter is the amount of SO_2 deposited on the metal surface, which depends on the concentration as well as the air velocity [91]. This may explain some of the contradicting opinions found in the literature.

The different corrosion mechanism at high SO_2 concentrations is also of importance in the so-called inert-gas systems which are used in order to reduce explosion hazards in cargo tanks of crude oil tankers. The gas usually contains $2.5–4.5\% O_2$, $12–14\% CO_2$, and $200–300$ ppm SO_2. It has been reported that corrosion inside cargo tanks on "inert-gas tankers" is lower than that in tankers without this system [114]. Systematic laboratory studies have shown that the main reason for this is the formation of a protective coating on the steel surface consisting of an inner layer containing FeS and $FeSO_3 \cdot 3H_2O$ [86,87].

5.5 WEATHERING STEEL

By alloying steel with small additions of elements such as copper, chromium, nickel, phosphorus, silicon, and manganese one improves the corrosion resistance in outdoor atmospheres. Such low-alloyed steels are called weathering steels.

5.5.1 Fundamentals

The composition and the crystallographic structure of rust on weathering steels are similar to the conditions on carbon steels. α-FeOOH, γ-FeOOH, and amorphous or noncrystalline matter are present in proportions similar to those on carbon steel [62,67]. The content of Fe_3O_4 remains low even after long exposure periods [115]. However, δ-FeOOH has been reported as a major constituent of the inner film created during long-term field exposure of weathering steel [63, 116].

In principle, the corrosion mechanism is similar for weathering and unalloyed carbon steels. On the former, however, the rust forms a more dense and compact layer which more effectively screens the steel surface from the corrosive components of the atmosphere. This layer may affect the corrosion process in several ways. The anodic reaction may be retarded by limiting the supply of water and corrosion-stimulating ions to the steel surface. The cathodic reaction may be affected by the low diffusion rate of oxygen, and, finally, the increased electrolytic resistance may also decrease the corrosion rate. It seems as though the sulfate nests decrease and ultimately disappear if the exposure conditions are favorable, that is, when frequent dry-to-wet alternations occur [117]. Such climatic changes are believed to favor bursting of the membranes around the sulfate nests and dissolution and washing away of their sulfate. The protective properties of the formed rust layer may also be monitored by measurements of the corrosion potential [76,89,118,119]. The corrosion potential of weathering steels reaches substantially nobler values than that of unalloyed carbon steels (Fig. 16).

Many investigations have been devoted to clarifying the beneficial influence of the alloying elements on the corrosion properties [120–123].

Among the individual elements, copper has the most pronounced effect in decreasing the corrosion rate. As illustrated in Fig. 23, an increase in the copper content from 0.01 to 0.04% decreases the corrosion rate by up to 70%. A further increase in the range from 0.20 to 0.50% copper improves the corrosion resistance only slightly [121]. Several mechanisms for the beneficial effects of copper have been proposed. According to one theory [124,125] copper ions dissolved from the base metal are able to precipitate sulfide ions originating either from sulfide inclusions in the steel or from the atmospheric pollution, and thus eliminate their detrimental effect. Another theory [3,126] states that the beneficial effect of copper is due to the formation of a surface coating of metallic copper, which either acts as protection in itself or promotes anodic passivation by supporting the cathodic reaction. The most likely explanation, however, is that copper forms basic sulfates with low solubility precipitating within the pores in the rust layer and thus decreasing the porosity [127]. Weathering steels usually contain 0.2 to 0.5% copper.

Chromium and nickel increase the corrosion resistance of weathering steels when they occur in combination with copper. Chromium is usually added in a content of 0.4–1.0% and nickel up to 0.65%. In this respect chromium seems to be more effective than nickel. Both elements also improve the mechanical properties of the steel. Nickel is supposed to act by forming insoluble basic sulfates in pores of

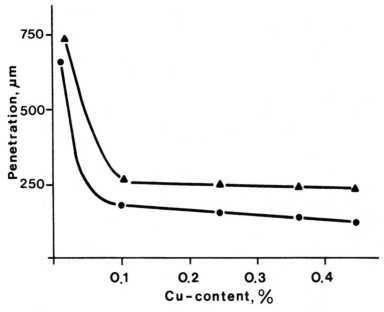

FIGURE 23 Effect of small alloying additions of copper on the atmospheric corrosion of carbon steel exposed for 15.5 years [121]. ▲ = Marine, ● = industrial.

the rust layer, whereas chromium is enriched in the inner rust layer together with copper and phosphorus [67,69]. They promote formation of a dense layer of amorphous FeOOH next to the steel surface [68]. This layer serves as a barrier to the transport of water, oxygen, and pollutants.

Phosphorus has a positive influence on the corrosion resistance of weathering steels [121,122]. An increase in the phosphorus content from less than 0.01 to 0.10% resulted in 20–30% improvement of the corrosion resistance of copper-bearing steels. According to one theory, phosphorus may form layers of insoluble phosphates in the rust, acting as transportation barriers in a similar way to the basic sulfates earlier mentioned.

5.5.2 Corrosion as a Function of Time and Environmental Parameters

The corrosion characteristics of weathering steels depend in a complex way on climatic parameters, pollution levels, the degree of sheltering, and the composition of the steel.

In the course of 3 or 4 years the corrosion rate is stabilized at a rather low value in most external atmospheres. In polluted atmospheres a dark brown to violet patina develops during this exposure. The formation of patina may, however, take a substantially longer time in rural areas. The rust formed is compact and contains no particles that can be wiped off the surface.

A key condition for the formation of a protective patina is a cyclical variation between wet and dry periods. Periodic flushing followed by drying is desirable. In rain-sheltered areas, the typical dark patina formed on open exposure is not obtained but the surface becomes coated with a layer of rust in lighter colors. Nevertheless, the rust layers are more compact and adherent than on carbon steel and the corrosion rate is usually so low that the absence of dark patina seems to be of only aesthetic significance [128]. An exception is atmospheres with high chloride contents, where accumulation of hygroscopic chlorides can give rise to very high corrosion rates [129]. On indoor exposure no systematic differences have been observed between weathering and plain carbon steels. The low corrosion rates observed in all cases were due to the low corrosivity of the environment and not to the composition of the steel [128]. Thus, there is no justification for using weathering steels in indoor conditions.

Under conditions with a very long time of wetness or permanent wetness, as on exposure in water or soil, the corrosion rate for weathering steels is usually about the same as that for carbon steel.

The corrosion rate for weathering steels also follows the power law valid for carbon steel, but with different values of the constants k and n. The constants vary with the environment as well as with the composition of the steel. Typical values of stabilized corrosion rates for atmospheres with different levels of SO_2 pollution are given in Table 7.

TABLE 7 Calculated Steady-State Corrosion Rates of Steels on Free Exposure in Different Types of Atmospheres [128]

SO_2 level ($\mu g/m^3$)	Type of atmosphere	Corrosion rate (μm/year)	
		Mild steel	Weathering steel
<20	Rural	5–10	2–5
20–115	Urban	10–30	2–6
>115	Industrial	>30	>6

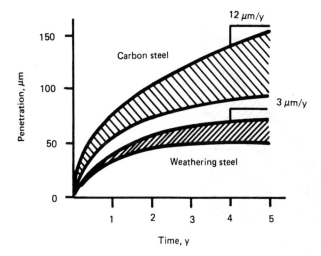

FIGURE 24 Corrosion losses of weathering steel (Atmofix 52 A) and unalloyed carbon steel in moderately SO_2-polluted atmosphere (limit of SO_2 deposition rate is 90 mg/m^2 day) [128].

In rural areas the corrosion rate is usually low; the time for developing a protective and nice-looking patina may be quite long. For carbon steel the corrosion rate is also rather low under these conditions.

In urban environments with SO_2 levels not exceeding about 115 $\mu g/m^3$ the weathering steels usually show stabilized corrosion rates, usually in the range 2–6 μm/year, i.e., only slightly higher values than in rural atmospheres. Mild steel shows markedly higher corrosion rate under these conditions (Fig. 24).

In more polluted industrial atmospheres with an SO_2 level exceeding 115 $\mu g/m^3$ significantly higher corrosion rates are also found for weathering steels (Fig. 25), indicating that the rust layer formed at very high SO_2 pollution is not very protective [128]. Although the surface coating may have a dark, pleasant appearance, it cannot be classified as a true patina. Under these conditions also loose rust particles are formed. The corrosion rate has, however, always been found to be less than that of carbon steel in these environments.

In marine environments heavily polluted with chlorides a protective patina does not develop and the corrosion rate may be high, especially close to the shore. This applies especially to rain-sheltered surfaces, where the corrosion rate may be very high due to accumulation of chlorides which are never washed away. Practical experience from Sweden shows, however, that at distances about 1 km from the shore the chloride deposition does not negatively affect patina formation [129].

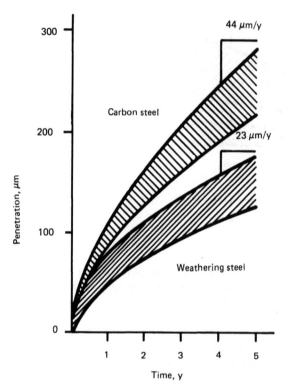

FIGURE 25 Corrosion losses of weathering steel (Atmofix 52 A) and
unalloyed carbon steel in heavily SO_2-polluted atmosphere [128].

5.5.3 Application Aspects

The protective rust layer formed on weathering steels is generally
considered a nice-looking, dark brown patina. So under urban and
rural conditions the weathering steels may be used without any additional protective coating of antirust paint, zinc, or aluminum. The
first applications were in the field of facade and roofing materials,
while in later applications heavy load-bearing structures dominated.
Power pylons and shells of self-supporting chimneys are examples of
structures successfully made of weathering steel. The reason for
this is that light structures require very careful design in order to
avoid corrosion problems. The main point in designing facades and
roofs is to avoid accumulations of water, corrosion products, and dirt.
This will prevent the occurrence of areas where the time of wetness
is very long. In accumulated rust and dirt, air pollutants also may
be deposited and give rise to very corrosive conditions.

During the first years of exposure, before the protective rust
layer has developed, rusting proceeds at a comparatively high rate.
Then rusty water is produced which may stain masonry, pavements,
etc. So precautions must be taken against detrimental staining effects. One can, for instance, cover the ground area exposed to
staining with an easily exchangeable material such as gravel, or one
can color the exposed masonry surfaces brown from the outset, so
that staining will not be visible. Staining may also be prevented by
careful draining of water from the roofs and facades.

In order to create optimal conditions for the formation of a protective patina, the designer must ensure open exposure of the surface
to the atmosphere, avoiding rain-sheltered areas. Facades should be
plane, and excessively deep recessing of, for example, windows
should be avoided. Otherwise there is a risk that patina will not
form and the coloring of the facade may become uneven.

5.6 ZINC

Zinc has a better resistance to atmospheric corrosion than carbon
steel and is extensively used for coating, or galvanizing, of ferrous
metal products. The atmospheric corrosion of zinc is influenced principally by the time of wetness and by the presence of air constituents
such as CO_2, SO_x, and Cl^-.

5.6.1 Fundamentals

The thermodynamic possibilities for reaction between zinc and the atmosphere with its various constituents will be surveyed in potential-
pH and concentration-pH diagrams [1]. Zinc is a relatively base metal.
The stability domains of various zinc-containing species in the system
$Zn-CO_2-H_2O$ at 25°C are shown in Fig. 26. The diagram is valid for a
total H_2CO_3 content of 10^{-5} mole/liter in the moisture film, that is, a
solution in equilibrium with the CO_2 content (about 0.03%) in outdoor
atmospheres. As shown by the diagram, there is a stability domain
for $ZnCO_3$ in the pH range 6 to 7. A supplementary diagram (Fig.
27a) shows how the width of the stability domain for $ZnCO_3$ varies with
the H_2CO_3. According to this diagram, based on equilibrium constants
selected by Hogfeldt and Sillén [55], there is no stability domain for
basic zinc carbonate. The dashed line representing equilibrium between $(H_2CO_3)_{tot}$ in the moisture film and the CO_2 present in outdoor
atmospheres indicates that $ZnCO_3$ is the stable corrosion product.
Grauer and Feitknecht [130], however, selected a somewhat higher
value for the solubility product of $ZnCO_3$. If this value is used, the
diagram (Fig. 27b) will also have a stability domain for basic zinc

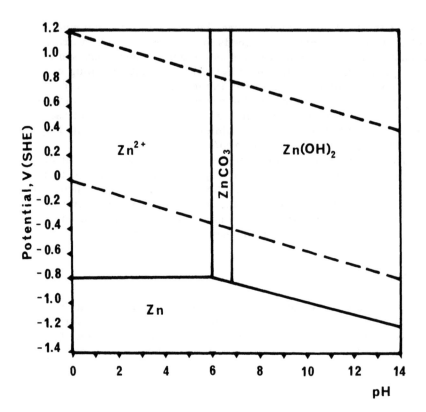

FIGURE 26 Potential-pH diagram, $Zn-CO_2-H_2O$; 25°C, 10^{-1} M Zn, 10^{-5} M H_2CO_3 [1].

carbonate $[Zn(OH)_{1.2}(CO_3)_{0.4}]$, and the dashed line indicates that this species is the stable corrosion product in contact with outdoor atmospheres. Figures 27a and 27b also differ with respect to the variant of $Zn(OH)_2$: in Fig. 27a ε-$Zn(OH)_2$ and in Fig. 27b amorphous $Zn(OH)_2$. The latter, preferred by Grauer and Feitknecht, has a higher solubility product than ε-$Zn(OH)_2$ and is evidently metastable. The stability domains of basic zinc sulfate and basic zinc chloride are shown in Figs. 27c and d.

The corrosion products found on zinc after outdoor exposure are largely in accordance with the thermodynamics described; zinc oxide (ZnO), zinc hydroxide [ε-$Zn(OH)_2$, β-$Zn(OH)_2$], basic zinc carbonates [$ZnOH(CO_3)_{0.5}$, $Zn(OH)_{1.2}(CO_3)_{0.4}$, $Zn(OH)_{1.5}(CO_3)_{0.25}$], and zinc carbonate have been found [7,130].

From this basis the corrosion process appears to proceed according to the following mechanism (Fig. 28). In moist outdoor conditions the zinc is oxidized with the formation of zinc hydroxide:

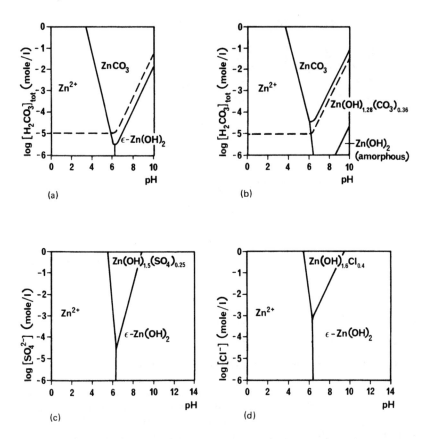

FIGURE 27 Stability domains of zinc carbonate and basic zinc carbonate (a) and (b), basic zinc sulfate (c) and basic zinc chloride (d) in aerated aqueous solutions with varying anion content and pH value; 25°C, 10^{-1} M Zn. The dashed line in (a) and (b) represents equilibrium conditions with outdoor atmospheres.

$$2Zn + H_2O + 1/2\ O_2 \rightarrow 2Zn(OH)_2$$

This reaction is of an electrochemical nature and involves cathodic reduction of oxygen and anodic oxidation of zinc.

The zinc hydroxide formed reacts with air constituents present, e.g., CO_2, SO_x, and Cl^-, with formation of the corresponding basic zinc salts at the hydroxide/air boundary, provided the pH value of the surface moisture is sufficiently high (see Fig. 27):

$$Zn(OH)_2 + 0.5CO_2 + H^+ \rightarrow ZnOH(CO_3)_{0.5} + H_2O$$

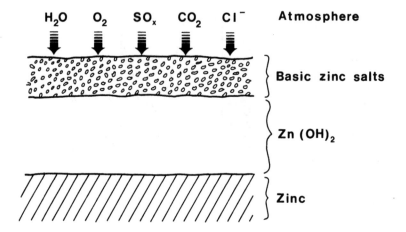

FIGURE 28 Corrosion products on zinc formed during atmospheric exposure.

$$Zn(OH)_2 + 0.25SO_2 + 0.25O_2 + 0.5\ H^+ \rightarrow$$

$$Zn(OH)_{1.5}(SO_4)_{0.25} + 0.5\ H_2O$$

$$Zn(OH)_2 + 0.6Cl^- + 0.6H^+ \rightarrow Zn(OH)_{1.4}Cl_{0.6} + 0.6H_2O$$

The zinc hydroxide and the basic zinc salts formed, the so-called zinc patina, protect the surface from further attack. If the surface moisture, however, reaches a low pH value permanently or occasionally, e.g., due to heavy air pollution with SO_x, then no zinc hydroxide or basic salts are formed; such deposits formed during earlier periods of higher pH values may even be dissolved:

$$Zn + SO_2 + O_2 \rightarrow ZnSO_4$$

$$Zn(OH)_2 + SO_2 + 1/2\ O_2 \rightarrow ZnSO_4 + H_2O$$

$$ZnOH(CO_3)_{0.5} + SO_2 + O_2 + 2H^+ \rightarrow ZnSO_4 + 1.5H_2O + 0.5CO_2$$

The $ZnSO_4$ is soluble in water. It may be washed away by rain and will then, of course, give no protection. In consequence, the corrosion rate will then be high. Abrasion and erosion may also contribute to the deterioration of the protective coating.

The transformations on zinc surfaces exposed to various types of outdoor atmospheres are surveyed in Table 8. Views of this general character have been presented by several authors [131–134], although in some cases with differing specific features. Thus Barton [133], assuming the formation of zinc sulfate from zinc hydroxide to

TABLE 8 Survey of Transformations on Zinc Surfaces in Various Types of Atmospheres

Rural atmosphere	$Zn \rightarrow Zn(OH)_2 \rightarrow ZnOH(CO_3)_{0.5}$
Urban or industrial atmosphere	$Zn \rightarrow Zn(OH)_2 \rightarrow ZnOH(CO_3)_{0.5}$
	$Zn \rightarrow Zn(OH)_2 \rightarrow ZnOH(CO_3)_{0.5} \rightarrow ZnSO_4$
	$Zn \rightarrow Zn(OH)_2 \rightarrow Zn(OH)_{1.5}(SO_4)_{0.25}$
	$Zn \rightarrow Zn(OH)_2 \rightarrow Zn(OH)_{1.5}(SO_4)_{0.25} \rightarrow ZnSO_4$
	$Zn \rightarrow ZnSO_4$
Marine atmosphere	$Zn \rightarrow Zn(OH)_2 \rightarrow Zn(OH)_{1.4}Cl_{0.6}$

be the rate-determining step, suggested that the sulfate and/or chloride bound at the corrosion product/air boundary acts as corrosion stimulator and the corrosion of zinc is proportional to the total amount of stimulator which comes into contact with the surface.

5.6.2 Corrosion as a Function of Time and Environmental Parameters

The following corrosion rates of zinc have been reported for various types of atmospheres [135,136]:

Atmosphere	Differential corrosion rate (μm/year)
Rural	0.2 to 2
Urban and industrial	2 to 16
Marine	0.5 to 8

In field tests low contents (<2%) of various alloying additions to zinc have not resulted in any significant improvement in the corrosion resistance. Larger additions of aluminum, however, have proved favorable. Thus, a zinc alloy containing 55% Al and 1.6% Si has shown two to six times better resistance than zinc when used as coating on carbon steel [137].

The penetration depth is generally reported to be an approximately linear function of the exposure time in rural and urban atmospheres

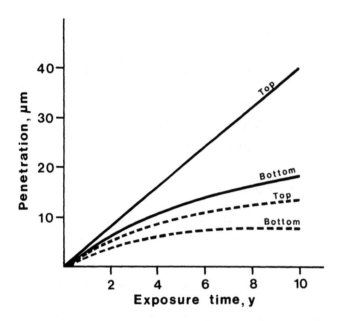

FIGURE 29 Predictive curves for atmospheric corrosion of galvanized
steel based on data obtained during 4 to 5 years of exposure in an
urban atmosphere (East Chicago) and a marine atmosphere (Kure
Beach, 245 m lot) [138]. —— = Urban, --- = marine.

[135]. Legault and Pearson [138], however, found that this is true
only for skyward surfaces exposed to urban atmospheres (Fig. 29).
Groundward surfaces in urban atmospheres and skyward as well as
groundward ones in marine atmospheres, show nonlinear relations
with time in accordance with the power law

$$p = kt^n$$

where the values of the constants k and n vary with the exposure
conditions. It may be noted that the corrosion rate was found to be
higher on the skyward surfaces than on the groundward ones. Further, it was concluded that with this equation one can reliably predict
the long-term atmospheric corrosion behavior of zinc on the basis of
two weight loss determinations during the initial 2-year period of exposure.

Several authors [139] have published equations which correlate the
atmospheric corrosion of zinc to the SO_2 content of the air, and in one
case also to the Cl^- deposition (Table 9). In most of these equations
the corrosion loss is a linear function of the SO_2 content (Fig. 30).

TABLE 9 Corrosion of Zinc and Galvanized Steel as Function of Environmental Parameters

Reference	Equation	Parameters
Barton et al. [100]	$K = 0.00076 t_w^{0.50} SO_2^{0.718}$	t_w = time of wetness, i.e., total time RH $\geq$ 80%, temp $\geq$ 0°C, hr/day SO_2 = SO_2 deposition, mg/m² day K = corrosion rate, g/m² day
Haynie and Upham [140]	$K = 0.001028(RH - 48.8)SO_2$	SO_2 = average SO_2 conc., μg/m³ RH = average relative humidity, % K = corrosion rate, μm/year
Guttman [141]	$K = 0.005461(t_w)^{0.8152}(SO_2 + 0.02889)$	t_w = electrochemically monitored time of wetness, hr SO_2 = SO_2 conc. when panels wet, ppm K = corrosion loss, g per 3 by 5 in. panel
Atteraas et al. [97]	$K = 0.22SO_2 + 6.0$ $K = 0.27Cl^- + 0.22SO_2 + 4.5$	SO_2 = SO_2 conc., μg/m³ Cl^- = chloride deposition, g/m² year K = corrosion rate, g/m² year
Hudson and Stanners [142]	$K = 0.16SO_2 + 6.32$	SO_2 = SO_2 conc., μg/m³ K = corrosion rate, g/m² year
Knotkova et al. [143]	$K = 0.17SO_2 + 3.6$	SO_2 = SO_2 deposition, mg/m² day K = corrosion rate, g/m² year

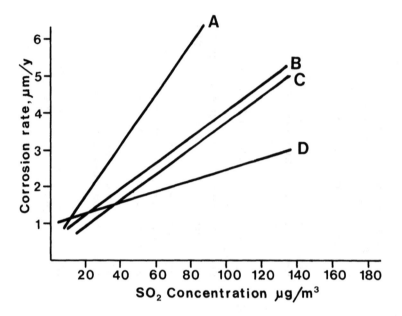

FIGURE 30 Corrosion rate of zinc as a function of SO_2 concentration
in the atmosphere [131]. Curve A is from Sweden's case study [154],
curve B from Hudson and Stanners [142], curve C from Haynie and
Upham [140], and curve D from Guttman [141].

The variation in constants among these equations might largely be due
to differences in time of wetness and temperature at the sites for which
the equations were determined. In some of the equations time of wetness has been included as an explicit parameter. The nomogram in
Fig. 31 represents a function suggested by Barton [78,101] and allows rapid estimation of the corrosion rate of zinc if the time of wetness and the SO_2 pollution are known.

As reported by Ellis [144], the corrosion rate of zinc surfaces is
very dependent on weather conditions during the early part of the exposure. Long-lasting rainfall or a relative humidity at or near 100%
during the first few days leads to a high corrosion rate, while drier
conditions lead to lower corrosion rates. After the initial period, however, the differential corrosion rate appears to be about the same in
the two cases (Fig. 32).

5.6.3 Application Aspects

A few decades back, galvanized steel was considered a corrosion-resistant material for structures in outdoor atmospheres. The corrosion

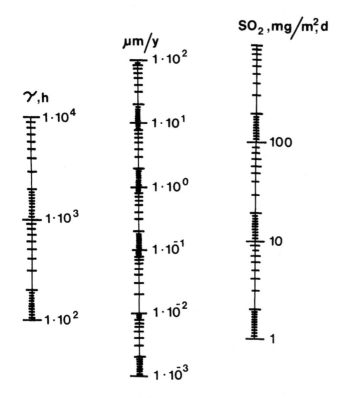

FIGURE 31 Nomogram for determination of the atmospheric corrosion rate (μm/year) of zinc from the SO_2 deposition (mg/m^2 day) and the time of wetness (hr/year) [78, 101].

rate was only 0.5 to 1.0 μm/year. With increasing pollution by SO_2 the corrosion rate of zinc has increased, in many urban or industrial areas up to 5 μm/year or more, and the life of galvanized steel structures has sometimes become unacceptably short. Today, it is often necessary to give galvanized steel additional protection, primarily by anticorrosion painting.

A high corrosion rate of zinc occurs when so-called white rust is formed in crevices where moisture collects, e.g., between stacked semiproducts like galvanized strips or sections. Then large amounts of bulky, loosely adherent, nonprotective corrosion products, white rust, are formed. It has not yet been established whether the corrosion products become nonprotective due to long-lasting moistening or to a poor supply of CO_2 so that protective basic zinc carbonate cannot form. When analyzed, presumably after exposure to ordinary

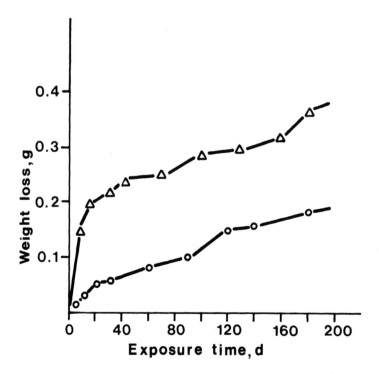

FIGURE 32 Memory effect on the atmospheric corrosion of zinc samples exposed on August 4, 1947 (△) and samples exposed on February
17, 1948 (○) [144].

CO_2-containing air, white rust contained considerable amounts of carbonate [145]. White rust is most frequent during rainy seasons. It
can also form on moist zinc surfaces in contact with mineral wool, which
prevents access of air and drying. This situation may occur in heat-
insulated wall structures enclosing galvanized steel members.

Zinc, being a less noble metal than iron, gives cathodic protection
to steel in the presence of a sufficiently conductive electrolyte. The
zinc coating on galvanized steel gives corrosion protection in the atmosphere to sheared edges and scratches where the steel is bare. The
protective action, however, is restricted to 1−3 mm at the most. Barton [133] has questioned whether this remote action should be ascribed
to electrochemical protection or to the spread of corrosion products of
zinc, which cause inactivation of the sulfate present at the iron/rust
boundary.

A coating of the Zn alloy containing 55% Al and 1.6% Si mentioned
above is increasingly used as a protective coating for carbon steel. In

addition to its good corrosion resistance in various types of atmospheres, it gives cathodic protection to bare steel surfaces at pores, cut edges, etc. In this respect the aluminum-zinc alloy coating is intermediate between zinc and aluminum coatings. This promising coating is being produced under the trade name Galvalume, Aluzink or Zincalum.

5.7 ALUMINUM

Aluminum has, in spite of its very base nature, proved to be a very useful metal for various structures exposed to indoor and outdoor atmospheres. For example, it has been extensively used for buildings, electric overhead conductors, and transportation vehicles.

5.7.1 Fundamentals

As shown by the potential-pH diagram (Fig. 33), aluminum is thermodynamically stable only at low potentials [1]. It can be used in the presence of water only because of its property of developing a protective coating of alumina. At an Al^{3+} activity of 10^{-1} M the stability range of the oxide coating extends down to pH 2.5.

The following species have been identified in corrosion products formed on atmospheric corrosion of aluminum under outdoor conditions: amorphous $Al(OH)_3$, α-$Al(OH)_3$ (bayerite), and γ-Al_2O_3, the latter with varying amounts of water in the lattice [7].

When a fresh aluminum surface is exposed to dry air it is immediately covered with a dense, thin amorphous oxide coating (100–200 Å thick) which is protective. In moist atmospheres the oxide coating grows thicker. It consists of one dense, protective barrier layer next to the metal and one outer more permeable bulk layer [145].

Anions such as SO_4^{2-} or Cl^- deposited on the oxide surface may react with the oxide with the formation of water-soluble salts, e.g., $Al_2(SO_4)_3$, and may also be incorporated in the lattice to form a variety of basic salts and complexes, about which little is known so far. The oxide coating is protective in urban atmospheres with SO_2 pollution, producing a relatively low pH value in the moisture film. High SO_2, however, causing a very low pH value in the moisture film, will lead to dissolution of the protective coating. In the presence of chloride, the oxide coating is more permeable to ions. The chloride ions are believed to migrate into the oxide layer and lower its resistance to outward migration of Al^{3+} [146].

In the presence of chloride ions, pitting may be initiated. In the propagation stage aluminum is dissolved anodically to Al^{3+} ions within the pit. The cathodic reaction takes place either outside the pit

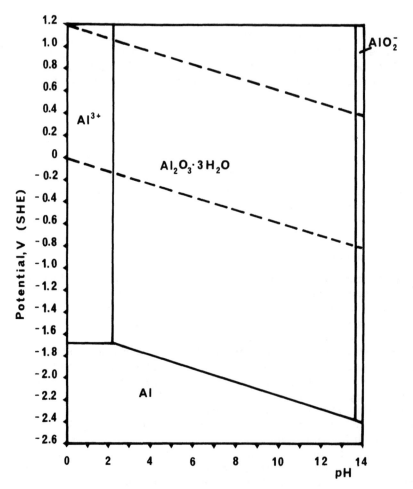

FIGURE 33 Potential-pH diagram, Al-H$_2$O; 25°C, 10^{-1} M Al [1].

close to its mouth or inside the pit and consists of the reduction of
oxygen or H$^+$ ions, respectively. In the former case it occurs preferentially at surface inclusions with a low oxygen overvoltage, such
as segregations of FeAl$_3$ or particles of deposited copper. The passivating oxide layer has low electronic conductivity, but the cathode reaction may locally destroy the protective oxide layer due to alkaliza-
tion, which lowers the electrode potential and may even make hydrogen liberation possible [4]. By hydrolysis of the Al^{3+} ions acid conditions are created within the pit and a cap of Al(OH)$_3$ and/or Al$_2$O$_3$
is formed over its mouth; the corrosion products finally block the operation of the pit.

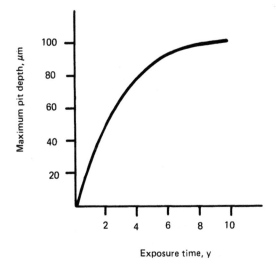

FIGURE 34 Maximum pit depth vs. exposure time for AlMn1.2 in the urban marine atmosphere of Gothenburg [147].

5.7.2 Corrosion as a Function of Time and Environmental Parameters

In clean outdoor atmospheres aluminum will retain its shiny appearance for years, even under tropical conditions. In polluted outdoor atmospheres small pits develop which are hardly visible to the naked eye. The pits become covered with crusts of aluminum oxide and hydroxide. The growth rate of the maximum pit depth is relatively high during the first few years of exposure, but decreases gradually so that the pit depth approaches a nearly constant value (Fig. 34) [147].

In clean or slightly polluted atmospheres most alloying constituents reduce the corrosion resistance of aluminum, although in such environments the corrosion rate is very low for unalloyed as well as for alloyed aluminum [78]. In industrial atmospheres, high in SO_2 pollution, the corrosion rate is largely the same for unalloyed aluminum and for most conventional aluminum alloys, i.e., alloys with Zn, Mg, Si, and/or Mn. Aluminum alloys with Cu, however, will generally show 4 to 20 times higher corrosion rates. The average penetration depth determined from the weight loss is very low and asymptotically approaches a limiting value. As shown in the following table, the mean corrosion rate for AlMn1.2 during 20 years of exposure does not exceed 1 μm/year even in polluted atmospheres; in clean atmospheres it is much lower [147]. The pitting is more significant, but in moderately polluted atmospheres the maximum pit depth

rarely exceeds 200 μm even after one or two decades of exposure
[148]:

Atmosphere	Mean corrosion rate (from weight loss) (μm/year)	Max. pit depth after 20 years exposure (μm)
Rural	<0.1	10 to 55
Urban	1	100 to 190
Marine	0.4 to 0.6	85 to 260

Rozenfeld [4] showed that the atmospheric corrosion at a relative
humidity of 98% is little affected by moderate SO_2 contents, probably
due to the low adsorption tendency for SO_2 on aluminum surfaces
(Fig. 6) [30]. Only at excessive SO_2 concentrations (above 0.01%
by volume) do severe corrosion effects occur [3]. In polluted environments, however, dust particles often play an important role in
accelerating corrosion if they are not washed away. This is the case
on sheltered exposures. There the corrosion rate, being low at first,
may after some time accelerate rapidly, eventually attaining a steady-
state value significantly higher than that in open-air exposure. This
behavior occurs even in semirural and urban atmospheres (Fig. 35)

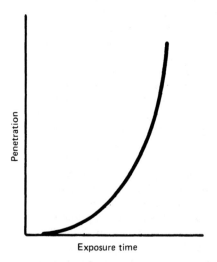

FIGURE 35 Sketch of corrosion-time curve for rain-sheltered aluminum
in a heavy industrial atmosphere [78].

[78]. Dust may accelerate corrosion by absorbing moisture and SO_x from the atmosphere, thus for long periods producing an acid medium on the surface; under such conditions the protective alumina coating is not stable (see Fig. 33). Further, carbonaceous dust may initiate pitting by galvanic action [4].

5.7.3 Application Aspects

Because of their good passivation properties in outdoor atmospheres, aluminum materials are commonly used for building purposes, e.g., for roofs, facades, and window frames. Of particular interest is the relatively good resistance of aluminum in SO_2-polluted atmospheres, where most other materials are susceptible to corrosion. Aluminum structures have been reported to be in excellent condition even after six to seven decades of exposure in urban atmospheres [147,149].

The low penetration rate and the shallow pitting do not generally influence the mechanical properties of aluminum structures, except in excessively polluted atmospheres. The shiny metal appearance will, however, gradually disappear and the surface will roughen under the formation of a grey patina of corrosion products. If the atmosphere contains much soot, this will be adsorbed by the corrosion products and give the patina a dark color. The shiny metal appearance may be retained by anodizing, that is, by anodic oxidation, which strengthens the oxide coating and improves its protective properties. An oxide coating with a thickness of 20 μm will generally preserve the original metal appearance for decades.

In rain-sheltered positions dust and other pollutants may collect and accelerate corrosion by disturbing the formation of a protective oxide coating. Such rain-sheltered areas will often show disturbing staining due to corrosion products and should, if possible, be avoided at the design stage. Existing sheltered areas should be cleaned regularly by washing.

In spite of the fact that the oxide coating is more permeable to ions in the presence of chloride, aluminum materials are also useful in marine atmospheres. Under these conditions, however, the aluminum is susceptible to bimetallic corrosion when in contact with a more noble metal such as carbon steel or copper [150]. In the presence of chloride, aluminum is also somewhat susceptible to crevice corrosion.

Aluminum is used as a protective coating on carbon steel. Because of the high electric resistance of the oxide layer, aluminum will not give cathodic protection to bare steel surfaces at sheared edges, scratches, etc., so bare steel surfaces may then become stained by rust, at least under rural or urban conditions. In marine atmospheres, however, where the oxide coating is more permeable to ions, aluminum will offer cathodic protection to carbon steel [155].

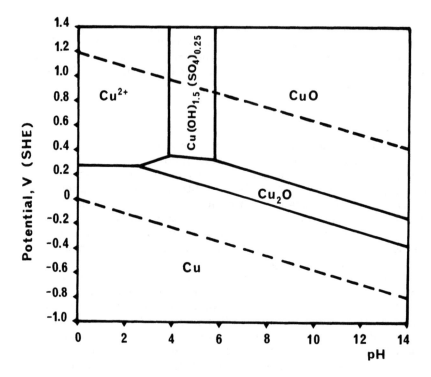

FIGURE 36 Potential-pH diagram, $Cu-SO_4^{2-}-H_2O$; 25°C, 10^{-1} M Cu, 10^{-3} M SO_4^{2-} [1].

5.8 COPPER

Copper has a long tradition as an architectural material. Because of its high electrical conductivity it is also used for electrical conductors of various kinds. Thus, large amounts of copper materials are used under atmospheric exposure in different types of climates all over the world.

5.8.1 Fundamentals

We may first consider the thermodynamic possibilities for the formation of copper species related to the various constituents in outdoor atmospheres [1]. The potential-pH diagram in Fig. 36 represents the system $Cu-SO_4^{2-}-H_2O$ at 25°C with a Cu^{2+} content of 10^{-1} mole/liter and an SO_4^{2-} content of 10^{-3} mole/liter. As can be seen, copper metal is stable in a great part of the stability region of water. This is consistent with copper being a noble metal. The diagram also shows a

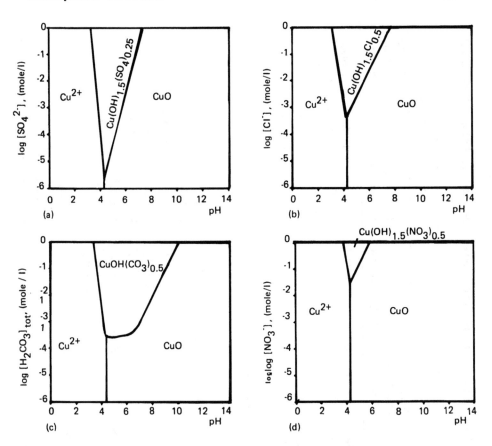

FIGURE 37 Stability domains of basic copper salts in aerated aqueous solutions with varying anion content and pH value; 25°C, 10^{-1} M Cu [1]. (a) Sulfate; (b) chloride; (c) carbonate; (d) nitrate.

stability domain for basic copper sulfate, $Cu(OH)_{1.5}(SO_4)_{0.25}$. The width of this domain depends on the SO_4^{2-} content; with decreasing content the width diminishes (Fig. 37a). The stability domains for $Cu(OH)_{1.5}Cl_{0.5}$, $CuOH(CO_3)_{0.5}$, and $Cu(OH)_{1.5}(NO_3)_{0.5}$ versus corresponding anion contents can also be seen in Fig. 37.

The coating of corrosion products or the so-called patina has a rather complex composition, varying from place to place (Table 10) [151]. The main components are generally copper(I) oxide and one or more basic copper salts [7,151]:

Basic sulfate: $Cu(OH)_{1.5}(SO_4)_{0.25}, Cu(OH)_{1.33}(SO_4)_{0.33}$

TABLE 10 Basic Copper Salts in Green Patina from Various Atmospheres, Defined by Anions Ranked with Respect to Content [151]

Reporter	Object	Time of exposure (years)	Country	Type of atmosphere			
				Rural	Urban or industrial	Marine	Mixed urban-marine
Vernon and Whitby	Copper roofs; copper conductor in marine atmosphere	12–300 13	UK	1. SO_4^{2-} 2. CO_3^{2-}	1. SO_4^{2-} 2. CO_3^{2-} 3. Cl^-	1. Cl^- 2. CO_3^{2-} 3. SO_4^{2-}	1. SO_4^{2-} 2. CO_3^{2-}, Cl^-
Vernon	Copper roof on church spire on the isle of Guernsey	33	UK			1. Cl^- 2. CO_3^{2-} 3. SO_4^{2-}	
Freeman, Jr.	Copper roofs	16–78	USA				1. SO_4^{2-} 2. CO_3^{2-}, Cl^-
Thompson, Tracy, and Freeman, Jr.	Copper panels from field test	20	USA	1. SO_4^{2-}, Cl^- 2. CO_3^{2-}	1. CO_3^{2-} 2. SO_4^{2-}	1. Cl^- 2. SO_4^{2-} 3. CO_3^{2-}	
Aoyama	Copper conductor for railway		Japan	1. NO_3^- 2. CO_3^{2-}	1. SO_4^{2-} 2. CO_3^{2-}	1. Cl^- 2. CO_3^{2-}	
Mattsson and Holm	Copper-base materials from field test	7	Sweden	1. SO_4^{2-} 2. NO_3^- 3. CO_3^{2-}	1. SO_4^{2-} 2. CO_3^{2-}	1. Cl^- 2. SO_4^{2-}	
Scholes and Jacob	Copper-base materials from field test	16	UK		1. SO_4^{2-} 2. Cl^-, CO_3^{2-}	1. Cl^- 2. SO_4^{2-}	

Basic chloride: $Cu(OH)_{1.5}Cl_{0.5}$

Basic carbonate: $CuOH(CO_3)_{0.5}$, $Cu(OH)_{1.33}(CO_3)_{0.33}$, $Cu(OH)_{0.67}(CO_3)_{0.67}$

Basic nitrate: $Cu(OH)_{1.5}(NO_3)_{0.5}$

In urban atmospheres basic sulfate is predominant, and in marine atmospheres basic chloride, while in rural atmospheres basic sulfate is usually the main component. This is in good agreement with the thermodynamics described. Unexpectedly, however, basic carbonate is sometimes found in practice. Figure 37c indicates that the conditions in the atmosphere would not favor the formation of this type of patina, for the H_2CO_3 content in the water film would only be about 10^{-5} mole/liter in equilibrium with the air. The presence of basic copper nitrate, also found in some locations, indicates that the water film on the metal surface may contain an appreciable amount of nitrate (Fig. 37d).

The corrosion products first formed are generally Cu_2O and CuO. The copper oxides react rather slowly with species in the air such as SO_x, Cl^-, CO_2, and NO_x with the formation of basic salts, provided the pH value of the surface moisture is sufficiently high:

$$Cu_2O + 0.5SO_2 + 1.5H_2O + 0.75O_2 \rightarrow 2\ Cu(OH)_{1.5}(SO_4)_{0.25}$$

$$Cu_2O + Cl^- + 1.5H_2O + 0.25O_2 \rightarrow 2\ Cu(OH)_{1.5}Cl_{0.5}$$

The corrosion products give some corrosion protection to the surface.

5.8.2 Corrosion as a Function of Time and Environmental Parameters

For copper in various types of atmospheres the following corrosion rates have been determined [102,151]:

Atmosphere	Corrosion rate (μm/year)
Rural	<1
Urban and industrial	1–3
Marine	1–2

The atmospheric corrosion rates for various types of copper alloys during long-term tests are shown in Fig. 38 [151]. The weight loss is approximately the same for low-alloyed coppers, brasses, nickel silvers, and tin bronzes. The reduction in mechanical properties due to general corrosion is generally small, mostly less than 5% in UTS and

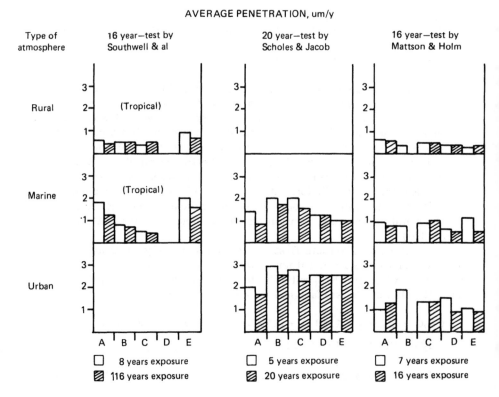

FIGURE 38 General corrosion rates for copper and copper alloys in the atmosphere, sloping surfaces. A = low-alloyed Cu; B = low-Zn brass; C = high-Zn brass; D = Ni silver; E = Sn bronze [152].

less than 10% in elongation [151]. For high-zinc brasses, however, greater changes occur because of dezincification [152]. Zinc is dissolved selectively from the alloy, which results in areas of porous copper with poor mechanical properties. The diagram in Fig. 39 shows that (α+β)-brasses are somewhat more prone to dezincification than pure β- and α-brasses, probably because of local cell action when α- and β-phases occur together in the structure. Small additions of As—about 0.02–0.04%—inhibit dezincification in brass, although only in α-brass.

High-zinc brasses, if under tensile stress, are also susceptible to stress corrosion cracking on atmospheric exposure outdoors [151]. The risk of stress corrosion in brasses is greatest in industrial and urban atmospheres characterized by high contents of sulfur dioxide

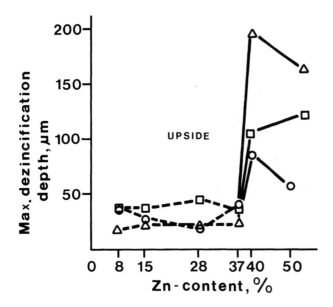

FIGURE 39 Maximum dezincification depth vs. zinc content in binary brasses after 16 years of exposure in different atmospheres [153]. △ = Urban, □ = marine, ○ = rural.

and ammonia. The stress corrosion susceptibility is markedly lower in marine atmospheres.

The corrosion rate of copper decreases somewhat during the first few years of exposure but soon approaches a largely constant value (Fig. 40) [152]. This behavior indicates the formation of a slightly protective layer of corrosion products. The protective action is probably due mainly to the presence of basic copper salts.

5.8.3 Application Aspects

Because of its low corrosion rate and its interesting patina formation, copper has long been used for building structures such as roofs, facades, and gutters. Many copper roofs on castles and other monumental buildings have lasted for several centuries.

If conditions are favorable, copper may after some years develop a characteristic blue-green patina in outdoor atmospheres. Sometimes a user is at first disappointed and complains when his newly laid copper roof shows an ugly, mottled appearance [153]. After 6 to 12 months, however, the surface has usually acquired a uniform dark-brown color. In general, the surface does not develop beyond this

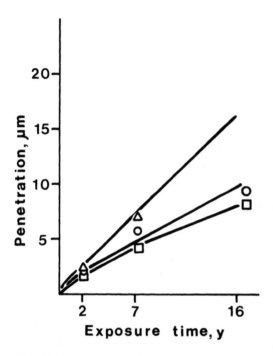

FIGURE 40 Penetration depth vs. exposure time for copper [153].
△ = Urban, ○ = marine, □ = rural.

stage for a number of years. After 5 to 10 years a green patina may begin to appear on sloping surfaces under urban or marine conditions. Vertical surfaces generally stay black much longer, as their time of wetness is shorter. In a marine atmosphere the surfaces facing the sea acquire a green patina sooner than the other surfaces. This is, of course, due to the greater supply of chlorides from the sea winds. If the atmosphere is little polluted, the patina may take an extremely long time to form (hundreds of years) due to the scarce supply of anions for the formation of basic copper salt. It may even fail to form at all if the temperature is low or the exposure conditions very dry. As shown by the potential-pH diagram, green patina will also fail to form under acidic conditions, e.g., near chimneys, where acid smoke strikes the roof. Under such conditions, the corrosion products are soluble.

Rainwater, running off the copper-bearing surfaces, has generally picked up traces of dissolved copper. The fraction of the corrosion products that leave the surface, dissolved in rainwater, is greatest in urban and industrial atmospheres highly polluted with SO_x [152].

Such water may cause blue staining on masonry, stonework, etc. Therefore, the rainwater from copper surfaces should be properly collected and drawn off through gutters and spouts.

5.9 CONCLUSIONS

Atmospheric corrosion proceeds in a relatively complicated system consisting of metal, corrosion products, surface electrolyte, and atmosphere. It is an electrochemical process, the electrolyte being a thin layer of moisture on the metal surface. The composition of the electrolyte depends primarily on the deposition rates of the air pollutants and fluctuates with wetting conditions. Because of the fluctuations accurate potential-pH diagrams cannot be calculated for the corroding systems. Nevertheless, approximate potential-pH diagrams are generally in accordance with the composition of the corrosion products and form a thermodynamic basis for understanding even the atmospheric corrosion processes. On some metals protective layers of corrosion products may be formed under favorable conditions. The presence of a layer of electrolyte, the composition of the electrolyte, and the properties of corrosion products on the surface determine the kinetics of the process. Thus, in practice the time of wetness and the deposition rate of pollutants, mainly SO_2 and chlorides in coastal regions, determine the atmospheric corrosion rate of the common structural metals. These parameters may also be used for classification of the corrosivity of the atmosphere for different metals.

The corrosion of carbon steel proceeds in local cells, which in SO_2-polluted atmospheres consist of sulfate nests surrounded by cathodic areas. The corrosion mechanism is in general quite well understood. The protective properties of the rust on carbon steel, which consists mainly of FeOOH, are poor. Thus the corrosion rate is high especially in polluted atmospheres. However, by alloying the steel with small amounts of, e.g., P, Cu, Cr, and Ni the protective properties of the rust may be improved. A dense layer of rust with an even distribution of sulfate provides good corrosion resistance and a nice-looking appearance to the weathering steels.

Zinc is usually subject to general corrosion in the atmosphere; the detailed mechanism of the anodic and cathodic processes is not very well known. In spite of the fact that zinc is a base metal, its corrosion rate is usually an order of magnitude lower than that of carbon steel. This is due to the formation of a protective coating of zinc carbonate or basic zinc salts. In SO_2-polluted atmospheres the pH of the surface moisture may decrease to values so low that the basic zinc salts are not stable. A protective zinc patina formed earlier is then transformed to soluble $ZnSO_4$, which may be washed away by rain.

On exposure to air aluminum is covered with a thin oxide or hydroxide layer which is stable down to pH 2.5. Anions such as SO_4^{2-} or Cl^- may locally react with the oxide with the formation of soluble salts. This occurs especially under heavy deposits and in rain-protected positions, where the surface electrolyte may be aggressive. The corrosion then proceeds as a localized attack of the pitting type. The corrosion products finally block the growth of pits, so during long-term exposure the pit depth approaches a nearly constant value. Thus aluminum also has very good corrosion resistance in polluted atmospheres.

Copper, like zinc, is subject to general corrosion in the atmosphere. In this case also the detailed mechanism of the anodic and cathodic reactions is not very well known. Copper is a rather noble metal. In the atmosphere it becomes covered by copper oxides, which ultimately may be transformed into basic copper salts, usually sulfate or, in marine atmospheres, chloride. The basic copper salts are the main components of the green patina. Under acidic conditions, e.g., in severely SO_2-polluted atmospheres, soluble corrosion products may be formed and the corrosion rate increases.

Even if a great number of investigations have been devoted to the mechanism of atmospheric corrosion and substantial knowledge has been gained, there are still gaps in our understanding. Nevertheless the present knowledge of the mechanism of atmospheric corrosion might be useful as a basis for the development of new short-term test methods, more representative of conditions in practice than some of the accelerated corrosion tests generally used today, e.g., the salt spray and Kesternich tests.

REFERENCES

1. E. Mattsson, NACE Plenary Lecture—1982, *Mater. Perform.*, *21*(7): 9 (1982).
2. K. Barton and Z. Bartonova, *Werkst. Korros.*, *20*: 216 (1969).
3. N. D. Tomashov, *Theory of Corrosion and Protection of Metals*, Macmillan, New York, 1966.
4. I. L. Rozenfeld, *Atmospheric Corrosion of Metals*, National Association of Corrosion Engineers, Houston, 1972.
5. A. Bukowiecki, *Schweiz. Arch.*, *32*: 42 (1966).
6. V. V. Skorcheletti and T. S. Tukaschinskij, *Zh. Prikl. Chim.*, *26*: 30 (1953).
7. K. Barton, *Schutz gegen atmosphärische Korrosion*, Verlag Chemie, Weinheim, 1972.
8. H. Schwarz, *Werkst. Korros.*, *23*: 648 (1972).
9. K. Bohnenkamp, *Werkst. Korros.*, *19*: 792 (1968).
10. G. Schikorr, *Werkst. Korros.*, *14*: 69 (1983).

11. K. Barton, Z. Bartonova, and E. Beranek, *Werkst. Korros.*, 25: 659 (1974).
12. D. Knotkova-Cermakova, J. Vlckova, and D. Kuchynka, *Werkst. Korros.*, 24: 684 (1973).
13. R. Ericsson and T. Sydberger, *Werkst. Korros.*, 31: 455 (1980).
14. P. V. Strekalov and Wu Din Wuy, *Zasch. Met.*, 18: 881 (1982) (in Russian).
15. V. Kucera, *Ambio*, 5: 243 (1976).
16. S. Haagenrud and B. Ottar, in *Proceedings of the 7th Scandinavian Corrosion Congress*, Trondheim, Norway, Royal Norwegian Council for Scientific and Industrial Research, Oslo, 1975.
17. J. M. Waldman et al., *Science*, 218: 677 (1982).
18. N. D. Tomashov, G. K. Berukshtis, and A. A. Lokotilov, *Zavod. Lab.*, 22: 345 (1956).
19. P. J. Sereda, *ASTM Bull.*, 228: 53 (1958).
20. V. Kucera and E. Mattsson, in *Corrosion in Natural Environments*, ASTM STP 558, American Society for Testing and Materials, Philadelphia, 1974, p. 239.
21. I. Jirovsky, I. Kokoska, and J. Prusek, *Werkst. Korros.*, 27: 16 (1976).
22. F. Mansfeld and J. V. Kenkel, *Corros. Sci.*, 16: 111 (1976).
23. V. Kucera and J. Gullman, in *Electrochemical Corrosion Testing*, ASTM STP 727, American Society for Testing and Materials, Philadelphia, 1981, p. 238.
24. Ju. N. Mikhailovskij et al., *Zasch. Met.*, 9: 264 (1973).
25. Ju. N. Mikhailovskij, P. V. Strekalov, and V. V. Agafonov, *Zasch. Met.*, 16: 308 (1980).
26. *Acidification Today and Tomorrow*, Swedish Ministry of Agriculture, Stockholm, 1982.
27. *Ecological Effects of Acid Deposition*, Expert Meeting I, 1982, Stockholm Conference on the Acidification of the Environment. National Swedish Environment Protection Board, Rept. 1636, Stockholm, 1983.
28. C. Brosset, *Acid Particulate Air Pollutants in Sweden*, Swedish Water and Air Pollution Research Laboratory, Gothenburg, 1975.
29. H. Dovland, E. Joranger, and A. Semb, *Wet and Dry Deposition of Pollutants in Norway*, Norwegian Institute for Air Research, Kjeller, 1976.
30. T. Sydberger and N.-G. Vannerberg, *Corros. Sci.*, 12:775 (1972).
31. W. H. J. Vernon, *Trans. Faraday Soc.*, 31: 1678 (1935).
32. R. Preston and B. Sanyal, *J. Appl. Chem.*, 6: 26 (1956).
33. J. B. Upham, *Air Pollut. Control Assoc.*, 17: 398 (1967).
34. F. H. Haynie and J. B. Upham, in *Corrosion in Natural Environments*, ASTM STP 558, American Society for Testing and Materials, Philadelphia, 1974.
35. A. Kutzelnigg, *Werkst. Korros.*, 8: 492 (1957).

36. D. W. Rice et al., *J. Electrochem. Soc.*, *128*: 275 (1981).
37. *Corrosion of Metals, Classification of Corrosion Aggressivity of Atmosphere*, Standard of Mutual Economic Aid, ST SEV 991-78.
38. ISO/DP 9223, *Corrosion of Metals and Alloys. Classification of Corrosivity Categories of Atmospheres*, 1986.
39. P. Grennfelt, *Nitrogen Oxides in Swedish and Norwegian Cities*, Swedish Water and Air Pollution Research Laboratory, Report for Project KHM EM 705, Gothenburg, 1982.
40. P. Grennfelt, *Nitrogen Oxides in the Atmosphere*, Rept. B 585, Swedish Water and Air Pollution Research Laboratory, Gothenburg, 1980.
41. P. Grennfelt, *Long- and Short-Range Transport of SO_2 and NO_2*, Rept. B 484, Swedish Water and Air Pollution Research Laboratory, Gothenburg, 1979.
42. I. Smith, *Nitrogen Oxides from Coal Combustion—Environmental Effects*, Rept. ICTIS/TR, IEA Coal Research, London, 1980.
43. K. Katoh et al., *Boshoku Gijutsu*, *30*: 337 (1981).
44. S. Luckat, *Quantitative Untersuchung des Einflusses von Luftverunreinigungen bei der Zerstörung von Naturstein*, Forschungsbericht (Rept.) 106 08 003/02, Zollern Institute, Dortmund, West Germany, 1981.
45. K. E. Johnson and J. F. Stanners, *The Characterisation of Corrosion Test Sites in the Community*, Rept. EUR 7433, Commission of the European Communities, Brussels, 1981.
46. B. Bonnarens and A. Bragard, *Recherche Collective sur la Corrosion Atmospherique des Aciers*, Rept. EUR 7400, Commission of the European Communities, Brussels, 1981.
47. H. Rodhe, Stockholm University, private communication.
48. D. Knotkova-Cermakova and J. Vlckova, *Werkst. Korros.*, *21*: 16 (1970).
49. P. J. Sereda, *Ind. Eng. Chem.*, *52*: 157 (1960).
50. G. K. Berukshtis and G. B. Klark, in *Corrosion of Metals and Alloys*, Part 5-7, Izd. AN SSSR, Moscow, 1965 (in Russian).
51. P. V. Strekalov, V. V. Agafonov, and Yu. N. Mikhailovskii, *Zasch. Met.*, *8*: 577 (1972).
52. A. I. Golubev, in *Corrosion and Protection of Constructional Materials*, Nauka, Moscow, 1966.
53. S. Haagenrud, V. Kucera, and L. Atteraas, in *Proceedings of the 9th Scandinavian Corrosion Congress*, Danish Corrosion Center, Copenhagen, 1983.
54. M. Pourbaix, *Atlas of Electrochemical Equilibria in Aqueous Solutions*, National Association of Corrosion Engineers, Houston, CEBELCOR, Brussels, 1974.
55. E. Hogfeldt and L. G. Sillén, *Tentative Equilibrium Constants for 25°C (on Infinite Dilution Scale)*, The Royal Institute of Technology, Stockholm, 1966.

56. K. Barton and Z. Bartonova, *Werkst. Korros.*, *21*: 85 (1970).
57. J. Dearden, *J. Iron Steel Inst.*, *186*: 241 (1948).
58. T. Sydberger, *Influence of Sulphur Pollution on the Atmospheric Corrosion of Steel*, thesis, Chalmers Univ. of Technology, Gothenburg, Sweden, 1976.
59. J. Detournay et al., *Corros. Sci.*, *15*: 295 (1975).
60. J. E. Hiller, *Werkst. Korros.*, *20*: 943 (1966).
61. S. Karaivanov and G. Gawrilov, *Werkst. Korros.*, *24*: 30 (1973).
62. R. Engelhardt et al., *Neue Huette, 16*: 593 (1971).
63. T. Misawa, *Corros. Sci.*, *13*: 659 (1973).
64. P. Keller, *Werkst. Korros.*, *20*: 102 (1969).
65. P. Keller, *Werkst. Korros.*, *22*: 32 (1971).
66. H. Schwarz, *Werkst. Korros.*, *23*: 648 (1972).
67. H. Baum et al., *Neue Huette, 19*: 423 (1974).
68. T. Misawa et al., *Corros. Sci.*, *11*: 35 (1971).
69. G. Becker et al., *Arch. Eisenhuettenwes.*, *40*: 341 (1969).
70. I. Suzuki, Y. Hisamatsu, and N. Masuko, *J. Electrochem. Soc.*, *127*: 2210 (1980).
71. S. J. Ali and G. C. Wood, *Br. Corros. J.*, *4*: 133 (1969).
72. H. Okada and H. Shimada, *Corrosion, 30*: 97 (1974).
73. K. A. Chandler and J. E. Stanners, in *Proceedings of the 2nd International Congress on Metallic Corrosion*, National Association of Corrosion Engineers, Houston, 1966.
74. H. Schwarz, *Werkst. Korros.*, *16*: 93 (1965).
75. J. Honzak and D. Kuchynka, *Werkst. Korros.*, *21*: 342 (1970).
76. I. Matsushima and T. Ueno, *Corros. Sci.*, *11*: 129 (1971).
77. K. Barton, D. Kuchynka, and E. Beranek, *Werkst. Korros.*, *29*: 199 (1978).
78. K. Barton, in *Air Pollution Control* (G. M. Bragg and W. Strauss, eds.), Wiley, New York, 1981.
79. J. F. Henriksen, *Corros. Sci.*, *9*: 573 (1969).
80. K. E. Heusler, *Z. Elektrochem.*, *62*: 582 (1958).
81. J. Bockris et al., *Electrochim. Acta, 4*: 325 (1961).
82. G. M. Florianovitch and J. M. Kolotyrkin, *Electrochim. Acta, 12*: 879 (1967).
83. A. G. Tanner, *Chem. Ind.*, *13*: 1027 (1964).
84. U. R. Evans and C. A. Taylor, *Corros. Sci.*, *12*: 227 (1972).
85. U. R. Evans, *Br. Corros. J.*, *7*: 10 (1972).
86. L.-G. Johansson, *SO_2-Induced Corrosion of Carbon Steel in Various Atmospheres and Dew Point Corrosion in Stack Gases*, thesis, Chalmers Univ. of Technology, Gothenburg, Sweden, 1982.
87. L.-G. Johansson and N.-G. Vannerberg, *Corros. Sci.*, *21*: 863 (1981).
88. D. A. Hegg and P. V. Hobbs, *Atmos. Environ.*, *12*: 241 (1978).
89. M. Pourbaix, *Corros. Sci.*, *12*: 161 (1972).

90. E. Langle, *Schweiz. Arch.*, *34*: 147 (1968).
91. T. Sydberger and R. Ericsson, *Werkst. Korros.*, *28*: 154 (1977).
92. G. Burgmann and D. Grimme, *Stahl Eisen*, *100*: 641 (1980).
93. M. Pourbaix, in *Proceedings of the 7th International Congress on Metallic Corrosion*, Associacao Brasileira de Corrosao, Rio de Janeiro, 1978.
94. K. Bohnenkamp, G. Burgmann, and W. Schwenk, *Galvano Organo*, *43*: 587 (1974).
95. H. Guttman and P. J. Sereda, in *Metal Corrosion in the Atmosphere*, ASTM STP 435, American Society for Testing and Materials, Philadelphia, 1968.
96. F. H. Haynie and J. B. Upham, in *Corrosion in Natural Environments*, ASTM STP 558, American Society for Testing and Materials, Philadelphia, 1974.
97. L. Atteraas et al., in *Proceedings of the 8th Scandinavian Corrosion Congress*, Helsinki University of Technology, Helsinki, Finland, 1978.
98. T. Hakkarainen and S. Ylasaari, in *Atmospheric Corrosion* (W. H. Ailor, ed.), Wiley, New York, 1982.
99. Ju. N. Mikhailovskij, *Zasch. Met.*, *16*: 395 (1980).
100. K. Barton et al., *Zasch. Met.*, *16*: 387 (1980).
101. Czechoslovak Standard CSN 03 8204, Determination of the Corrosion Aggressivity of Atmospheres for Metals and Metallic Coatings, 1978.
102. *Effects of Sulphur Compounds on Materials, Including Historic and Cultural Monuments*, in Airborne Sulphur Pollution. Effects and Control, United Nations, New York, 1984.
103. H. R. Ambler and A. Bain, *J. Appl. Chem.*, *5*: 437 (1955).
104. F. H. Haynie et al., *Mater. Perform.*, *15*: 48 (1976).
105. *Pictorial Surface Preparation Standards for Painting Steel Surfaces*, Swedish Standard SIS 05 59 00, 1967.
106. L. Igetoft, Swedish Corrosion Institute, private communication.
107. B. Gronvall, L. Thureson, and V. Victor, *Blasted Steel Surfaces—Surface Roughness, Cleanliness and Cold Working*, Bull. 67, Swedish Corrosion Institute, Stockholm, 1971.
108. C. Calabrese and J. R. Allen, *Corrosion*, *34*: 331 (1978).
109. J. E. O. Mayne, *J. Appl. Chem.*, *9*: 673 (1959).
110. K. Kesternich, *Metalloberflaeche*, *29*: 52 (1975).
111. K. Barton and E. Beranek, *Werkst. Korros.*, *11*: 348 (1960).
112. K. Barton, in *First International Congress on Metallic Corrosion*, Butterworths, London, 1961, p. 685.
113. D. Cermakova and K. Barton, *Zasch. Met.*, *3*: 145 (1967).
114. C. F. Day et al., *Trans. Inst. Mar. Eng.*, *84*: 33 (1972).
115. H. Okada et al., *Corrosion*, *26*: 429 (1970).

116. J. T. Keiser, C. W. Brown, and R. H. Heidersbach. *Corros. Sci.*, *23*: 251 (1983).
117. H. Schwitter and H. Bohni, *J. Electrochem. Soc.*, *127*: 15 (1980).
118. R. Bruno et al., *Corrosion*, *29*: 95 (1973).
119. M. Pourbaix, in *Passivity of Metals*, Electrochemical Society, Princeton, 1978.
120. *ASTM Proc.*, *53*: 110 (1953).
121. C. P. Larrabee and S. K. Coburn, in *First International Congress on Metallic Corrosion*, Butterworths, London, 1961.
122. J. C. Hudson and J. F. Stanners, *J. Iron Steel Inst.*, *180*: 271 (1955).
123. D. M. Buck, *ASTM Proc.*, *19*: 224 (1919).
124. G. Wranglén, *Corros. Sci.*, *9*: 585 (1969).
125. D. Fyfe et al., *Corros. Sci.*, *10*: 817 (1970).
126. C. Carius, *Z. Metallk.*, *22*: 237 (1930).
127. H. R. Copson, *ASTM Proc.*, *52*: 1005 (1952).
128. D. Knotkova et al., in *Atmospheric Corrosion of Metals*, ASTM STP 767, Philadelphia, 1982, p. 7.
129. J. Gullman et al., *Corrosion Resistance of Weathering Steels—Typical Cases of Damage in Building Context and Their Prevention*, Bull. 94, Swedish Corrosion Institute, Stockholm, 1985.
130. R. Grauer and W. Feitknecht, *Corros. Sci.*, *7*: 629 (1967).
131. R. W. Bailey and H. G. Ridge, *Chem. Ind.*, *78*: 1222 (1957).
132. T. Biestek and J. Niemec, *Proc. Inst. Mech. Precyz*, *14*(2): 38 (1966).
133. K. B. Barton, in *Intergalva '70* (Zinc Development Association, ed.), Industrial Newspaper Ltd., London, 1971, p. 199.
134. J. A. van Oeteren, in *Korrosionsschutz durch Beschichtungsstoffe*, vol. 1, Carl Hauser Verlag, München, 1980, pp. 91, 126.
135. C. J. Slunder and W. K. Boyd, *Zinc: Its Corrosion Resistance*, Zinc Institute Inc., New York, 1971.
136. E. Mattsson, *Tek. Tidskr.*, *98*: 767 (1968).
137. T. Johnsson and V. Kucera, *Proceedings of the Thirteenth International Galvanizing Conference*, Portcullis Press, London, 1982, p. 47/1.
138. R. A. Legault and V. P. Pearson, in *Atmospheric Factors Affecting the Corrosion of Engineering Metals*, ASTM STP 646, American Society for Testing and Materials, Philadelphia, 1978, p. 83.
139. *The Costs and Benefits of Sulphur Oxide Control. A Methodological Study*, OECD, Paris, 1981.
140. F. H. Haynie and J. B. Upham, *Mater. Prot. Perform.*, *9*(8): 35 (1970).
141. H. Guttman, in *Metal Corrosion in the Atmosphere*, ASTM STP 435, American Society for Testing and Materials, Philadelphia, 1968, p. 235.

142. J. C. Hudson and J. F. Stanners, *J. Appl. Chem.*, 3: 86 (1953).
143. D. Knotkova et al., in *9th International Congress on Metallic Corrosion*, National Research Council Canada, Ottawa, 1984.
144. O. B. Ellis, *Proc. ASTM*, 49: 152 (1949).
145. C. E. Bird and F. J. Strauss, *Mater. Perform.*, 15(11): 27 (1976).
146. H. Kaesche, *Localized Corrosion*, National Association of Corrosion Engineers, Houston, 1974, p. 516.
147. E. Mattsson and S. Lindgren, *Metal Corrosion in the Atmosphere*, ASTM STP 435, American Society for Testing and Materials, Philadelphia, 1968, p. 240.
148. E. Mattsson, *Tek. Tidskr.*, 98: 767 (1968).
149. H. P. Godard, W. B. Jepson, M. R. Bothwell, and R. L. Kane, *The Corrosion of Light Metals*, Wiley, New York, 1967, p. 92.
150. V. Kucera and E. Mattsson, in *Atmospheric Corrosion* (W. H. Ailor, ed.), Wiley, New York, 1982, p. 561.
151. E. Mattsson and R. Holm, in *Atmospheric Corrosion* (W. H. Ailor, ed.), Wiley, New York, 1982, p. 365.
152. R. Holm and E. Mattsson, in *Atmospheric Corrosion of Metals*, ASTM STP 767, American Society for Testing and Materials, Philadelphia, 1982, p. 85.
153. E. Mattsson and R. Holm, *Sheet Met. Ind.*, 45: 270 (1968).
154. *Air Pollution Across National Boundaries. The Impact on the Environment of Sulfur in Air and Precipitation*, Sweden's case study for the U.N. Conference on the Human Environment, Royal Ministry of Agriculture, Stockholm, 1971.
155. E. Mattsson and W. Ericsson, *Aluminum and Zinc Coatings on Steel—A Comparison of Corrosion Resistance in the Atmosphere*, Research Rept. 1274, Svenska Metallverken, Västerås, Sweden, 1959.

6

LOCALIZED CORROSION

HANS BÖHNI

Institute for Materials, Chemistry, and Corrosion, Federal Institute of Technology, Zurich, Switzerland

6.1 INTRODUCTION

Generally, localized corrosion includes various types of corrosion phenomena such as pitting, crevice corrosion, and intergranular attack as well as stress corrosion cracking and corrosion fatigue. The latter two forms of corrosion are discussed in other chapters of this book and will therefore not be considered here. Since the book is entirely devoted to corrosion mechanisms the scope of this chapter will be limited to pitting corrosion, where several detailed mechanisms have been proposed, in contrast to crevice corrosion, where only fragmentary contributions are available at present. It is the author's opinion that for a more rigorous treatment of crevice corrosion, still deeper insight into the corrosion mechanisms of pitting is required. With respect to intergranular attack, mechanical aspects would certainly have to be included, which is beyond the scope of this contribution.

Furthermore, the discussion of comprehensive mechanisms of localized corrosion is mainly restricted to contributions from the past decade. Since most of the earlier published models have already been extensively discussed in several review articles and in proceedings of conferences on localized corrosion, they will not be discussed in detail here again. Interested readers are referred to the original literature.

6.2 INITIATION MECHANISMS

During the past two decades a number of models have been proposed to describe initiation processes leading to passive film breakdown and the subsequent localized attack. In the review articles that have been published [1−8] the models are usually divided into three groups: (a) adsorption and adsorption-induced mechanisms, where the adsorption of aggressive anions not only is necessary, as is the case for most of the mechanisms proposed, but also plays a major role in pit initiation, (b) ion migration or penetration models, and (c) mechanical film breakdown theories. In the discussion of mechanically induced pit initiation processes, breakdown due to the introduction of externally applied stresses will be excluded.

Certainly a strict differentiation of the various existing theories on pit initiation is difficult, because many of them have common features. A classification different from the one used here may therefore be equally suitable. Furthermore, statements concerning prerequisites necessary for breakdown to occur will be omitted here. It is the author's opinion that these often very specific requirements, which are usually postulated in advance for certain corrosion phenomena without sufficient knowledge of the corrosion mechanisms involved, narrow rather than broaden the scope of progress in corrosion research.

6.2.1 Adsorption and Adsorption-Induced Mechanisms

Most of the pit initiation mechanisms proposed so far in the literature include the adsorption of aggressive anions at energetically preferred places as a necessary step in the whole nucleation process. Nevertheless, the significance of this step may differ quite markedly depending on the type of mechanism considered. Therefore only those mechanisms are summarized in this section in which the adsorption process has a decisive influence on pit initiation but is not necessarily the rate-determining step.

Several authors, including Uhlig and co-workers [9−11], Kolotyrkin [12], and Hoar and Jacob [5], proposed concepts based on either competitive adsorption or surface complex ion formation. In the case of competitive adsorption Cl^- anions as well as passivating agents are adsorbed simultaneously. Above a critical potential Cl^- adsorption is favored and breakdown of passivity occurs. Support for this concept was gained by experimental results on the inhibiting action of various anions. It was suggested that there were critical Cl^-/inhibitor concentration ratios, depending on the potential, above but not below which pitting was observed [10−13,99]. The mechanism may be consistent with localized attack, since adsorption and displacement will probably occur at discrete sites such as lattice defects and dislocations, but the occurrence of induction times varying with passive film thickness cannot readily be explained [14].

The formation on the film/solution interface of transitional Cl^--containing complexes which are much more soluble than the complexes formed in the absence of halides, leading to a locally thinned passive layer, was originally proposed by Hoar and Jacob [5]. Heusler and Fischer [15,16] and Strehblow and co-workers [18,19] have considered this mechanism in detail and with more experimental support.

The results obtained by Heusler and Fischer [15] for iron in Cl^--containing borate and phthalate buffer solutions, using a rotating ring disk electrode, show that after addition of chloride to the solution an irregularly oscillating Fe^{3+} dissolution occurs, whereas the total current density remains constant. Finally, a steep increase of the Fe^{2+} as well as of the total current is observed, indicating the end of the induction period and the beginning of pit growth. Heusler and Fischer assumed that the chloride ions adsorbed on the oxide surface form two-dimensional clusters or islands fluctuating on the surface and strongly accelerate the Fe^{3+} transfer. As the oxide dissolves currentless during the induction period, they further assumed that the chloride catalyzes the transfer of oxygen ions from the oxide to the solution even more strongly. Therefore cathodic polarization, which is necessary for cathodic oxygen transfer, is small and has only a negligible effect on the Fe^{3+} transfer. Furthermore, Heusler and Fischer pointed out that the currentless oxide dissolution stops as soon as a new steady state is established, corresponding to a thinner oxide film. Obviously, pit initiation cannot be explained entirely by two-dimensional nucleation; a three-dimensional oxide layer must be removed. The authors therefore supposed that during the lifetime of chloride clusters new nuclei appear beneath them, leading to a further thinning of the oxide film, until finally complete local removal is achieved, as shown schematically in Fig. 1. Some evidence for the existence of chloride islands even at potentials below the critical pitting potential was obtained by Janik-Czachor et al. [20]. Since the thickness of the locally formed salt layers observed in these experiments was estimated to be 6—8 nm, precipitation of corrosion products cannot be excluded. Furthermore, Heusler and Fischer observed a potential dependence of the induction time analogous to the one obtained for the formation of two-dimensional nuclei as a function of overvoltage [21], using the potential difference $\Delta\epsilon = \epsilon - \epsilon_p$ between the actual electrode potential and the critical pitting potential. These results imply the existence of a critical nucleation potential as a necessary requirement, a concept which is discussed in more detail in the next section.

Strehblow et al. [18,19] investigated the general attack of passive iron by hydrogen fluoride. They proposed this as a simpler model and one experimentally more accessible than the local breakdown of passive iron by the other halides. The results of this investigation using rotating ring disk electrodes and surface analytical methods such as X-ray photoelectron spectroscopy (XPS) and low-energy ion scattering

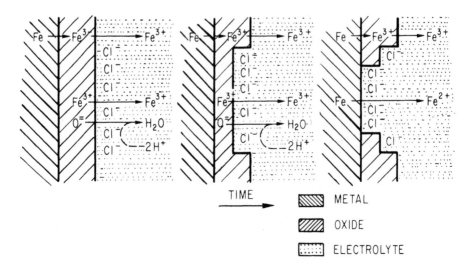

FIGURE 1 Pit initiation on passive iron due to formation of two-dimensional Cl^- clusters according to Heusler and Fischer [15]. (After Kaesche [7].)

(ISS) indicate that the breakdown process occurs with complete removal of the passivating oxide layer. Hydrogen fluoride catalyzes the transfer of Fe^{3+} and Ni^{2+} ions from the oxide into the electrolyte in a manner similar to the influence of chloride as proposed by Heusler and Fischer.

In case of iron [18] the following reactions are suggested at the film/solution interface:

$$Fe_{ox}^{3+} + HF_{aq} \rightleftarrows FeF_{ad}^{2+} + H^+ \tag{1}$$

$$FeF_{ad}^{2+} \longrightarrow FeF_{aq}^{2+} \tag{2}$$

$$FeF_{aq}^{2+} + 4HF_{aq} + H_2O \rightleftarrows FeF_5 \cdot H_2O_{aq}^{2-} + 5H^+ \tag{3}$$

Reaction (1) is assumed to be fast enough that the electrosorption equilibrium is not disturbed by the rate-determining charge transfer reaction (2). Finally, a fast complexing reaction forms the stable and soluble fluoro complex. Strehblow and co-workers concluded from the XPS and ISS analyses that the passive layer is thinned during this stage and does not contain any fluoride.

In Fig. 2 the depth profiles of two passivated iron specimens are shown, one of which was also exposed to hydrogen fluoride. The larger Fe 2P 3/2 iron metal peak of the nonsputtered specimen at

E_B = 706.8 eV after HF exposure and the decreasing oxide peak during sputtering are explained by a thinned passive film. Similarly,
the O 1s peak at E_B = 529.9 eV of the specimen exposed to HF disappears within half the sputtering time of the specimen that has not
been in contact with HF. This also indicates thinning of the oxide
layer. In the final stage the formation of Fe^{2+} ions is detected at
the rotating ring disk electrode, indicating removal of the passive
film and the onset of a general attack.

The main difference between these contributions, which are important for a better understanding of pit nucleation processes, is the
different behavior of Cl^- and F^- anions. Whereas Cl^- ions in the suggested model form only two-dimensional clusters leading to localized
thinning of the passive layer, the F^- ions supposedly adsorb homog-
eneously on the oxide surface and therefore cause a general attack.
In this mechanism, based on the catalytic effect of adsorbed halide
ions on the dissolution kinetics of the passive film, both groups of
authors do not take into account structural parameters such as point
defects, dislocations, and inclusions. This aspect must be included
in further investigations extending these mechanisms. It may then
be possible to gain deeper insight into questions concerning the different behavior of the F^- ions compared to the other halides.

To explain the effect of adsorbed halides, especially Cl^- ions, Sato
[22] proposes a somewhat different mechanism for breakdown of passive films. His theoretical concept is based on the potential-dependent transpassive dissolution depending on the electronic properties
of the passive film. In the case of a thin stoichiometric oxide film
having no localized electron levels in the forbidden energy gap, the
Fermi level of electrons in the film is located within the forbidden
band gap and is equilibrated with the Fermi level in the metal. On
anodic polarization of the electrode, band bending occurs. If the
anodic polarization is large enough the Fermi level finally becomes
lower than the valence band edge E_V and electron tunneling occurs,
as shown in Fig. 3 schematically for an n-type passive oxide. The
charge accumulation at the film/electrolyte interface leads to an increase of the potential difference across the Helmholtz layer and
therefore to a potential-dependent transpassive dissolution, determined by the band structure of electron levels. In the presence of
more electronegative or lower-valence anions than the matrix oxide
ion, such as Cl^- ions, electron acceptor levels appear in the lower
part of the forbidden energy gap, as also shown in Fig. 3. Therefore, the critical potential above which potential-dependent dissolution of the film occurs will be less noble at the sites of chloride ion
adsorption or incorporation than at the sites without chloride. As a
result of the increased dissolution rate above the critical potential,
local thinning of the passive film occurs until a new steady state is
reached. Sato points out that the local thinning of the oxide film as

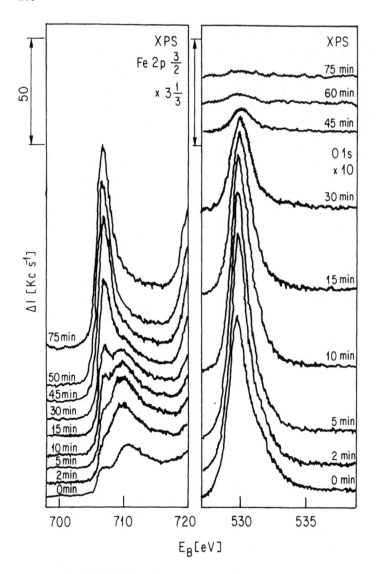

FIGURE 2 X-ray photoelectron spectroscopy (XPS)-depth profile of iron: (a) passivated of $\varepsilon = 1.0$ V in 1 M $HClO_4$. (After Löchel and Strehblow [18].)

a mechanism for pit initiation must be seen in view of an increased probability of electrocapillary film breakdown, which will be discussed in a later section. Furthermore, Sato also includes the effect of dislocations in his theoretical approach, which create additional electron

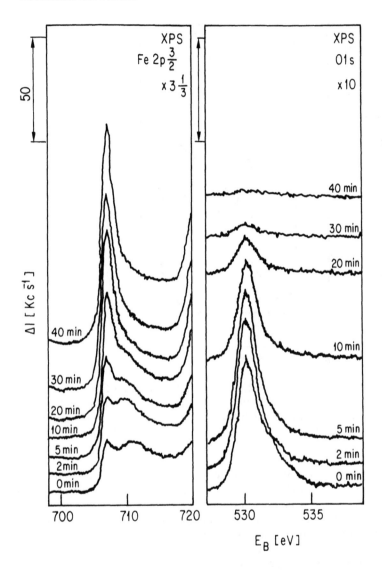

FIGURE 2 (continued) X-ray photoelectron spectroscopy (XPS)-depth profile of iron: (b) additionally exposed to 1 M $HClO_4$ + 0.1 M HF for 72 sec at $\varepsilon = 1.00$ V. (After Löchel and Strehblow [18].)

levels within the forbidden gap, similar to the influence of Cl^- ions facilitating localized dissolution.

So far, detailed experimental support has not been reported. Adequate knowledge of the electronic properties of passive films is lacking

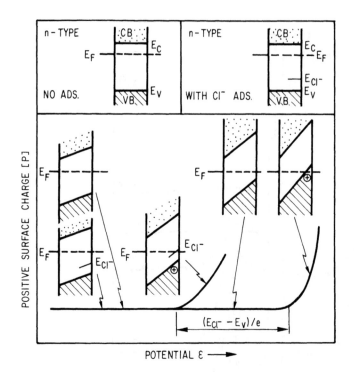

FIGURE 3 Electron energy level diagrams of thin anodic films of n-type oxide with and without adsorbed chloride ions as function of potential. E_{Cl^-} is an acceptor level introduced by the adsorbed chloride ions. (After Sato [22].)

at the present stage. However, Sato concludes from the proposed mechanism that the electrochemical stability of a passive film depends strongly on the electron energy band structure in the film. Thus n-type passive oxide films, where the Fermi level is in the upper part of the forbidden energy gap, are more stable than p-type films, where the Fermi level is situated near the valence band, which is in qualitative agreement with results for iron and nickel as shown in Fig. 4 [22–25].

6.2.2 Penetration and Ion Migration Models

The theoretical models summarized in this section require either penetration of damaging anions from the oxide/electrolyte interface to the metal/oxide interface or migration of cations or their respective vacancies as the decisive process.

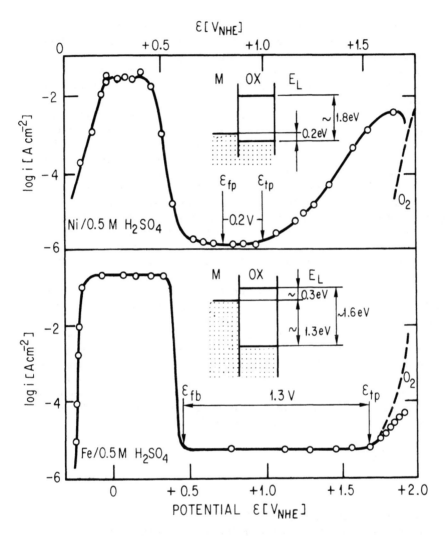

FIGURE 4 Anodic polarization curves of iron and nickel in sulfuric acid solutions and estimated electron energy band structure of the passive films; ε_{fb} is the flat band potential, ε_{tp} the transpassivation potential. (After Sato [22].)

According to the penetration theory originally presented by Hoar et al. [26], the aggressive anions adsorbed on the oxide film enter and penetrate the film when the electrostatic field across the film/solution interface reaches a critical value corresponding to the critical breakdown potential. Thus a "contaminated oxide" film is produced which is a much better ion conductor than the original passive layer. Rapid cation egress occurs and pitting can proceed.

Experimental results, including induction time measurements on iron and nickel in Cl^--containing electrolytes, show clearly that the observed induction periods are smaller by orders of magnitude than the estimated values, based on high-field-strength transport mechanisms for the migration of aggressive anions through the passivating oxide film [2]. Furthermore, it is difficult to explain pitting of iron by SO_4^{2-} or ClO_4^- [27] or of aluminum in NO_3^--containing solutions [29] with this theory. For these ions, penetration through an oxide lattice is even less probable because of their large diameters, but nevertheless small induction times have been observed [27]. Surface analytical studies of iron by Auger spectroscopy do not support the penetration theory [28, 30]. These results indicate that chlorides are not incorporated in the oxide film, which is in contrast to the experimental observations of Augustinsky and co-workers [31–33].

A very extensive theoretical model to explain the growth kinetics of a passive film developed by Lin, Chao, and Macdonald [34,35] has been applied by the same authors to account for chemical breakdown. It involves transport of both anions and cations or their respective vacancies. As a consequence of the diffusion of metal cations from the metal/film to the film/solution interface, metal vacancies may accumulate, forming voids at the metal/film interface if the submergence into metal occurs at a lower rate. When the voids grow to a critical size the passive film will collapse, leading to pit growth. The criterion for pit initiation is expressed as follows:

$$(J_{ca} - J_m)(t - \tau) > \xi \tag{4}$$

where

J_{ca} = cation diffusion rate in the film
J_m = rate of submergence of metal holes into the metal
ξ = critical amount of metal holes
t = time to accumulate metal holes to the critical amount (incubation time)
τ = constant

J_{ca} may be calculated for thin passive films (10–40 Å) by using the proposed point defect model for anodic films in a simplified form according to:

$$J_{ca} = \chi K D_{V_M^{\chi'}} c_{V_M^{\chi'}} (f/s) \tag{5}$$

where

$D_{V_M^{X'}}$ = diffusivity of cation vacancy

$c_{V_{M}^{X'}}(f/s)$ = concentration of cation vacancies at film/solution interface

X = charge of cation

Assuming further local equilibrium of Schottky-pair reaction at the film/solution interface:

$$\left[c_{V_O^{··}}(f/s)\right]^{X/2}\left[c_{V_M^{X'}}(f/s)\right] = \text{const} \tag{5a}$$

the following general relationship between the cation diffusion rate J_{ca} and the concentration of oxygen vacancies is obtained:

$$J_{ca} = J^0\left[c_{V_O^{··}}(f/s)\right]^{-X/2} \tag{6}$$

In the presence of a chloride-containing solution, the chloride ions are incorporated at the film/solution interface, occupying oxygen vacancies, $Cl_O^{··}$.

$$V_O^{··} + Cl_{aq}^- \rightleftarrows Cl_O^{··} \tag{6a}$$

Again, local equilibrium is proposed. Thus $c_{V_O^{··}}$ will decrease with increasing chloride concentration as well as potential according to the following equation:

$$c_{V_O^{··}}a_{Cl^-} = u\ \exp\left(\frac{F\alpha\varepsilon_{app}}{RT}\right) \tag{7}$$

Combining Eqs. (4), (6), and (7), the critical pitting potential ε_p as a function of the chloride activity is obtained:

$$J^0 u^{-X/2}\ \exp\left(\frac{XF\alpha\varepsilon_p}{2RT}\right)a_{Cl^-}^{X/2} = J_m \tag{8}$$

Equation (8) is easily rearranged to yield

$$\varepsilon_p = A - B\ \log\ a_{Cl^-} \tag{9}$$

with

$$A = \frac{4.606RT}{\chi F\alpha} \log \frac{J_m}{J^0_u{}^{-\chi/2}}$$

and

$$B = 2.303RT/\alpha F$$

Equation (9) corresponds to the commonly observed relationship between critical pitting potential and chloride concentration in pitting experiments. Inserting the experimentally determined α values for iron and nickel in Eq. (9), Macdonald and co-workers [35] obtained reasonably good agreement with the experimental values measured by various authors. However, it should be mentioned that the B values cited in the literature are not always suitable for proving validity of a mechanism, since the experimental techniques used in pitting experiments usually do not permit differentiating between pit initiation and limiting pit growth potentials.

The defect model can also be used to calculate incubation times. The authors derived the following simplified mathematical expression:

$$t = \xi'\left[\exp\left(-\frac{\chi F\alpha \, \Delta\varepsilon}{2RT}\right)\right] + \tau \tag{10}$$

where ξ' is a constant for a given system, and τ is the time over which transient diffusion, not considered in this model, is significant compared with steady-state diffusion. Equation (10), which is valid only for $\Delta\varepsilon = \varepsilon - \varepsilon_p > 0.050$ V, gives a linear relationship between $\log$th t and $\Delta\varepsilon$ up to $\Delta\varepsilon$ values of approximately 0.2 V with a slope of -18.8 V^{-1} for iron. This is in excellent agreement with experimental values obtained by Heusler and Fischer [15], as shown in Fig. 5. Like the adsorption-induced mechanism of Heusler and Fischer, the point defect model requires the occurrence of a critical pit initiation potential, which still must be questioned. Results [41] from electrochemical noise measurements on aluminum below the pit growth-limiting potential do not support the existence of a limiting pit initiation potential; instead, a continuously decreasing nucleation rate is observed with decreasing potential. In addition, the measured induction times usually show a large scatter [18], which makes it difficult to achieve more than qualitative agreement. Nevertheless, further experiments using other metals and environments would be very valuable, especially if the experimental confirmation could be extended to more model-sensitive parameters than the slopes in Eqs. (9) and (10). The point defect mechanism, however, is one of the most detailed models that

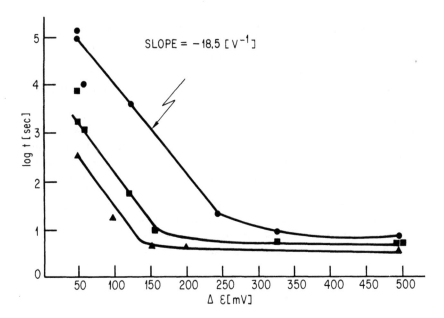

FIGURE 5 Times t for incubation of pitting on passive iron in borate solution, 25°C, as a function of the potential difference $\Delta \varepsilon = \varepsilon - \varepsilon_p$ according to Lin et al. [35].

has been proposed. Furthermore, surface structural effects such as dislocations, grain boundaries, and inclusions, which have a pronounced influence on pit initiation, may readily be included.

6.2.3 Mechanical Film Breakdown Theories

Mechanical breakdown theories for passive films have been discussed in the literature either as an additional process combined with other mechanisms or as the principal step in pit initiation, giving direct access of the electrolyte to the base metal within the crack. In this section only mechanisms in which the mechanical breakdown process is of primary importance will be discussed.

In 1971 Sato [37] proposed a breakdown mechanism for anodic films, starting from thermodynamic considerations. He showed that thin films always contain a significant film pressure mainly due to electrostriction:

$$P - P_0 = \frac{\varepsilon(\varepsilon - 1)E^2}{8\pi} - \frac{\gamma}{L} \tag{11}$$

where P is the film pressure, P_0 the atmospheric pressure, ε the dielectric constant within the film, E the electric field strength, γ the surface tension of the film, and L the film thickness. An approximate calculation of the electrostriction pressure in anodic films with an electric field strength of $\sim 10^6$ V/cm gave pressure values larger than 1000 kg/cm^2, so that metal oxides or hydroxides with a critical compressive stress for breakdown of 100–1000 kg/cm^2 could easily deform or break. According to Sato's electrostriction hypothesis, both the surface tension of the film and the film thickness have a pronounced influence on film pressure. A decrease in the interfacial tension increases the film pressure and therefore facilitates breakdown. Sato also proposed that adsorption of chloride ions, depending on their concentration, greatly reduces the surface tension. When P exceeds a critical value the film breaks down. For these critical conditions Sato derived the following simplified relation between the breakdown potential and the chloride concentration:

$$\frac{d\varepsilon_p}{d \ln a_{MCl}} = -\frac{8\pi kT \Gamma^*_{Cl^-}}{\varepsilon(\varepsilon + 1)E} \tag{12}$$

where $\Gamma^*_{Cl^-}$ is the critical adsorption density of Cl$^-$ at ε_p. This is in qualitative agreement with the experimentally obtained relationship. It is interesting to note that the term on the right in Eq. (12) again does not contain model-sensitive parameters, as mentioned in the preceding section, which makes proof of this theory rather difficult.

In 1982 Sato [22] postulated a somewhat different mechanism for mechanical breakdown. He studied stability against breakdown in aqueous solution in terms of electrocapillary energetics, including the role of aggressive anions. Using a simple model for electrical breakdown of lipid membranes [38], Sato derived the following equation for the work A_b required to form a cylindrical breakthrough pore in the passive film:

$$A_b = [2\pi rL\gamma + \pi r^2(\gamma_m - \gamma)] - 0.5\pi r^2 \Delta C_d \varepsilon^2 \tag{13}$$

where r is the pore radius, L the film thickness, γ the surface tension of the film/electrolyte interface, γ_m the surface tension of the metal/electrolyte interface, ε the potential difference between metal and electrolyte, and ΔC_d the difference in electrode capacitance between the passivated metal and the film-free metal in the solution.

Figure 6 shows the calculated work for pore formation as a function of the pore radius for different potential differences. The remaining parameters are kept constant. Sato suggested that for nulei to grow to macroscopic size a critical radius r* corresponding to

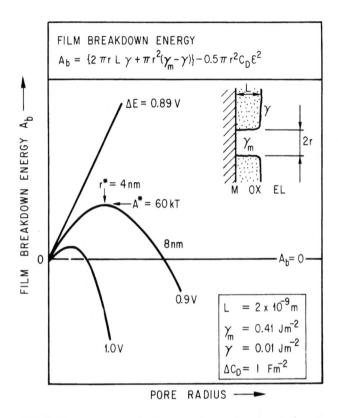

FIGURE 6 Electrocapillary energy for the formation of a breakthrough pore in a thin surface oxide film on metals as a function of the pore radius. $\Delta\varepsilon = \varepsilon - \varepsilon_{pzc}$, where ε_{pzc} is the potential of zero charge of the film-free metal. (After Sato [22].)

a critical pore formation energy A_b^* must be exceeded. Furthermore, Sato calculated A_b^* and r^* as function of the potential:

$$A_b^* = \frac{\pi(L\sigma)^2}{\gamma - \gamma_m + 0.5th\,\Delta C_d\varepsilon^2} \qquad (14)$$

$$r^* = \frac{L\sigma}{\gamma - \gamma_m + 0.5th\,\Delta C_d\varepsilon^2} \qquad (15)$$

From Eqs. (14) and (15) it may be concluded that A_b^* and r^* increase steeply with decreasing ε and a critical value ε_p may be obtained

$$\varepsilon_p = \sqrt{\frac{\gamma_m - \gamma}{0.5 \text{th} \, \Delta C_d}} \qquad (16)$$

below which the passive film is stable against electrocapillary break-down. As in the electrostriction theory, a decrease in surface tension, which may arise from chloride ions on the metal surface, will shift the breakdown potential in the active direction. A limitation of the model in its present stage of development, with experimental proof still lacking, is that structural parameters which influence pit initiation processes have not been taken into account.

In concluding this section on pit initiation, an approach which promises to yield new insights into these processes must be mentioned. Studies of passive film breakdown by electrochemical noise analysis [36,41] have shown that the onset of pitting can be detected by a large increase in current noise. Although no comprehensive mechanism has emerged yet from measurements of this type, the results indicate clearly that in case of aluminum as well as iron-chromium-nickel alloys the breakdown consists of a dynamic breakdown-repair process rather than a steady deterioration of the properties of the passive film. This was suggested by Videm [43] several years ago.

6.3 PIT GROWTH KINETICS

In discussions of corrosion mechanisms leading to a localized attack on passive metals, pit initiation processes are usually distinguished from the subsequent growth stage. These stages may easily be defined in terms of theoretical considerations, where the initiation process is restricted to the local destruction of the passive film. Unfortunately, many of the investigations of pit initiation reported in the literature, such as the electrochemical noise measurements mentioned above, include pit growth processes as well. To define the scope of the following section, only stable pit growth leading to open pits of macroscopic size is considered here. This limitation seems justified, since the mechanisms proposed for stable pit growth are generally much more accessible to experiment than is the very early growth stage, where experimental proof is usually rather difficult to obtain. Furthermore, the discussion will again be limited to contributions of a comprehensive nature. Special attention is given to mechanisms which include the time dependence of pit growth, a topic somewhat neglected in previous contributions.

Theoretical models of pit growth providing quantitative results are usually restricted to a simplified pit geometry such as cylindrical

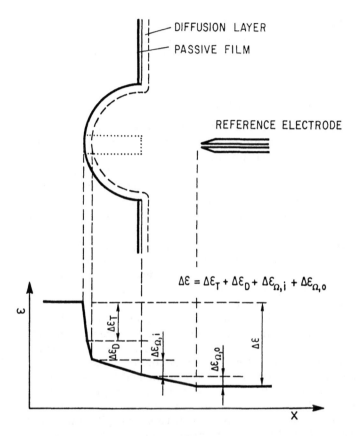

FIGURE 7 Schematic representation of the potential contributions in and outside a hemispherical pit (see text).

pores, where only the pit bottom dissolves and the walls are protected by a passive film. In the case of hemispherical pits, cylindrical volumes above the center of the pit bottom are usually considered, not allowing for interactions with the surroundings (unidirectional pit model). Therefore these theoretical models cannot easily be transformed to more realistic conditions. The mechanisms are grouped according to the processes which are decisive for pit growth, such as diffusion or ohmic transport phenomena. In Fig. 7 this aspect is shown schematically in terms of potential contributions in and outside the pit. The potential difference between metal and reference electrode consists of a charge transfer overvoltage, $\Delta\varepsilon_T$, a diffusional contribution $\Delta\varepsilon_D$, the ohmic potential drop $\Delta\varepsilon_{\Omega,i}$ within the pit, and the ohmic potential drop $\Delta\varepsilon_{\Omega,o}$ outside the pit. It should be remembered that $\Delta\varepsilon_{\Omega,o}$ does not depend only on the current of a single pit; other pits growing simultaneously on the same surface may also contribute to $\Delta\varepsilon_{\Omega,o}$.

6.3.1 Charge-Transfer Influence During Initial Growth

Vetter and Strehblow [27,39] concluded from theoretical considerations that during the early stage of growth of a hemispherical pit the metal chloride concentration within the pit does not increase sufficiently for precipitation to occur. They transformed the potential and concentration distribution, readily available for a convex hemispherical surface, to the concave conditions that occur in a hemispherical pit by introducing a correction term. For instance, for an initial current density of 9 A/cm^2 on iron in a chloride-containing sulfuric acid solution, a concentration difference of approximately 0.9 mol FeCl$_2$/liter was obtained, which is far below the saturation concentration (~ 4.4 mol/liter). Similarly, they estimated the ohmic potential drop within a hemispherical pit by applying the same mathematical equations as in the case of diffusion. For iron a $\Delta\varepsilon_{\Omega,i}$ value of ~ 18 mV was obtained, which again cannot explain the stability of growing pits. Instead, an ion-conducting salt layer present on the metal surface due to adsorption of aggressive anions was proposed.

Experimental support for Vetters and Strehblow's view of the stability of growing pits was obtained by Strehblow and Wenners [44] with their investigations of the early-stage dissolution processes on iron and nickel at high current densities. In buffered and moderately concentrated Cl$^-$ solutions, linear growth of the pit radius with time was observed. From the experimentally determined growth rates, assuming a hemispherical geometry of the growing pits, the metal dissolution current densities were calculated. The current densities varied from 6 to 20 A/cm^2 and showed approximately a semilogarithmic potential dependence. Galvanostatic pulse measurements with a microelectrode in solutions of high chloride concentration with current densities in the range 1–1000 A/cm^2 gave a similar potential dependence. In addition, a transition time could be determined before the electrode potential increased by some volts. Strehblow and Wenners [44] explained this transition time as the time required to form a saturated solution at the metal surface and therefore applied the equation of Sand [45] for nonstationary diffusion, extended to include migration effects. Reasonably good agreement with the experimental results was obtained. The authors concluded that in the presence of chlorides (>1 mol/liter) an adsorption layer or a thin poreless salt film is formed at the beginning to prevent repassivation. After the transition time this chloride layer grows in thickness and finally a porous layer appears by precipitation from a saturated solution, which accounts for the observed ohmic potential increase (Figs. 8 and 9).

PORELESS   POROUS

METAL   SALT LAYER   ELECTROLYTE

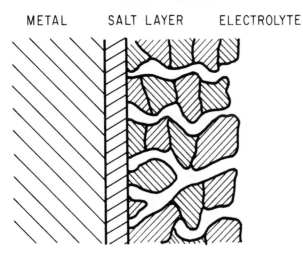

FIGURE 8 Schematic representation of the salt layers formed during dissolution at high current densities. (After Strehblow and Wenners [44].)

This mechanism for the early stage of pit growth is based on the high current densities, up to 100 A/cm^2, that occur during this period. Such high current densities cannot be obtained in simple potentiodynamic experiments with moderately concentrated electrolytes because of the formation of passive films. Since no significant potential differences within the pits were estimated, a different electrode state (pitting state) consisting of an adsorption layer or a thin poreless salt film, as mentioned above, was proposed, which had to prevent repassivation of the metal surface at these high potentials. Whether this pitting state is also present at potentials close to the pitting potential, where lower current densities are attained, should also be studied. Furthermore, the current potential curves showing large scatter did not permit definite conclusions concerning the rate-determining step. Finally, it should be pointed out that Vetter and Strehblow [27,39,44] as well as Strehblow and Wenners assumed for their calculation a constant current density within the pit, as previously suggested by other authors [39,42]. A constant current density, which is consistent with a charge transfer-controlled step, seems reasonable for the initial growth period, but must be questioned for longer growth times.

Popov et al. [46–55] proposed an extended theoretical model for the experimental results obtained by Strehblow and Wenners [44] for iron and nickel. They first considered processes occurring at the

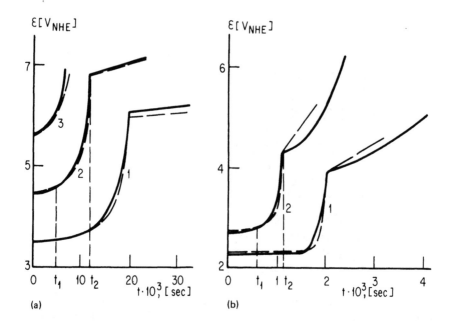

FIGURE 9 Galvanostatic potential transient; solid line = experimental values [43]; dashed line = theoretical values [50]. (a) Fe in 1 M KCl; 1, 31.6 A/cm^2; 2, 42.1 A/cm^2; 3, 52.6 A/cm^2; (b) Ni in 4 M KCl; 1, 72 A/cm^2; 2, 90 A/cm^2. (After Popov et al. [51].)

bottom of the hemispherical pit, such as adsorption, charge transfer, and dissolution of complexes formed at the surface [47,51]. Subsequently, they discussed the transport of the components in and outside the pit, generated by the primary processes or already present, considering diffusion, migration, and convection.

Analyzing the results of the galvanostatic pulse experiments obtained by Strehblow and Wenners, as shown in Fig. 9, Popov et al. distinguished several successive stages during pit growth. In the early stage, $0 < t < t_1$, corresponding approximately to the transition time discussed above, the metal ions passing into the solution at high rates do not influence the electrochemical kinetics. Even the continuously decreasing concentration of free solvent molecules (water) due to hydration of the metal ions has no effect during this stage. Popov et al. suggested that at high Cl$^-$ concentrations, corresponding to high Cl$^-$ coverage, the metal dissolution is charge transfer-controlled [51]. At $t \sim t_1$ the decreasing free solvent concentration is supposed to influence the electrochemical kinetics significantly, and finally a dehydrated boundary layer at the pit bottom is formed, consisting of hydrated metal ions and Cl$^-$ ions.

In the following stage, $t > t_2$, the thickness of the dehydrated layer, called the resistive layer, grows until the whole volume within the pit is filled. Popov et al. estimated from their mathematical evaluation that the growth rate of the resistive layer is larger by orders of magnitude than the growth rate of the pit. Consequently, the resistive layer will occupy the whole volume and finally even grow into the solution adjacent to the pit, filling a hemispherical volume until a critical radius is reached and stationary conditions are achieved [52,54,55].

Experimental support for this extended and detailed theoretical treatment of Strehblow and Wenners' results, only briefly outlined in this section, is limited to the very beginning of pit growth. From their theoretical considerations Popov et al. calculated the number of Cl⁻ ions and water molecules taking part in the primary dissolution process occurring at the pit bottom. In the case of nickel they proposed the following reaction as the rate-determining step for $t < t_1$ [47,51]:

$$[NiCl(H_2O)_4]_{ad}^{\sigma-1} \rightarrow [NiCl(H_2O)_4]_{aq}^{+1} + (2-\sigma)e^- \tag{17}$$

where σ is the amount of charge involved in a preceding chemisorption process in which the surface complex for the subsequent dissolution process is formed. Concerning the existence of a resistive layer, no direct experimental evidence exists so far. Popov et al. assumed that in the resistive layer the current flow is achieved by migration of metal ions in the presence of an electric field, whereas the negative ions are stationary. Further experimental support is certainly needed to show the validity of this interesting theoretical contribution.

6.3.2 Diffusion-Related Mechanisms

The presence of salt layers on the surface of pits grown on passive metals has also been recognized by other authors. For example, Engell [3], in a survey on stability and breakdown phenomena of passivating films, demonstrated that sufficient changes in concentration can occur adjacent to an active metal surface to cause precipitation of a metal salt having considerable solubility. Strehblow et al. [56] and Rosenfeld et al. [57] also contributed arguments in favor of salt film formation. Beck and Alkire [58,59] proposed salt film formation at very early stages of pit growth by applying simple transport laws.

Based on Tafel kinetics and on considerations of ohmic resistance within the solution, Beck and Alkire estimated that the current density at a pit nucleus with a diameter of approximately the thickness

of the passivating film would be in the range 10^3 to 10^6 A/cm^2, leading to salt film precipitation with 10^{-4} to 10^{-8} sec. In view of these
short transition times necessary to form salt films, they considered
the growth rate of hemispherical pits in which the corrosion current
is controlled by the rate of dissolution of this metal salt film. On the
basis of simple diffusion processes, Beck and Alkire [58] calculated
the dissolution rate in hemispherical pits in a simplified form according to Eq. (18):

$$i_L = \frac{zFDc_s}{d_p} \qquad (18)$$

where i_L is the limiting current dissolution, c_s the saturation concentration, d_p the pit radius, D the diffusivity, z the valence, and F
the Faraday constant. Since geometry and migration exert opposite
and nearly equal influences on the dissolution rate, they are neglected in Eq. (18). Under these conditions the following growth rate
of hemispherical pits is obtained:

$$\frac{d_p}{dt} = \frac{i_L M}{zF\rho} = \frac{Dc_s M}{\rho d_p} \qquad (19)$$

(where ρ is density and M molecular weight), which on integration
yields

$$d_p = \left(r_1^2 + \frac{2Dc_s M}{\rho} t\right)^{1/2} \qquad (20)$$

$$i_p = \frac{zFDc_s}{[r_1^2 + (2DMc_s/\rho)t]^{1/2}} \qquad (21)$$

where r_1 is the radius of the initial pit nucleus corresponding approximately to the thickness of the passivating oxide film.
    In addition, the same authors calculated the thickness of the salt
films during pit growth, assuming high field conduction. The results
obtained for titanium showed that even for larger pits the thickness
was always below 200 Å and therefore not visible by eye. Furthermore, they concluded that the occurrence of polished hemispherical
pits, often observed at later stages of pit growth, cannot be explained
by these thin barrier-type films; subsequently, thicker films having
different properties must be formed. Considering Eqs. (20) and (21),
a parabolic growth law is proposed for pit growth. Experimental proof
of this model for the early stage growth could not be attained yet.

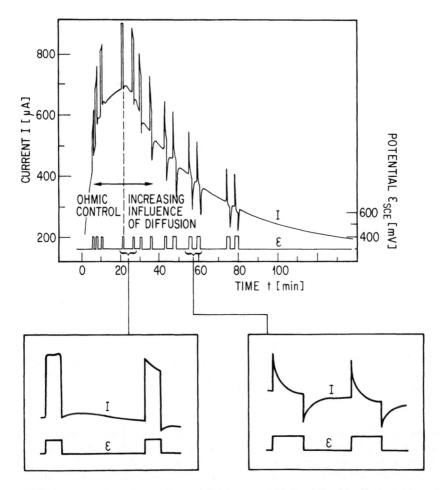

FIGURE 10 Time dependence of pit current for Ni wire electrode under potentiostatic conditions with superimposed potential pulses; wire diameter 0.5 mm, 0.1 M NaCl, pH 11. (After Heimgartner and Böhni [62].)

Usually much longer transition periods, where diffusion processes start to control metal dissolution, have been observed experimentally, as demonstrated for nickel in Cl⁻-containing solutions [60,62] or copper under pitting-like electropolishing conditions [61]. In Fig. 10 the time dependence of the pit current for a nickel wire electrode under potentiostatic control with superimposed potential pulses is shown [62]. The shape of the current transients obtained at the current maximum indicate that a transition from mixed charge transfer/ohmic resistance

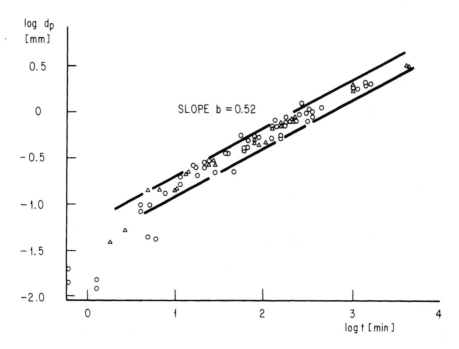

FIGURE 11 Time dependence of actual pit depth for Ni wire electrodes at various potentials and for various wire diameters; 0.1 M NaCl, pH 11; $d_p = at^b$; $\circ$ = Ni 99.0; $\triangle$ = Ni 99.98. (After Heimgartner [60].)

control to diffusion occurs. Since the potentials applied in these experiments were much closer to the pitting potentials than those considered in the theoretical calculations of Beck and Alkire [58], appreciably lower initial current densities were obtained, which readily explains the differences in transition times.

At later stages a parabolic rate law is observed. In Fig. 11 the time dependence of the actual pit depth is shown for various applied potentials. Except for the transition period mentioned above, potential-independent pit growth over an extended period of time corresponding to a diffusion-controlled parabolic rate law is obtained.

Furthermore, Hunkeler [63], evaluating pit growth data from the literature, showed that growth rates obeying a parabolic time law over an extended period are frequently observed. Results for nickel, stainless steels, and iron under polarized conditions as well as at the rest potential have been reported [64–66,68]. Whether these parabolic rate laws can be explained only by limiting diffusion processes is not evident at the present stage. Pit growth measurements on aluminum [63,69,70] have clearly shown that ohmic control of pit growth also yields a parabolic rate law, as will be discussed in more detail in the next section.

Beck and Chan [71] showed that mass transfer as well as ohmic resistance can be important in pit growth. In a contribution on the effect of hydrodynamics on pit growth, using a flow channel, they demonstrated that either a mass transport or an ohmic limited current density is attained, depending on the flow velocity. In the case of diffusion-limited current density, Eq. (18) is valid, as discussed before, whereas for ohmic control the current density is described by

$$i_\Omega = \frac{\sigma \, \text{th} \, \Delta\varepsilon_\Omega}{a d_p} \qquad (22)$$

in which σ is the conductivity, $\Delta\varepsilon_\Omega$ is the ohmic overpotential, d_p is the pit radius, and a is a geometric constant to correct for the electrolyte resistance from the convex hemisphere to the concave pit surface. The factor a has been estimated to be about 3 [27]. In Fig. 12 the effect of flow velocity on current density in a single artificial pit is shown. In addition, the diffusion limiting current density given by Eq. (18) and the ohmic limit according to Eq. (22) for the various conductivities tested are included. The experiments confirm that for solutions where $i_L < i_\Omega$ the current densities increased with flow rate as predicted by Eqs. (18) and (22), finally reaching the ohmic limit. For solutions where $i_\Omega < i_L$ the current drops and again approaches the ohmic limit of the bulk concentration, as would be expected when the reaction products formed in and near the pit are removed at higher flow rates. These results obtained on iron with artificial pits can not be transformed to actual pitting conditions, as already pointed out by the authors. Nevertheless, the results show convincingly that diffusion processes connected with salt precipitation may control pit growth if the ohmic control is suppressed.

6.3.3 Ohmic Resistance-Controlled Growth

As pointed out by Beck [72], pitting on titanium and aluminum occurs at a high ohmic-limited current density. Generation of large amounts of hydrogen bubbles within the pit increases the mass transport rate so that solid salt does not form and mass transport-limited current densities are not attained. Bubbles generated within the diffusion layer enormously increase the mass transfer coefficient [73]. Therefore fluid flow of the bulk solution has little effect on pit growth [74] under such conditions.

In contributions on the mechanism of pit growth on aluminum, Hunkeler and Böhni [63,69,70] obtained further and more detailed support for ohmic resistance-controlled pit growth. Using a metal

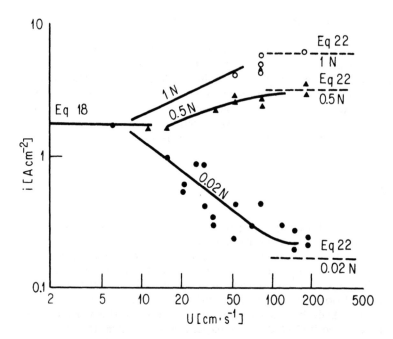

FIGURE 12 Effect of flow velocity on current density in artificial pit on iron in NaCl solutions at $\varepsilon = 0.3$ V_{SCE} compared to predictions of Eqs. (18) and (22). (After Beck and Chan [71].)

foil technique, they showed that the pit growth rate is not only time-dependent but also markedly influenced by the applied potential. In Fig. 13 the pit depth d_p and the pit current density i_p are shown as functions of time. Increasing the potential yields higher growth rates. For all potentials and chloride concentrations tested a parabolic pit growth law was found:

$$d_p = at^b \tag{23}$$

where t is the time, a is a constant depending on various parameters such as the potential, the temperature, and the composition of the electrolyte, and b is a system-independent constant. The value of b was approximately 0.5 for all systems tested. With Eq. (23) the pit growth rate can be calculated and readily transformed into current density:

$$i_p = i_p^0 t^{b-1} \tag{24}$$

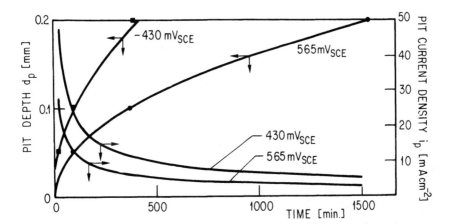

FIGURE 13 Influence of potential on the pit growth (pit depth and pit current density); Al, 10^{-2} M NaCl, pH 11. (After Hunkeler and Böhni [69,70].)

where i_p is the current density at the pit bottom and i_p^0 is the initial pit current density at $t = 1$. Additions of foreign ions may either accelerate or inhibit pit growth, depending on the type and concentrations of the ions added. In the absence of inhibition, Hunkeler and Böhni found that the accelerating effect depends directly on the conductivity of the bulk solution. In Fig. 14 normalized pit growth times to reach a given pit depth (0.2 mm) in different electrolytes are shown as a function of normalized electrolyte resistance. The linear relationship obtained clearly indicates ohmic solution resistance-controlled pit growth, neglecting (at present) an ohmic potential drop $\Delta \varepsilon_{\Omega,o}$ outside the pit. The ohmic potential within the pit $\Delta \varepsilon_{\Omega,i}$, considering a cylindrical volume of length d_p (pit depth) above the pit bottom, can be calculated according to

$$\Delta \varepsilon_{\Omega,i} = \frac{i_p d_p}{\sigma} \qquad (25)$$

From Eq. (25) a linear relation between the pit growth parameter $i_p d_p$ and the ohmic potential drop is to be expected. In recent contributions [75] direct experimental evidence is given, as demonstrated in Fig. 15. The results also show that the conductivity values obtained from the slopes in Fig. 15 correspond reasonably well with the values for the bulk solution, at least for dilute or moderately concentrated solutions. The authors pointed out that because of the vigorous

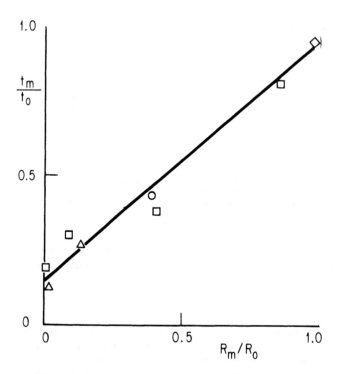

FIGURE 14 Normalized pit growth times to reach a constant pit depth of 0.2 mm as a function of the normalized resistance of the bulk electrolyte; Al, $\varepsilon = -0.48\ V_{SCE}$, 10^{-2} M NaCl, and 10^{-4} M Na_2SO_4 (reference solution). t_m, R_m = pit growth time and electrolytic resistance in presence of various amounts of additional foreign anions; t_0, R_0 = pit growth time and electrolytic resistance of reference solution; $\square$ = $+SO_4^{2-}$; $\triangle$ = $+ClO_4^{-}$; $\circ$ = $+CrO_4^{2-}$; $\diamond$ = $+NO_3^{-}$. (After Hunkeler [63].)

generation of hydrogen during pitting of aluminum no significant change in the composition of the electrolyte within the pit occurs, in contrast to situations where diffusion processes control pit growth.

Equation (25) is readily integrated, assuming that $\Delta\varepsilon_{\Omega,i}$ is constant under potentiostatic conditions. With

$$i_p = d_p \frac{zF\rho}{M} \tag{26}$$

a parabolic rate law is obtained:

$$d_p = \left(\frac{\Delta\varepsilon_{\Omega,i} M\sigma}{zF\rho} t \right)^{1/2} \tag{27}$$

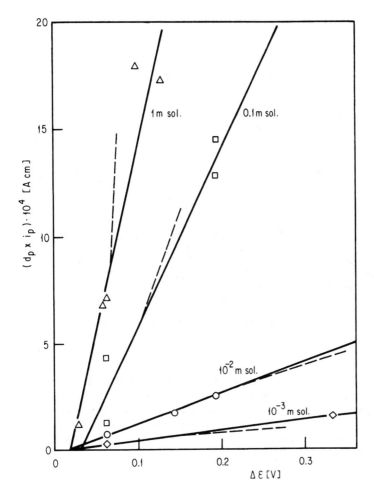

FIGURE 15 Pit growth parameter $d_p \times i_p$ as a function of the potential difference $\Delta \varepsilon = \varepsilon - \varepsilon_p$ for various solutions of different conductivities; solid lines = experimental values, dashed lines = values calculated according to conductivity of bulk electrolyte. (After Hunkeler and Böhni [75].)

Experimental proof for ohmic resistance-controlled pit growth on aluminum was also obtained for open-circuit conditions. Literature data from long-term corrosion tests on aluminum immersed in various tap waters were evaluated by the same authors [75]. From this study the following relationship between the mean value $\bar{a}$ [see Eq. (23)] and the conductivity, in accordance with Eq. (27), could be obtained:

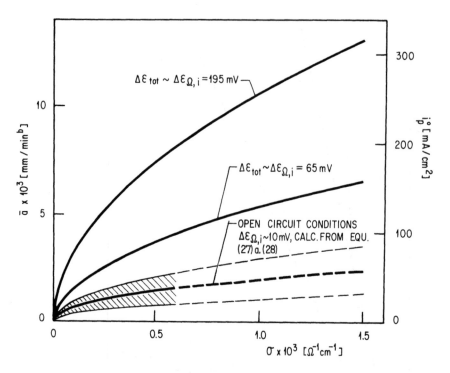

FIGURE 16 Plot of mean value $\bar{a}$ [see Eq. (23)] or initial pit current density i_p^0 as a function of the bulk electrolyte conductivity for polarized and open-circuit conditions on Al according to Eqs. (27) and (28). Values for open-circuit conditions were evaluated from literature data for long-term immersion tests (shadowed area). (After Hunkeler and Böhni [75].)

$$\bar{a} = 0.063 \pm 0.028\sigma^{0.5} \tag{28}$$

In Fig. 16 the mean values $\bar{a}$ are shown as a function of the conductivity for a polarized specimen, using the metal foil technique referred to above [69, 70], as well as for simple immersion tests under open-circuit conditions. From Eqs. (27) and (28) $\Delta\varepsilon_{\Omega,i}$ may be calculated. For the immersion tests a value of ~ 10 mV was obtained, as indicated in Fig. 16, which seems quite reasonable since pitting corrosion usually proceeds at or slightly above the pitting potential. These results clearly show that ohmic resistance-controlled pit growth on aluminum also occurs under open-circuit conditions. Therefore experimental results obtained from short-time potentiostatic measurements may easily be transformed to more practical conditions. In addition, the results indicate that in the case of aluminum potential contributions due to charge transfer may be neglected, which is in agreement with experimental results obtained by various authors [75,97,98].

The ohmic potential drop during the growth of pits will now be considered in more detail. Newman et al. [76] and Melville [77] suggested from their theoretical considerations that the ohmic potential drop is mainly restricted to the solution within the pit and $\Delta\varepsilon_{\Omega,o}$ can be neglected. This assumption may certainly be valid in case of single-pit growth, but it must be questioned if a large number of pits grow simultaneously on a metal surface. Assuming that the current I_p flowing out of a single pit is distributed uniformly over the whole remaining surface, as will occur in the case of a conducting passive layer, $\Delta\varepsilon_{\Omega,o}$ is calculated according to

$$\Delta\varepsilon_{\Omega,o} = R_{\Omega} N 2\pi d_p^2 ip \tag{29}$$

where R_{Ω} is the ohmic resistance of the volume of the electrolyte between metal surface and Haber-Luggin capillary and N is the number of pits growing. Equation [29] was used by Rosenfeld and Danilov [67] to study pit growth kinetics on stainless steels. Hunkeler and Böhni [75] applied Eq. (29) for the estimation of $\Delta\varepsilon_{\Omega,o}$ during pit growth on aluminum. The results indicate that for moderate chloride concentrations and slight deviations from the pit growth limiting potentials, pit densities below 10 pits/cm^2 do not give rise to a significant potential drop outside the pits, whereas in dilute electrolytes and for high potentials even smaller pit densities yield appreciable values of $\Delta\varepsilon_{\Omega,o}$. For those conditions both contributions, $\Delta\varepsilon_{\Omega,i} + \Delta\varepsilon_{\Omega,o} = \Delta\varepsilon_{\Omega,tot}$, must be considered. Combining Eqs. (25), (26), and (20), the time dependence of pit growth can be derived:

$$\frac{c}{2\sigma} d_p^2 + \frac{R_{\Omega} N 2\pi c}{3} d_p^3 = \Delta\varepsilon_{tot} t \tag{30}$$

where c is a conversion factor (= $ZF\rho/M$). Equation (30) indicates that for $c/2\sigma \gg (R_{\Omega}N2\pi c/3)d_p$, when the ohmic potential drop within the pit dominates the pit growth, a parabolic rate law is obtained (b $\sim$ 0.5) as expected, whereas for the opposite conditions, $(R_{\Omega}N2\pi c/3)d_p \gg c/2\sigma$, a b value of 1/3 is attained, indicating slower pit growth due to ohmic contributions from neighboring pits.

6.4 STABILITY OF LOCALIZED CORROSION PROCESSES

Most models of stability of pitting presented up to the present have been extensively discussed in several review articles published during the past decade [3,6,7]. As no new fundamental theories have been proposed since then, the scope of the following section will be restricted to aspects of stability in connection with the mechanisms of pit growth discussed before. Since important knowledge on the growth kinetics of localized corrosion processes, especially with respect to crevice corrosion, is still lacking, only fragmentary contributions on the stability of such processes are presently available. More detailed insight into the growth kinetics is needed for a better understanding of the stability of localized corrosion.

6.4.1 Critical Concentrations and Salt Film Formation

Various authors [12,26,40,78–81] suggested that the concentration of the ionic species of the electrolyte within a pit is different from that in the bulk electrolyte. It has also been proposed that critical concentrations must be exceeded for stable pit growth to occur. Galvele [6,82,83] considered localized acidification as the main reason for pit initiation as well as for stable pit growth. Using a unidirectional pit model similar to the one developed by Pickering and Frankenthal [84], the concentrations of Me^{2+}, $MeOH^-$, and H^+ ions as a function of pit depth × current density were calculated for various metals. Galvele assumed that the metal ions hydrolyze inside the pits and that the aggressive anion salt acts as a supporting electrolyte, which considerably simplifies the transport equations, since only transport by diffusion remains important. From this calculation concentration diagrams for ions inside the pits were obtained, as shown in Fig. 17 for iron. In contrast to Vetter and Strehblow [27], who questioned the mechanism of pitting based on local changes in the composition of the solution at least during the initial stage of pit growth, Galvele found that although the chloride ion changes were not significant, the pH changes due to hydrolyses of metal ions were important (see Fig. 17). In addition, the critical H^+ concentration, above which pits would grow, was determined, assuming that the critical concentration is equal to the pH at which the passivating oxide film is in equilibrium with a concentration of 10^{-6} mol/liter of the metal ion, as used by Pourbaix for his diagrams [85].

The calculations for iron, as indicated in Fig. 17, showed that the critical acidification was reached for $d_p i_p$ values approximately equal to 10^{-6} A/cm, d_p being the pit depth and i_p the current density in the pit. For current densities of the order of 1 A/cm^2, which seems reasonable for various metals [27,40,86,87] at least for the early stage, a pit size of 10^{-6} cm will be large enough so that the critical acidification is attained at the pit bottom. Similar results were obtained for Zn, Ni, Co, Al, and Cr [82]. Furthermore, Galvele extended his model to metal/electrolyte systems with additional components present in the solution such as inhibiting anions of weak acids.

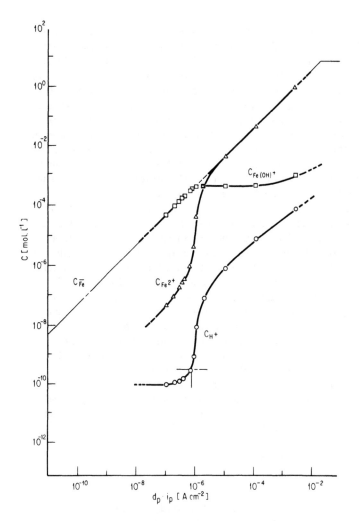

FIGURE 17 Concentrations of Fe^{2+}, $Fe(OH)^+$, and H^+ as a function of the product of the pit depth d_p and the current density i_p in a uni-directional pit. $Fe = Fe^{2+} + 2e$; $Fe^{2+} + H_2O \rightleftarrows Fe(OH)^+ + H^+$; $*k_1 \approx 10^{-7}$. (After Galvele [61].)

These calculations yield a shift of the H^+ concentration to larger $d_p i_p$ values due to the effect of the buffer ion concentration. Assuming that the relation between potential and current density inside the pit follows a logarithmic law, he predicted an increase of the pitting potential with the buffer concentration according to a semilogarithmic relationship, as is usually observed in presence of inhibitors.

In another contribution [83], Galvele extended his calculations to take full hydrolysis and precipitation into account. In this case no arbitrary assumptions must be made to define the critical acidification, since the critical value will be that above which most of the corrosion products are soluble. Galvele assumed that under this condition the passive film will be undermined and pitting will occur. From calculations based on full hydrolysis the pitting potentials of bivalent metals such as iron, nickel, cadmium, and zinc were found to be pH-independent up to pH values of approximately 9 to 10, but pH-dependent for higher pH values. In the case of trivalent metal ions such as aluminum the condition for pitting is not affected by the external pH over a wide range of pH values, as is indeed observed experimentally.

Considering the stability of pitting corrosion in terms of local acidification, it should be remembered that in all cases where hydrogen generation occurs in the pits a critical concentration buildup within the pit cannot be considered the important requirement for stable pit growth. Because of the rigorous convection, no significant concentration gradients can develop. It must also be pointed out that in the case of iron and nickel passivity is achieved even at pH 0, so acidification by hydrolysis may not be used to explain the stability of pits. In neutral or alkaline solutions local acidification may certainly have an important influence. However, additional parameters stabilizing the active state, such as halide accumulation or the ohmic potential drop, must be considered simultaneously.

The stability of pitting dissolution of metals based on critical ion concentrations was also examined by Sato [88,89]. He suggested that a critical concentration buildup Δc^* must be attained. The criterion determining the stability of pitting is then given by

$\Delta c > \Delta c^*$ for stable pit growth

$\Delta c < \Delta c^*$ for repassivation

Δc^* may be calculated using Eq. (18), as already applied by other authors [27,58,90]. According to Eq. (18), Δc is proportional to the product of the pit dissolution current density i_p and the pit size d_p. The critical pit size d_p^* is therefore a function of i_p and hence of the metal electrode potential ε. Assuming a Tafel equation for the dissolution current, the following relationship is obtained:

$$\varepsilon = - a \log d_p^* + b' \tag{31}$$

where a is a constant and $b' = \log(zFD \, \Delta c^*/2) + b$. Equation (31) indicates that for large defects less noble potentials are required to start pit growth. Since the b' value also contains the critical concentration buildup Δc^*, smaller pits may grow in systems with lower critical concentration buildup.

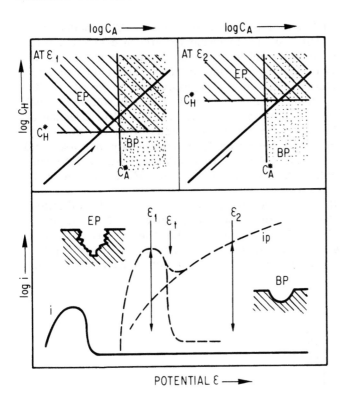

FIGURE 18 Diagrams showing local buildup of hydrogen ion concentration C_H and aggressive anion concentration C_A in a pit electrolyte which leads to etching pits at ε_1 and brightening pits at ε_2; $C_H{}^*$ and $C_A{}^*$ are critical ion concentrations for stable etching and brightening pits, respectively; ε_t is the boundary potential between etching pits and brightening pits. (After Sato [88].)

Sato points out, considering the physical significance of Δc^*, that two models of metal dissolution must be distinguished: active or etching dissolution, which takes place at lower potentials, and transpassive or brightening dissolution, which occurs at relatively noble potentials. He further suggests that for the stability of etching pits a critical hydrogen ion concentration $C_H{}^*$ must be reached, whereas for the stability of brightening pits a critical concentration of aggressive anions $C_A{}^*$ is required. Since the concentration of both hydrogen ions and aggressive anions within the pit increases with increasing metal dissolution, the type of pitting that occurs depends on whether $C_H{}^*$ or $C_A{}^*$ is reached first, as shown schematically in Fig. 18.

The stability criteria proposed by Sato can be applied only if concentration gradients develop due to diffusion. Therefore the same

objections must be made as for the model of Galvele. This model cannot be generalized. With regard to ohmic resistance-controlled pit
growth, it should be pointed out that similar relationships are often
obtained. Using the model of Sato, for instance, Eq. (31) predicts
a semilogarithmic relation between the repassivation or growth-limiting potential and the chloride concentration in the bulk solution, which
is also obtained experimentally on aluminum [60,70]. But because of
the large amount of hydrogen bubbles generated within the pits during pitting of aluminum, the mass transfer is enormously increased and
therefore the concept of critical concentrations proposed by Sato cannot be applied to explain the chloride dependence of the growth-limiting potential on aluminum.

   The effect of salt films formed within the pit during the initiation
or growth process has been discussed frequently [7,27,46—55,58,
91,92]. Beck and Chan [71] demonstrated that the formation of salt
films on stainless steel is decisive for the stability of growing pits.
In Fig. 19 the effect of flow velocity on current density in a pit is

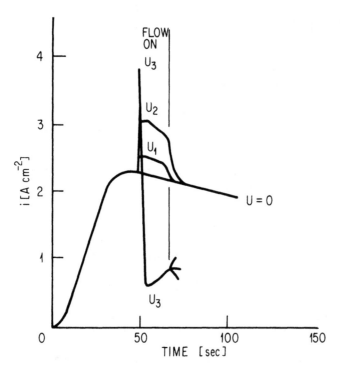

FIGURE 19 Effect of flow velocity U on current density in an artificial
pit of type 304 stainless steel; 1 N NaCl $\varepsilon = 0.5$ V_{SCE}, $U_1 < U_2 < U_3$.
(After Beck and Chan [71].)

shown. The current increases with flow velocity at low values, as expected, for diffusion-controlled growth. At the highest velocities, the current immediately peaks at a value below the ohmic limit and then drops to a lower value. The authors suggested that repassivation occurs as soon as the salt film is removed. From their experiments they further assumed that the salt film is more important in maintaining pitting than a low pH. The rather high and still increasing current density after repassivation was explained by crevice effects, which could not be entirely suppressed in these experiments.

6.4.2 Potential Drop Within the Pit

Large potential variations within pits were reported several years ago [78,79,93]. Differences of 1 V or more were measured between the bulk electrolyte and the electrolyte at the bottom of the pit. According to Pickering and Frankenthal [84,94—96], only a small fraction of this potential drop can be accounted for as IR drop through an unobstructed column of electrolyte within a pit. From their calculations of the concentration and potential gradients based on diffusion and migration processes in the pit they obtained values up to 0.1 V as indicated in Fig. 20. Potential measurements on iron, using a microprobe which was moved into the pit, gave excellent agreement with the theoretical calculations for the outer part of the pit up to about

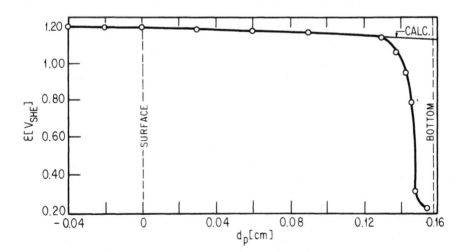

FIGURE 20 Plot of measured potential as a function of depth x into the pit for a single-pit iron specimen; experimental values are compared with calculated values. (After Pickering and Frankenthal [84].)

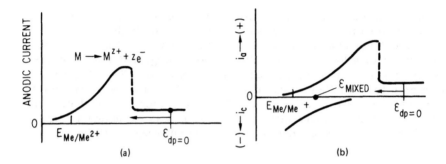

FIGURE 21 Schematic illustrating the limiting electrode potential as a shift of the electrode potential from the value at the external surface $\varepsilon_{dp=0}$ to that at the bottom of the pit ε_{dp}: (a) in the case of a single current-producing reaction the limiting potential corresponds to the equilibrium potential $E_{Me/Me}{}^{2+}$, $\varepsilon_{LIM} = E_{Me/Me}{}^{2+}$; (b) in the case of two reactions of opposite signs occurring the limiting potential corresponds to the mixed potential ε_{MIXED}, $\varepsilon_{LIM} = \varepsilon_{MIXED}$. (After Pickering [96].)

0.02 cm from the pit bottom. In this region a potential change of approximately 1 V in the less noble direction was observed (see Fig. 20), in contrast to the calculated values. It was concluded that these data were inconsistent with the existing models of pit growth. Pickering and Frankenthal presented a modified model, where pit growth is explained by active metal dissolution including a high-resistance path due to constriction caused by a hydrogen bubble at the pit bottom. Pickering [96] extended this model to discuss the limiting electrode potential which may be attained at the pit bottom. In the case of a single current-producing reaction it is the equilibrium potential of this reaction. When two reactions of opposite sign occur in the pit it will be the mixed potential of these reactions. Figure 21 illustrates schematically the mechanism of Pickering and Frankenthal as well as the concept of the limiting electrode potentials at the pit bottom.

The hydrogen formation observed at applied potentials above the thermodynamic values may certainly be explained by the large potential drop measured on single artificial pits on iron. However, further experimental support showing whether a high-resistance path due to constrictions caused by hydrogen gas bubbles is the only possible explanation would be valuable. Moreover, most of the experiments of Pickering and Frankenthal were performed on iron, which does not passivate under these conditions. Additional tests with passive materials should be carried out.

REFERENCES

1. J. Kruger, in *Passivity and Its Breakdown on Iron and Iron Base Alloys*, National Association of Corrosion Engineers, Houston, 1976, p. 91.
2. H.-H. Strehblow, *Werkst. Korros.*, 27: 792 (1976).
3. H.-J. Engell, *Electrochim. Acta*, 22: 987 (1977).
4. T. P. Hoar, *Corros. Sci.*, 7: 341 (1967).
5. T. P. Hoar and W. R. Jacob, *Nature*, 216: 1299 (1967).
6. J. R. Galvele, in *Passivity of Metals* (R. P. Frankenthal and J. Kruger, eds.), The Electrochemical Society, Princeton, N.J., 1978, p. 285.
7. H. Kaesche, in *Die Korrosion der Metalle*, Springer-Verlag, New York, 1979, p. 249.
8. S. Szklarska-Smialowska, in *Localized Corrosion*, National Association of Corrosion Engineers, Houston, 1974, p. 312.
9. H. H. Uhlig, *J. Electrochem. Soc.*, 97: 215c (1950).
10. H. Böhni and H. H. Uhlig, *J. Electrochem. Soc.*, 116: 906 (1969).
11. H. P. Leckie and H. H. Uhlig, *J. Electrochem. Soc.*, 113: 1262 (1966).
12. Ya. M. Kolotyrkin, *J. Electrochem. Soc.*, 108: 209 (1961); *Corrosion*, 19: 261t (1963).
13. H.-H. Strehblow and B. Titze, *Corros. Sci.*, 17: 461 (1977).
14. C. L. McBee and J. Kruger, in *Localized Corrosion*, National Association of Corrosion Engineers, Houston, 1984, p. 252.
15. K. E. Heusler and L. Fischer, *Werkst. Korros.*, 27: 551 (1976).
16. K. E. Heusler and L. Fischer, *Werkst. Korros.*, 27: 788 (1976).
17. H.-H. Strehblow and M. B. Ives, *Corros. Sci.*, 16: 317 (1976).
18. B. Löchel and H.-H. Strehblow, *Electrochim. Acta*, 28: 565 (1983).
19. H.-H. Strehblow, B. Titze, and B. P. Löchel, *Corros. Sci.*, 19: 1047 (1979).
20. M. Janik-Czachor, A. Szummer, and Z. Szklarska-Smialowska, *Corros. Sci.*, 15: 775 (1975).
21. T. Erdey-Gruz and M. Vollmer, *Z. Pyys. Chem. Abt. A*, 157: 165 (1931).
22. N. Sato, *J. Electrochem. Soc.*, 129: 255 (1982).
23. U. Stimming and J. W. Schultze, *Electrochim. Acta*, 24: 859 (1979).
24. U. Stimming and J. W. Schultze, *Ber. Bunsenges. Phys. Chem.*, 80: 1297 (1976).

25. J. H. Kennedy and K. W. Frese, Jr., *J. Electrochem. Soc.*, *125*: 723 (1978).

26. T. P. Hoar, D. C. Mears, and G. P. Rothwell, *Corros. Sci.*, *5*: 279 (1965).

27. K. J. Vetter and H.-H. Strehblow, *Ber. Bunsenges. Phys. Chem.*, *74*: 1024 (1970); *75*: 449 (1970).

28. Z. Szklarska-Smialowska, H. Viefhaus, and M. Janik-Czachor, *Corros. Sci.*, *16*: 649 (1976).

29. J. R. Galvele, S. M. de De Micheli, I. L. Muller, S. B. de Wexler, and I. L. Alanis, in *Localized Corrosion*, National Association of Corrosion Engineers, Houston, 1974, p. 580.

30. M. Janik-Czachor and S. Kaszczyszyn, *Werkst. Korros.*, *33*: 500 (1982).

31. J. Painot and J. Augustynski, *Electrochim. Acta*, *20*: 747 (1975).

32. J. Augustynski and J. Painot, *J. Electrochem. Soc.*, *123*: 841 (1976).

33. M. Koudelkova and J. Augustynski, *J. Electrochem. Soc.*, *124*: 1165 (1977).

34. C. Y. Chao, L. F. Lin, and D. D. Macdonald, *J. Electrochem. Soc.*, *128*: 1187 (1981).

35. L. F. Lin, C. Y. Chao, and D. D. Macdonald, *J. Electrochem. Soc.*, *128*: 1194 (1981).

36. U. Bertocci, in *Proceedings of the 7th International Congress on Metallic Corrosion*, Rio de Janeiro, 1978, p. 2010; *J. Electrochem. Soc.*, *128*: 520 (1981); *Surf. Sci.*, *101*: 608 (1980).

37. N. Sato, *Electrochim. Acta*, *16*: 1683 (1971).

38. I. G. Abidor, V. B. Arakelyan, L. V. Chernomordik, Y. A. Chizmadzkev, V. F. Pastushenko, and M. R. Tarasevic, *J. Electroanal. Chem. Interfacial Electrochem.*, *104*: 37 (1979).

39. K. J. Vetter and H.-H. Strehblow, in *Localized Corrosion*, National Association of Corrosion Engineers, Houston, 1974, p. 240.

40. H. Kaesche, *Z. Phys. Chem.*, *34*: 87 (1962).

41. H. Böhni and Ph. Corboz, in *Corrosion and Corrosion Protection* (R. P. Frankenthal and F. Mansfeld, eds.), The Electrochemical Society, Pennington, N.J., 1981, p. 92.

42. H.-J. Engell and N. D. Stolica, *Z. Phys. Chem.*, *20*: 113 (1959); *Arch. Eisenhuettenwes.*, *30*: 239 (1959).

43. K. Videm, Kjeller Rept. 149, Institut for Atomenergi, Kjeller, Norway, 1974.

44. H.-H. Strehblow and J. Wenners, *Z. Phys. Chem.*, *98*: 199 (1975); *Electrochim. Acta*, *22*: 421 (1977); H.-H. Strehblow and M. B. Ives, *Corros. Sci.*, *16*: 317 (1976).

45. H. J. Sand, *Z. Phys. Chem.*, *35*: 641 (1900).

46. Yu. A. Popov, Yu. V. Alekseev, and Ya. M. Kolotyrkin, *Elektrokhimiya, 14:* 1447 (1978).
47. Yu. A. Popov, Yu. V. Alekseev, and Ya. M. Kolotyrkin, *Elektrokhimiya, 14:* 1601 (1978).
48. Yu. A. Popov, Yu. V. Alekseev, and Ya. M. Kolotyrkin, *Elektrokhimiya, 15:* 403 (1979).
49. Yu. A. Popov, Yu. V. Alekseev, and Ya. M. Kolotyrkin, *Elektrokhimiya, 15:* 533 (1979).
50. Yu. A. Popov, Yu. V. Alekseev, and Ya. M. Kolotyrkin, *Elektrokhimiya, 15:* 665 (1979).
51. Yu. A. Popov, Yu. V. Alekseev, and Ya. M. Kolotyrkin, *Elektrokhimiya, 15:* 669 (1979).
52. Yu. A. Popov, Yu. V. Alekseev, and Ya. M. Kolotyrkin, *Elektrokhimiya, 15:* 894 (1979).
53. Yu. A. Popov, Yu. V. Alekseev, and Ya. M. Kolotyrkin, *Elektrokhimiya, 15:* 898 (1979).
54. Yu. A. Popov, Yu. V. Alekseev, and Ya. M. Kolotyrkin, *Elektrokhimiya, 15:* 1071 (1979).
55. Yu. A. Popov, *Dokl. Akad. Nauk SSSR, 240:* 373 (1978); *243:* 1487 (1978).
56. H.-H. Strehblow, K. J. Vetter, and A. Willigallis, *Ber. Bunsenges. Phys. Chem.,* 75: 822 (1971).
57. I. L. Rosenfeld, I. S. Danilov, and R. N. Oranskays, *J. Electrochem. Soc.,* 125: 1729 (1978).
58. Th. R. Beck and R. C. Alkire, *J. Electrochem. Soc.,* 126: 1662 (1979).
59. R. Alkire, D. Ernsberger, and T. R. Beck, *J. Electrochem. Soc.,* 125: 1382 (1978).
60. P. Heimgartner, in *Proceedings of the 8th International Congress on Metallic Corrosion,* 1981, Deutsche Gesellschaft für chemisches Apparatewesen, vol. I, p. 12. P. Heimgartner and H. Böhni, *Corrosion,* 41: 715 (1985).
61. R. Kirchheim, K. Maier, and G. Tölg, *J. Electrochem. Soc.,* 128: 1027 (1981).
62. P. Heimgartner, dissertation No. 7519, ETH Zürich, (1984).
63. F. Hunkeler, dissertation No. 6663, ETH Zürich, 1980.
64. Z. Szklarska-Smialowska, *Werkst. Korros.,* 22, 780 (1971).
65. N. Sato, T. Nakagawa, K. Kudo, and M. Sakashita, *Trans. Jpn. Inst. Met.,* 13: 103 (1972).
66. Y. Hisamatsu, T. Yoshii, and Y. Matsumura, in *Localized Corrosion,* National Association of Corrosion Engineers, Houston, 1974, p. 427.
67. I. L. Rosenfeld and I. S. Danilov, *Corros. Sci.,* 7: 129 (1967).
68. G. Butler, P. Stretton, and J. G. Beyon, *Br. Corros. J.,* 7: 168 (1972).

69. F. Hunkeler and H. Böhni, *Werkst. Korros.*, *32*: 129 (1981).
70. F. Hunkeler and H. Böhni, *Corrosion*, 37: 645 (1981).
71. T. R. Beck and S. G. Chan, *Corrosion*, *11*: 665 (1981).
72. T. R. Beck, *Corrosion*, *33*: 9 (1977).
73. N. Ibl, *Chem. Ing. Tech.*, *35*: 353 (1963).
74. C. Edeleann, *J. Inst. Met.*, *89*: 90 (1960/61).
75. F. Hunkeler and H. Böhni, *Werkst. Korros.*, *34*: 593 (1983), *Corrosion*, *40*: 534 (1984).
76. J. Newman, D. N. Hanson, and K. Vetter, *Electrochim. Acta*, *22*: 829 (1977).
77. P. H. Melville, *J. Electrochem. Soc.*, *127*: 864 (1980).
78. G. Herbsleb and H.-J. Engell, *Werkst. Korros.*, *17*: 365 (1966).
79. G. Herbsleb and H.-J. Engell, *Z. Elektrochem.*, *65*: 881 (1961).
80. M. Pourbaix, *Corrosion*, *26*: 431 (1970).
81. H.-H. Uhlig, *Trans. AIME*, *140*: 411 (1940).
82. J. R. Galvele, *J. Electrochem. Soc.*, *123*: 464 (1976).
83. J. R. Galvele, *Corros. Sci.*, *21*: 551 (1981).
84. H. W. Pickering and R. P. Frankenthal, *J. Electrochem. Soc.*, *119*: 1297 (1972).
85. M. Pourbaix, in *Atlas of Electrochemical Equilibria in Aqueous Solutions*, Pergamon, Oxford, 1966, p. 70.
86. Z. Szklarska-Smialowska and M. Janik-Czachor, *Br. Corros. J.*, *4*: 138 (1969).
87. W. Schwenk, *Corrosion*, *20*: 129t (1964).
88. N. Sato, in *Corrosion and Corrosion Protection* (R. P. Frankenthal and F. Mansfeld, eds.), The Electrochemical Society, Pennington, N.J., 1981, p. 101.
89. N. Sato, *J. Electrochem. Soc.*, *129*: 260 (1982).
90. Y. Hisamatsu, in *Passivity and Its Breakdown on Iron and Iron Base Alloys*, National Association of Corrosion Engineers, Houston, 1976, p. 99.
91. T. R. Beck, *J. Electrochem. Soc.*, *120*: 1317 (1973).
92. J. W. Tester and H. S. Jsaacs, *J. Electrochem. Soc.*, *122*: 1438 (1975).
93. C. M. Chen, F. H. Beck, and M. G. Fontana, *Corrosion*, *27*: 234 (1971).
94. R. P.. Frankenthal and H. W. Pickering, *J. Electrochem. Soc.*, *119*: 1304 (1972).
95. H. W. Pickering and R. P. Frankenthal, in *Localized Corrosion*, National Association of Corrosion Engineers, Houston, 1974, p. 261.
96. H. W. Pickering, in *Corrosion and Corrosion Protection* (R. P. Frankenthal and F. Mansfeld, eds.), The Electrochemical Society, Pennington, N.J., 1981, p. 85.

97. P. L. Joseph, V. Balasubramanian, and B. A. Shenoi, *J. Electrochem. Soc. Jpn.*, *35*: 8 (1967).
98. G. Sussek, M. Kesten, and H.-G. Feller, *Metall.*, *33*: 1031 (1979).
99. F. Hunkeler and H. Böhni, *Werkst. Korros.*, *34*: 68 (1983).

7

THE EFFECT OF HYDROGEN ON METALS

M. R. LOUTHAN, JR.

Department of Materials Engineering, College of Engineering, Virginia Polytechnic Institute and State University, Blacksburg, Virginia

7.1 INTRODUCTION

Hydrogen embrittlement and/or hydriding has led to failures of fuel cladding in nuclear reactors [1], breakage of aircraft components [2], leakage from gas-filled pressure vessels used by NASA [3], delayed failure in numerous high-strength steels [4], reductions in mechanical properties of nuclear materials [5], and blisters or fisheyes in copper [6], aluminum [7], and steel parts [8]. The role of hydrogen in stress corrosion cracking (SCC) of titanium [9], aluminum [10], and both austenitic [11] and high-strength [12] steels has also been emphasized. Because of these numerous adverse effects, the process of hydrogen embrittlement has received widespread study.

The adverse effects of hydrogen on the mechanical properties of metals and alloys have been known since W. H. Johnson wrote his classic paper, "On Some Remarkable Changes Produced in Iron and Steel by the Action of Hydrogen and Acids." This paper was published in the *Proceedings of the Royal Society of London* (Vol. 23) in 1875. By the early 1940s many hundreds of papers had been written on the subject, and by 1960 the volume of available literature was overwhelming. This situation has not improved with time, and today any attempt to present a comprehensive review of the literature on hydrogen metals would require a lifetime effort.

The effects of hydrogen on metals depend on several factors including:

Hydrogen solubility and diffusivity
The possibility of reaction to form a hydride
The possibility of reaction between hydrogen and impurity and/or
   alloying elements

Although there are several proposed mechanisms by which hydrogen may affect the load-bearing capabilities of metals, most require
the development of a critical hydrogen concentration within the metal
lattice. This critical concentration is proposed to lower lattice cohesion [13], cause localized plasticity [14], restrict glide [15], precipitate as gas bubbles [16], or cause embrittlement by a combination of
these and other mechanisms [17–27]. Regardless of the mechanism,
however, the fact that hydrogen degrades mechanical properties is
well known because the process has been extensively studied. These
studies have produced a degree of understanding which often permits
design engineers to choose materials to minimize the probability of
adverse hydrogen effects. The objective of this chapter is to emphasize the understanding which has been developed and to provide
some insight into the process of hydrogen embrittlement.

7.2 HYDROGEN EFFECTS

7.2.1 Blistering

Hydrogen can be absorbed by a metal during casting, forging, heat
treatment, welding, and finishing or during service. If the external
hydrogen pressure is reduced after absorption, hydrogen diffusion
from the metal is required for the system to obtain thermodynamic
equilibrium. The internal hydrogen concentration (C) in equilibrium
with an external hydrogen pressure (p) over a hydride-free metal is

$$C = C_0 p^{1/2} \exp(-\Delta H_s/RT) \text{ cc gas (NTP)/cc metal}$$

Therefore, the equilibrium hydrogen in any metal is a sensitive function of temperature. Hydrogen diffusion from the metal is required
for equilibrium if the temperature is reduced. If the temperature
change is rapid, diffusion may be too slow to maintain equilibrium and
the metal lattice may become supersaturated with hydrogen.
   Local equilibrium may be maintained within the metal lattice, however, by precipitation of molecular hydrogen at internal cavities (or
defects). The hydrogen pressure at these cavities will depend on
the initial concentration and the amount of the temperature decrease.
This hydrogen pressure at an internal defect can exceed the pressure required to strain the metal plastically. Such strain opens mi-
crocavities and causes blistering. For practical purposes, however,
blisters typically do not originate because of exposure to hydrogen gas

but are developed by reactions between the metal and water vapor during melting and heat treatment, or from hydrogen introduced during finishing operations such as electropolishing and acid cleaning.

Iron and aluminum will react with atmospheric moisture at elevated temperatures according to the equations

$$Fe + H_2O \rightarrow FeO + 2H$$

$$2Al + 3H_2O \rightarrow Al_2O_3 + 6H$$

The metal lattice becomes supersaturated with hydrogen because the effective atomic hydrogen pressure during the reaction is many times the equilibrium pressure. Subsequent precipitation of molecular hydrogen leads to the development of high pressures and blisters (Fig. 1a).

If hydrogen is absorbed while the metal is liquid, gas bubbles (blow holes) may develop because of the decrease in hydrogen solubility during solidification (Fig. 1b). Precipitation of hydrogen gas at internal defects has also led to blistering after acid pickling, electroplating, and other finishing operations.

The cause of blistering is well known; a change in environment (temperature, hydrogen overpressure) leads to precipitation of hydrogen gas from a supersaturated metal lattice. Because this mechanism is well known, handling and finishing techniques have been developed to minimize this form of embrittlement. Vacuum melting and degassing minimize the quantity of hydrogen in the metal. Acid pickling and other such processes which may introduce hydrogen are avoided when practical, and possible moisture sources, such as the coatings of welding electrodes, are carefully considered before use. Melts are degassed by bubbling an active or inert gas through the liquid. Chlorination of aluminum is an effective method of removing hydrogen, and hydrogen is removed from magnesium by holding the metal at temperatures just above the melting temperature to allow for hydrogen outgassing before casting.

Inclusions and other defects are typical sites for hydrogen precipitation. These sites can be controlled, to a limited extent, by processing techniques. For any given hydrogen content, the extent of blistering decreases when the number of precipitation sites is increased. Thus a homogeneous distribution of very small inclusions will minimize susceptibility to embrittlement.

The observation that a homogeneous distribution of inclusions may be beneficial in designing for resistance to hydrogen-induced blistering has also found application in the development of high-strength steels which are resistant to other forms of hydrogen embrittlement. For example, as exploration for new sources of oil and gas intensifies, it becomes increasingly necessary to use high-strength steels because

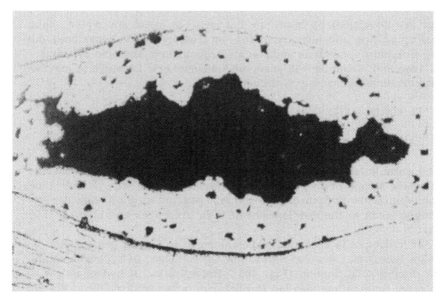

(a)

FIGURE 1 a) Hydrogen blister in aluminum. Blister formed by exposing 1100 Al to room air in muffle furnace at 773 K. b) Hydrogen
blow holes in iron casting. Iron was melted and cast in 1 atm hydrogen; porosity was caused by hydrogen precipitation due to a large
drop in solubility on solidification. (From Fig. 33 in Ref. [6].)

well depths exceed 4500 m. At these depths hydrogen sulfide gas,
H_2S, is frequently encountered. This gas causes extensive damage
to high-strength steels (the damage is termed sulfide stress cracking, SSC) and the damage process is known to be related to hydrogen embrittlement. Consequently, research on SSC-resistant steels
has concentrated on designing compositions to ensure fine, homogeneous microstructures. Other techniques to develop steels to resist
sour gas environments include the use of inclusion shape control and
the use of steels with very low inclusion levels [28]. In addition to
these indirect effects of inclusions on embrittlement, hydrogen may
diffuse into a metal and react directly with inclusions or alloy phases
[29a−c] to form a gaseous product. This phenomenon is often termed
hydrogen sickness.

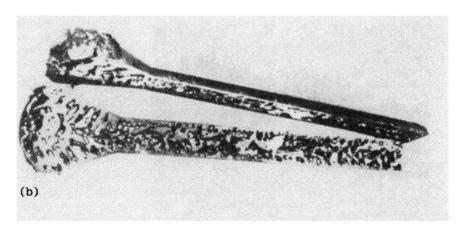

(b)

7.2.2 Hydrogen Sickness

Hydrogen sickness in copper is more correctly described by the term steam embrittlement and is observed in copper which contains oxygen. It is caused by hydrogen reaction with oxygen and the formation of water. The problem was encountered originally when hydrogen gas was used to maintain a reducing atmosphere during the heat treatment of copper. Hydrogen diffused into the metal during such anneals and reacted with oxide particles to form water. At temperatures above the critical temperature for water, the steam pressure generated by the reaction often exceeded the strength of the copper and caused plastic deformation and/or tearing. Examination of steam-embrittled copper frequently reveals grain boundary cracks or cavities (Fig. 2). However, cavities are also found within the grains and along prior grain boundaries in alloys that have received thermomechanical treatments after oxide precipitation. Problems related to the occurrence of steam embrittlement can be avoided if:

Alloys chosen for hydrogen service are free of oxide inclusions.
Alloys containing oxide are not annealed in hydrogen.

Silver is also susceptible to steam embrittlement; however, in this case, oxide reduction, per se, may not occur. Blisters have been found on high-purity (99.999+%) silver annealed in hydrogen [29b]. Hydrogen diffuses into the silver and reacts with the dissolved oxygen to form water vapor, which is nearly insoluble and therefore

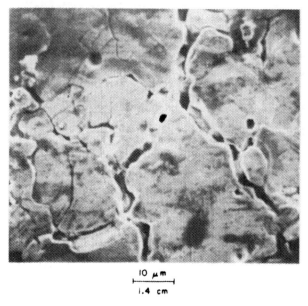

10 μm
├────────┤
1.4 cm

(a)

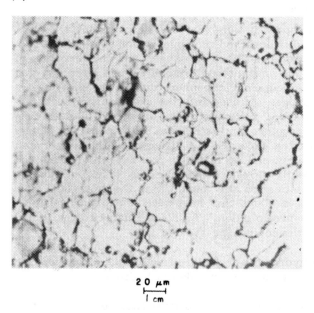

2 0 μm
├───┤
1 cm

(b)

FIGURE 2 Pores and cracks along grain boundaries of ETP copper exposed to hydrogen at 500°C.

precipitates. Detailed metallographic studies indicate that grain boundary precipitation is predominant. Individual bubbles continue to grow as the reaction proceeds until bubble agglomeration leads to grain boundary cracking.

Often blistering is also observed in steels used in the chemical and petrochemical industry because of hydrogen reaction with carbide inclusions and carbon in solid solution [19]. The reaction is termed hydrogen attack, and high pressures are developed because of methane formation. The formation of methane reduces the mechanical properties because in addition to the formation of high-pressure gas bubbles, the carbides and dissolved carbon are eliminated, thus lowering the strength. The susceptibility to hydrogen attack can be reduced by the use of steels containing titanium, vanadium, or other additions to react with carbon to form carbides that are not reduced by hydrogen.

Hydrogen attack was first recognized as a major problem in the petrochemical industry in 1940 and in 1949. G. A. Nelson published his classic paper on that subject, "Hydrogenation Plant Steel," in the *Proceedings of the American Petroleum Institute, Refining Division.* The purpose of Nelson's study was to define practical limits for plant operations based on the operating hydrogen pressure and the temperature of service. These operating limits were based on service experience and the resulting empirical plots separated the service conditions into safe and unsafe areas. This separation is shown schematically in Fig. 3a. Such curves are now termed Nelson diagrams and are developed by the American Petroleum Institute (API) using input data from almost all of the oil companies. The input data result in different curves for each steel used. The curve is drawn so that it lies below all unsatisfactory data points; therefore the accumulation of additional data can only shift the existing curves down and to the left. An example of an actual Nelson curve for a carbon steel with trace alloying elements is shown in Fig. 3b. The open symbols in the graph signify satisfactory industrial experience, while the closed symbols indicate unsatisfactory experience. It should also be noted that the curves are developed on the basis of input data from at least nine different oil companies. The conversion of Cr and V content to molybdenum equivalents should also be noted. This conversion is empirical and is used simply for convenience in placing the data on a Nelson curve.

A second form of hydrogen-induced damage to carbon steels during elevated-temperature service is decarburization. Thermodynamic calculations show that hydrogen should react with carbon in solution and/or metal carbides to form methane. The methane reaction with a carbide can be written as

$$2H_2 + M_XC \rightarrow CH_4 + XM$$

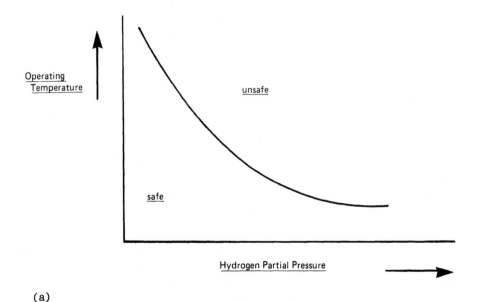

(a)

FIGURE 3 a) Schematic of safe operating limits for steel in hydrogen.
b) Operating limits for steels in hydrogen service showing effect of
trace alloying elements. (From Ref. [30a].)

If this reaction removes carbides from the surface of a steel and the
reaction between hydrogen and dissolved carbon

$$2H_2 + [C] \rightarrow CH_4$$

also takes place, the surface carbon content in the steel will be reduced to zero. This reduction will cause dissolution of carbides at
elevated temperatures. The carbon content at the surface will be
maintained at zero and carbides remote from the surface will be dissolved. A schematic of the carbon distribution which develops because of the surface hydrogen-carbon reaction is shown in Fig. 4a,
and photomicrographs illustrating the decarburized (single-phase region) zone and the two-phase region are shown in Fig. 4b. The decarburization will weaken the structure and can, if undetected, cause
failure of the system.

   Hydrogen sickness, hydrogen attack, blistering, decarburization, and other forms of damage caused by hydrogen reaction with
alloy or impurity elements are generally described as irreversible hydrogen embrittlement. Although some of the mechanical properties

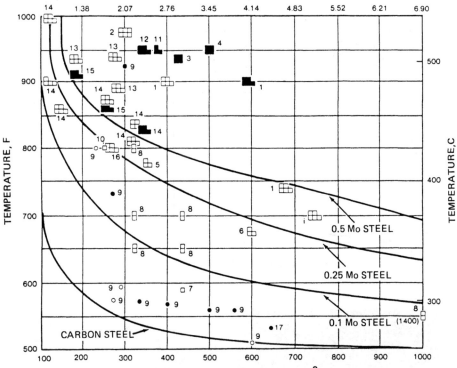

HYDROGEN PARTIAL PRESSURE, MPa abs

TEMPERATURE, F

TEMPERATURE, C

0.5 Mo STEEL

0.25 Mo STEEL

0.1 Mo STEEL (1400)

CARBON STEEL

HYDROGEN PARTIAL PRESSURE, Lb/in^2 abs

References	CR	Analysis MO	V	Mo Equivalent
1. Shell Oil Co.*		0.50		0.50
2. Weld Deposits, D. J. Bergman*	0.79	0.39		0.59
3. Weld Deposits, D. J. Bergman*	0.80	0.15		0.35
4. Weld Deposits, D. J. Bergman*	0.50	0.25		0.37
5. Continental Oil Co.*		0.25		0.25
6. Standard Oil Co. of California*		0.27		0.27
7. Standard Oil Co. of California*	0.50	0.06	0.08	
8. A. O. Smith Corp.*			0.13-0.18	
9. Shell Development Co. Drawing No. VT 659-2				
10. AMOCO Oil Co.*	0.04			0.02
11. R. W. Manuel, *Corrosion*, 17[9], pp. 103-4, Sept., 1961	0.27	0.15		0.22
12. The Standard Oil Co. of Ohio*	0.11	0.43		0.50
13. Exxon Corp.*				
14. Union Oil of California*				
15. Amoco Oil Co.*				
16. Standard Oil Co. of California*				
17. Gulf Oil Corp.*				

NOTES:
1. Mo has four times the resistance of Cr to H_2 attack.
2. Mo is equivalent to V, Ti, or Nb up to 0.1 percent.
3. Si, Ni, Cu, P, and S do not increase resistance.

*Private communication to Subcommittee on Corrosion

(b)

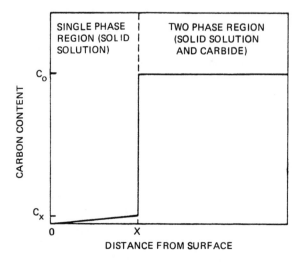

(a)

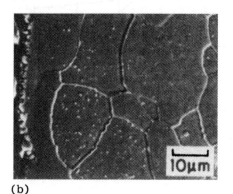 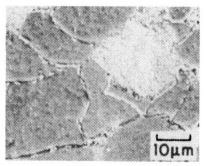

(b)

FIGURE 4 a) Schematic carbon distribution of partially decarburized
steel c_0 = original carbon content, c_x = carbon content in solid solution of ferrite. b) Microstructure of 1Cr-0.5Mo steel after 3000 hr
of hydrogen exposure; etched in 4% picric acid and 1% nitric acid
methanol solution. (From Ref. [30b].)

may be restored by heat treatments, the removal of strengthening
phases, such as carbides, causes strength losses which are not generally recovered. Furthermore, it is doubtful that the microvoids,
pores, blisters, or fissures developed during the attack can be healed
effectively by thermomechanical treatments. Therefore, for design or
fabrication considerations, prevention is a much more effective measure than subsequent attempts to recover strength and ductility
losses.

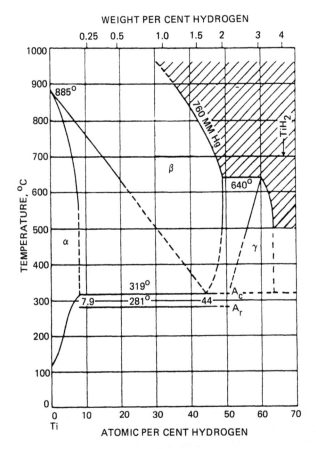

FIGURE 5 Titanium-hydrogen phase diagram for hydrogen at 1 atm. (From Ref. [57].)

7.3 HYDRIDE EMBRITTLEMENT

Zirconium, titanium, niobium, vanadium, uranium, and other such metals and their alloys form hydrides if the external hydrogen pressure is high enough. The titanium-hydrogen phase diagram (Fig. 5) is typical of titanium, zirconium, and hafnium, and shows that below the eutectoid temperature, hydrogen solubility decreases rapidly with decreasing temperature and is practically nil at room temperature. The other hydride formers show similar solubility decreases. Because of this decrease in hydrogen solubility, hydrides often precipitate even when the hydrogen content of the metal is very low. In some respects this is similar to blistering; a hydrogen-rich phase is formed, and this phase may affect the mechanical properties of the metal [17]. Under

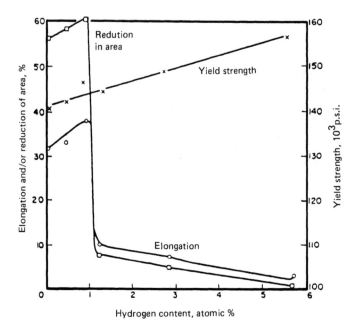

FIGURE 6   Tensile properties of Ti-8Mn as a function of hydrogen
content.

severe hydrogen charging conditions, the strains developed by precipitation of the less dense hydride phase can produce stresses sufficient to cause failure even when no load is applied [19].

Susceptibility to hydride embrittlement is dependent on numerous
variables, including hydrogen content, hydride distribution and morphology, temperature, and strain rate. The metal hydrides are not
only brittle and less dense than the metal matrix, but precipitate at
dislocations [31], at grain and twin boundaries [32], and at other
defects.   This combination of hydride-matrix properties influences
both the initiation and growth stages of fracture [32].

Hydride embrittlement generally increases with increasing hydrogen content (Fig. 6).   This increased susceptibility is due to an
increase in the number of hydrides per unit volume, a corresponding
decrease in the interhydride spacing, and a change in actual hydride
distribution [33]. Hydride distribution (orientation), which also affects interhydride spacing, is of such major importance in some cases
that the role of hydrogen content is suppressed.   Tensile specimens
of Zircaloy-2 containing hydrides oriented parallel to the stress axis
failed after 10% elongation, whereas comparable specimens containing
hydrides perpendicular to the stress axis, but with only 0.05 times
the total hydrogen, exhibited no macroscopic ductility [34].

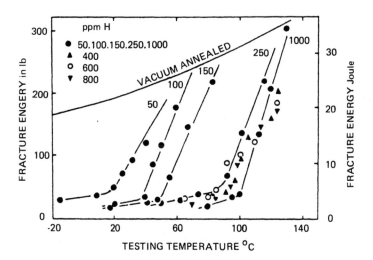

FIGURE 7 Ductile-to-brittle transitions in hydrided zirconium. (From Ref. [33].)

Increasing the temperature decreases the susceptibility to hydride embrittlement because, for any given hydrogen content, the volume fraction of hydride decreases. However, there is some evidence that above a critical temperature, the hydride may become ductile [35] and thus be less likely to enhance crack initiation. Ductile-to-brittle transitions have been observed in hydrided zirconium alloys [33], and the transition temperature has been shown to increase with increasing hydrogen content (Fig. 7). This phenomenon was shown to be consistent with variations in interhydride spacing and led to the conclusion that the embrittling effect of hydride precipitation is due to cracking along the hydride [33]. The shape and distribution of the hydrides are strongly dependent on the heat treatment prior to precipitation and on the cooling rate during precipitation [36]. Because cracks propagate along metal-hydride interfaces (Fig. 8), cooling rates also influence embrittlement. Slow cooling apparently promotes the formation of thin platelets primarily at grain boundaries and on specific crystallographic planes, whereas rapid quenching precipitates highly dispersed particles. Correspondingly, the toughness (resistance to hydride embrittlement) is higher in hydrided titanium when the hydrides are compact than when thin hydride platelets are present [37].

The existence of hydride precipitates before testing is not a necessary condition for hydride embrittlement. Vanadium alloys with hydrogen contents less than the apparent terminal solid solubility are embrittled because the test stresses (strains) cause hydride precipitation and subsequent embrittlement [38]. In such cases, the apparent solubility may be decreased by an applied stress because precipitation

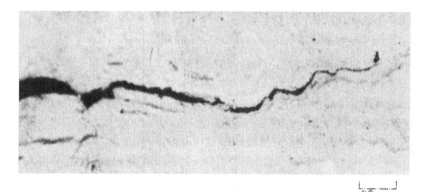

$\overline{?C\,\mu m}$

FIGURE 8 Cracking along metal-hydride interface. Cracks developed during tensile tests of hydrided Zircaloy-2. (From Ref. [58].)

of the low-density hydride phase is accompanied by nucleation of numerous dislocations at hydride-matrix interfaces [36]. Test stresses aid hydride nucleation and thus lower the apparent solubility. This effect is markedly different from the thermodynamically predicted increase in actual hydrogen solubility through the application of elastic tensile stresses [39].

Surface hydrides also affect the mechanical properties of hydride-forming metals. Failure of tensile specimens with surface hydride layers is initiated in the hydride and occurs at very low strains. Most of the conditions for surface hydride formation are met by exposure of hydride-forming metals to gaseous hydrogen, even at very low pressures. For example, calculations show that for alpha-phase titanium at room temperature, the equilibrium pressure above which hydrides will form is 5×10^{-14} torr. However, titanium alloys can be exposed to much higher hydrogen pressures with no adverse effects because of the protection afforded by the oxide film typically present on such alloys [40]. Surface hydride formation can be predicted from a knowledge of hydrogen transport rates in both the oxide film and the metal matrix [40]. Furthermore, surface hydride formation, per se, was shown to degrade the mechanical properties and may (without the application of stress) cause disintegration of the specimen (Fig. 9).

Hydride embrittlement, in most cases, is a maximum at high strain rates. Studies with hydrided titanium alloys [41a–c] show that impact and notch specimen testing are effective methods for determining the susceptibility to hydride embrittlement (Fig. 10).

Stress effects on hydride precipitation can result in the time-dependent development of hydride embrittlement. The single parameter having the greatest effect on the strength and ductility of many

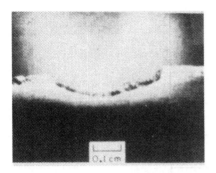

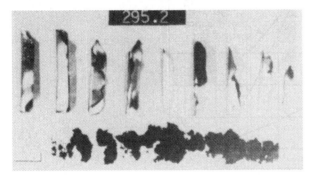

FIGURE 9 Disintegration of Ti-5Al-2.5Sn specimen exposed to hydrogen at 71°C. (From Ref. [40].)

hydride-forming metals and alloys is the hydride orientation [41b]. The orientation of the hydride platelets is influenced by both applied and residual stresses, and platelets tend to orient so that they are parallel to a tensile stress and perpendicular to a compressive stress (Fig. 11). The samples shown in Fig. 9 were machined from the same tabular section of Zircaloy-2 (a zirconium-tin alloy) and stressed as shown during hydride precipitation. The only difference between the two samples was the applied stress during precipitation, and when the experiment was repeated with the stress conditions reversed, the hydride orientation reversed. Because of this tendency and the tendency for the brittle hydrides to precipitate in regions of high tensile stresses, time-dependent fracture of hydrided zirconium, niobium, and vanadium alloys has been observed. This time-dependent fracture is modeled in Fig. 12. Hydride precipitation is induced in a region of stress concentration by the application of a tensile stress. In addition to this stress-induced precipitation, reorientation of hydrides may occur [41c] because of the resolution of hydrides which have unfavorable orientations relative

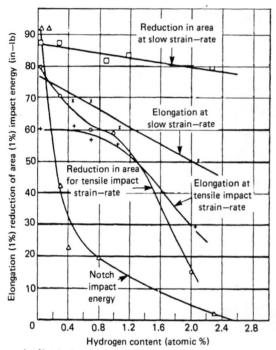

△   Notch—impact energy (in.—lb)
□   Reduction in area at a slow strain—rate (%)
o   Reduction in area at tensile impact strain—rate (%)
x   Elongation at a slow strain—rate (%)
+   Elongation at impact strain—rate (%)

FIGURE 10 Effect of testing techniques in revealing susceptibility of hydrided titanium to hydrogen embrittlement. (From Ref. [41].)

to the applied stress. The hydrides which are perpendicular to the crack-opening stress then provide low ductility and low-energy sites for crack growth.

7.4 GENERALIZED HYDROGEN EMBRITTLEMENT

7.4.1 Failure Under Static Conditions

Delayed failure (static fatigue) is a time-dependent form of hydrogen embrittlement that occurs primarily in high-strength martensitic steels [2]. This form of embrittlement is distinguished from the time-dependent hydride-induced fracture modeled in Fig. 10 because hydride formation does not occur. For many years, investigators thought that delayed failure hydrogen embrittlement was restricted to body-centered-

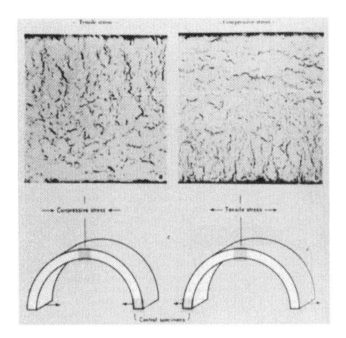

FIGURE 11 Hydride-orientation in stressed circumferential bend specimens of Zircaloy-2.

cubic and body-centered-tetragonal structures [42a], but recent studies have shown delayed failure, sustained load crack growth in a variety of austenitic alloys [42b]. The delayed failure process takes
place because of time-dependent damage accumulated in hydrogen-ex-
posed or hydrogen-charged specimens, loaded to stresses less than
the normal fracture stress. The samples fail after an incubation time,
which depends on temperature, hydrogen content, and other test or
exposure variables. This phenomenon has been extensively studied
with notched and precracked specimens and is perhaps most simply
illustrated by Fig. 13. The upper critical stress is the normal fracture stress below which failure will not occur. The incubation period
is the time required for the initial crack formation or the onset of irreversible damage, while the fracture time is the time to complete rupture. The lower critical stress and the failure time increase rapidly
with increasing notch root radius [43] and are sensitive functions of
temperature, hydrogen content, and specimen strength. The incubation time is reversible, and measurements of the kinetics of reversibility show an activation energy of approximately 9000 cal/mole [44].
This activation energy has been used to support and/or discredit
several proposed mechanisms for hydrogen embrittlement. However,
because the kinetics of hydrogen transport in steels are severely
complicated by trapping [45], use of this type of kinetic data is
questionable [14,22].

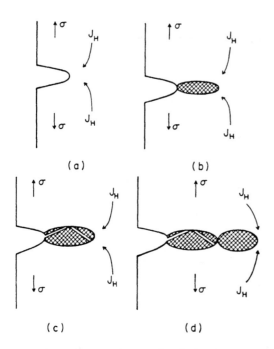

FIGURE 12 Schematic showing the mechanism of hydride embrittlement by stress-induced hydride formation. a) Flux of hydrogen to
the crack tip due to reduction of the hydrogen chemical potential in
the tensile stress field; b) formation of the hydride due to reduction
of the hydride chemical potential by the applied stress; c) cleavage
of the hydride along its cleavage plane resulting in crack advance;
d) process repeats the various steps. (From Ref. [41c].)

This concern also applies to the analysis of crack growth rate data
from fracture mechanics specimens. Hydrogen transport kinetics are
greatly affected by plastic deformation, and until these effects are
quantifiable, support of embrittlement models on the basis of activation energy determinations is tentative at best.

Delayed failure is one of the most insidious forms of hydrogen embrittlement because it can occur without warning in parts which are
not being exposed to hydrogen. The hydrogen absorbed during a
pickling or finishing operation can cause delayed failure during subsequent use, even when nominal test procedures fail to indicate hydrogen damage. For this reason, tests to determine susceptibility to
delayed failure were developed (ASTM Standard E-8-65T). Numerous
testing techniques have been reported for establishing delayed failure

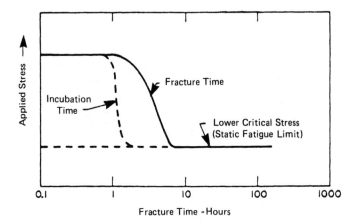

FIGURE 13 Schematic representation of delayed failure characteristics of a hydrogenated steel. (From Ref. [44].)

susceptibility; however, the results of the different techniques are not necessarily comparable. The difference in results shows that qualification of a particular plating process, finishing operation, or pickling bath may be as dependent on the test technique as on the process itself. For this reason, it is desirable that the susceptibility of a part to delayed failure or the qualification of a particular finishing operation be determined through test procedures recommended in the ASTM Standards rather than through tests described in ASTM STPs but not yet accepted as standard techniques. The role of test technique in determining the susceptibility to hydrogen damage is important in any test to qualify a material for hydrogen service. This fact is perhaps best illustrated by the observation that type 304 stainless steel is generally not susceptible to hydrogen-induced sustained load crack growth even when tested in 210-MPa hydrogen gas at stress intensities approaching those necessary to cause crack growth in the absence of hydrogen [42a]. This type of test will therefore suggest that hydrogen embrittlement of type 304 stainless steel is unlikely. Unfortunately, if this same steel is tested in tension in 70-MPa hydrogen gas [21] severe hydrogen-induced reductions in strength and ductility are observed. Furthermore, the fracture mode is changed from microvoid coalescence to faceted fracture. These results suggest that the steel is very susceptible to embrittlement and, in fact, indicate that hydrogen-induced sustained load crack growth should be observed.

Support for the idea that hydrogen induced slow crack growth was obtained in tests of cathodically charged type 304L stainless steel [59]

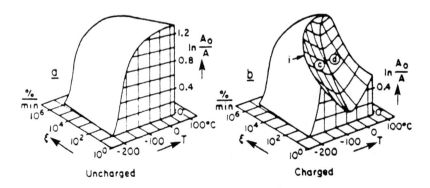

FIGURE 14 Effect of strain rate and temperature on the susceptibility of mild steel to hydrogen embrittlement. (From Ref. [46].)

and tests with other more stable austenitic steels [60]. One of the early examples of hydrogen embrittlement [46] shows the temperature and strain-rate dependence of ductility losses in cathodically charged mild steel (Fig. 14). These and similar data show that the degree of embrittlement

   Increases with increasing hydrogen content [47]
   Is a maximum at an intermediate temperature [46]
   Decreases with increasing strain rates [46]

Embrittlement in hydrogen-charged austenitic steels [16], nickel [48], and high-strength steels [2] shows similar characteristics. Furthermore, studies of the effects of temperature, strain rate, and hydrogen content on hydrogen transport by dislocations [49] suggest that such transport plays a key role in hydrogen embrittlement processes. This conclusion led to the following phenomenological description [50a, b] of hydrogen embrittlement in non-hydride-forming metals: "The deleterious effects of hydrogen on the tensile properties of metals are caused by the association and movement of hydrogen with dislocations. Hydrogen-dislocation interactions modify plastic deformation processes by stabilizing microcracks, by changing the work-hardening rate, and by solid-solution hardening."

This description, coupled with the concept of hydrogen binding to dislocations, provides a basis for explaining the temperature, strain rate, and hydrogen content dependence of hydrogen embrittlement. The amount of hydrogen transported (C_T) is related to dislocation movement into any region of a specimen during a dynamic test through the equations

$$C_T \propto \rho_m C_{\perp} \bar{V}$$

and

$$C_{\perp} = C \exp(-G_B/RT)$$

where C is the nominal hydrogen content, $C_{\perp}$ the hydrogen content
at a dislocation core, ρ_m the mobile dislocation density, $\bar{V}$ the average dislocation velocity, and G_B the hydrogen-dislocation binding
energy, which is negative when binding occurs.

Susceptibility to hydrogen embrittlement is a maximum at some intermediate temperature and apparently vanishes at elevated temperatures, even under conditions of high-temperature creep and stress
rupture [51], but embrittlement has been observed in tests at temperatures as low as 77 K [44].

The decreasing susceptibility with increasing temperature at any
given hydrogen concentration is caused by an exponential decrease
of the ratio $C_{\perp}/C$ with temperature. The ratio approaches 1 as T increases, and for a given amount of dislocation motion less hydrogen
is transported at the higher temperature. On the other hand, the
amount of hydrogen transported increases with decreasing temperature until the dislocations become saturated (i.e., $C_{\perp} = 1$) and/or
until the lower temperature reduces hydrogen mobility to the point
where dislocations must break away from their hydrogen atmospheres
in order to move.

Most proposed embrittlement mechanisms require that (a) a critical,
or threshold, amount of hydrogen be localized, and (b) the amount
of embrittlement be proportional to the amount of hydrogen localized.
Therefore, for a given strain rate and hydrogen content, C_T decreases as T increases at any temperature above the dislocation saturation temperature. Below this temperature C_T is limited by the
diffusivity of hydrogen and decreases as T decreases. Thus, a maximum in the amount of hydrogen transported and localized is obtained
at an intermediate temperature, and hence a maximum in embrittlement
is observed. The temperature of this maximum should increase with
increasing strain rate and increasing hydrogen content, as has been
observed in experimental studies [46].

The amount of hydrogen transported per unit strain decreases as
the strain rate increases (Fig. 15 and Ref. [49]). This dependence,
coupled with the dislocation-transport model for embrittlement, predicts the observed decrease in embrittlement susceptibility with increased strain rate.

For any given amount of dislocation motion, the amount of hydrogen transported is directly proportional to the hydrogen content until the dislocations become saturated with hydrogen. This

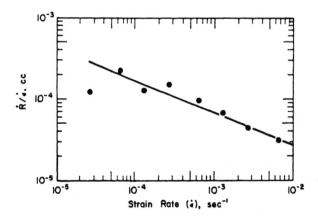

FIGURE 15 Strain rate dependence of tritium release from Armco iron deformed in tension at 21°C. (From Ref. [49].)

proportionality indicates that the strain-to-fracture should decrease linearly with hydrogen content until the dislocations are saturated. Further increases in hydrogen content above this saturation value should have little effect on hydrogen embrittlement. This behavior has been experimentally observed (Fig. 16 and Ref. [47]).

In addition to the effects of hydrogen content, temperature, and strain rate on the susceptibility to hydrogen embrittlement, studies have also shown that embrittlement is promoted by coplanar disloca-tion motion [5] and that susceptibility to embrittlement is greater for high-strength alloys than for low-strength alloys of similar structures (Fig. 17 and Ref. [25]). These embrittlement studies provide some simple criteria for alloy selection to minimize embrittlement susceptibil-ity. Because a critical quantity of hydrogen is apparently required for embrittlement, a low hydrogen solubility is desirable. For ex-ample, neither copper nor aluminum is susceptible to dynamic em-brittlement, and their hydrogen solubilities are much lower than those of readily embrittled metals such as alloys of iron and nickel. Em-brittlement is promoted by low stacking-fault energies, high negative dislocation-hydrogen binding energies, and high strengths; thus, al-loys having these general properties should be avoided. Alloy design can also minimize embrittlement. For example, fine dispersions of in-coherent precipitates have been effective in eliminating some adverse hydrogen effects in nickel [52], and prestraining in air before hy-drogen exposures has minimized embrittlement in other alloys [53]. The qualitative model for a role of dislocation transport in hydrogen embrittlement processes has been expanded into a semiquantitative

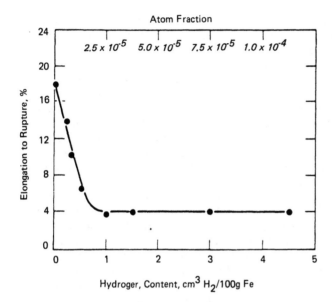

FIGURE 16 Variation in elongation to rupture of zone refined iron with hydrogen content. (From Ref. [47].)

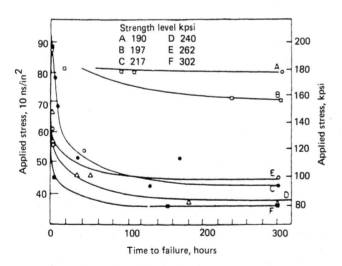

FIGURE 17 Effect of strength level on susceptibility to hydrogen embrittlement. Data are for static tests; however, similar dependence has been shown [5] for dynamic tests. (From Ref. [2].)

model [55] which shows that plastic zones can be locally enriched in
hydrogen whenever a mobile dislocation loses its atmosphere. This
type of argument has been used to explain observations such as those
described for the 304 stainless steel.

Sustained load crack growth studies are in effect static tests, and
little or no dislocation transport (and hence local enrichment of hydrogen) can occur. Because of this lack of local enrichment, no embrittlement takes place. Dynamic tests, such as the tensile test in
hydrogen, cause local enrichment and hence cause embrittlement.
Such interpretations thus provide some explanation for the observed
importance of test technique and indicate the importance of testing in
a hydrogen environment. This fact has led some investigators to
classify hydrogen environment embrittlement as either internal hydrogen embrittlement or environmental hydrogen embrittlement. Hydrogen environment embrittlement is embrittlement which results when
an initially hydrogen-free metal is deformed in a hydrogen environment. This phenomenon was first described in the mid-1950s [53,54]
in studies which showed that ductility of many metals tested in high-
pressure gaseous hydrogen was significantly less than expected. Subsequent studies [55] indicated that hydrogen is supplied to the root of
the propagating crack, both from the metal and directly from the gas
phase. Other investigators [3,4] confirmed these results and suggested that hydrogen environment embrittlement is distinctly different from other embrittlement modes. This distinction probably developed because tests in high-pressure hydrogen were often found
to be independent of exposure time [21]. This independence was interpreted as showing that diffusion and hence absorption are not vital
to hydrogen environment effects. However, the fact that plastic deformation during hydrogen exposure greatly affects absorption, and
that both the depth of significant absorption and the amount of hydrogen absorbed are significantly greater than predicted by diffusion theory, probably discounts the need for a distinction between
hydrogen environment and internal hydrogen embrittlement. Absorption readily occurs during environmental embrittlement, and
there is little need to invoke a different environmental embrittlement mode for this phenomenon. Furthermore, the effects of temperature, strain rate, and hydrogen content (pressure) on susceptibility to hydrogen embrittlement have been shown to be the same
in both internal (precharged) and environmental tests [16]. These
results suggest that a single process accounts for both forms of embrittlement. Nevertheless, environmental hydrogen embrittlement is
as well-studied phenomenon.

Hydrogen environments also affect slow crack growth rates for
precracked specimens [56] and the fatigue life [2] of many metals.
Crack growth in H-11 steel has been shown to be a sensitive function
of environment. The stress intensity for unstable crack growth was

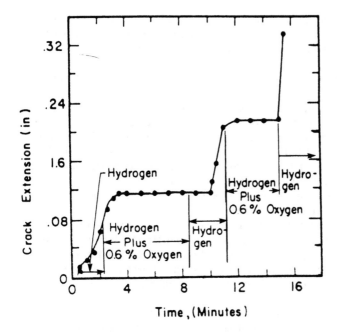

FIGURE 18 Subcritical crack growth in H-11 steel exposed to hydrogen and a hydrogen-0.6% oxygen environment. (From Ref. [56].)

∿11 kpsi/in. in hydrogen and was ∿40 kpsi/in. in dry argon. Furthermore, hydrogen-induced slow crack growth in compact tension specimens could be stopped if 0.6% oxygen was added to the atmospheric hydrogen environment (Fig. 18). This shows that oxygen is a remarkable protective agent against hydrogen-induced cracking. Other studies [3] have shown that oxygen impurities will inhibit cracking of steels in high-pressure hydrogen. The mechanism for this inhibition (or protection) is not established, but is probably related to oxygen inhibition of hydrogen absorption [50b].

7.4.2 Proposed Mechanisms for Generalized Embrittlement

The general mechanisms proposed to explain hydrogen embrittlement are numerous. However, many of the models are related, and mechanisms which have received considerable support include:

Precipitation of hydrogen as a gas at internal defects [27a]. The pressure developed by this precipitation is added to the applied stress and thus lowers the apparent fracture stress. Evidence to support this early theory continues to be developed, particularly for hydrogen-assisted cracking in hydrogen sulfide gas, where crack formation by pressurization involves hydrogen precipitation as molecular hydrogen at inclusion/matrix interfaces [61].

Interaction of dissolved hydrogen to reduce the cohesive strength of the lattice.

Adsorption of hydrogen to reduce the surface energy required to form a crack and thus lower the fracture stress.

Adsorption of hydrogen to increase the ease of dislocation motion or generation, or both. This idea differs in general from the previous models in that hydrogen is assumed to locally enhance plasticity rather than embrittle the lattice.

Formation of a hydrogen-rich phase whose mechanical properties differ from those of the matrix.

Association of hydrogen with dislocations to provide localized hydrogen accumulations and thereby embrittle the lattice. The dislocation-hydrogen association should not be considered as an embrittlement mechanism but is simply a method of accumulation of a high local hydrogen concentration.

Each of these proposed mechanisms was developed to explain experimental observations and/or industrial experience; thus, each is consistent with some portion of the available data for hydrogen effects on metals.

The internal pressure theory is probably the oldest, and until the mid 1960s it was the most widely accepted embrittlement model. The model assumes that supersaturated hydrogen atoms precipitate from solution as molecular hydrogen gas. This precipitation takes place at the interface between nonmetallic inclusions and the metal matrix. Many of the experimental observations used to support the pressure model were made on cathodically charged specimens and on samples quenched to room temperature after hydrogen exposures at elevated temperatures. In either case, the hydrogen fugacity in the charged sample corresponds to a very high hydrogen pressure. Thus, if local equilibrium is maintained and internal voids, microvoids, or interfaces exist, high hydrogen pressures will develop and can cause deformation and failure, even in the absence of an applied stress.

The pressure theory of embrittlement lost favor with the increased testing of precracked specimens in low-pressure hydrogen environments. For example, the statement that the occurrence of crack growth in low-pressure hydrogen is not compatible with a pressure mechanism of brittleness is typical of the criticisms of this model. However, whenever cathodic charging or hydrogen uptake through reactions with gases such as H_2S is possible, high hydrogen pressures can be developed as seen in Ref. [61].

Each of the above embrittlement models can be supported or criticized by selection of the data to be discussed [62]. Investigators advocating the hydrogen-induced loss in cohesion typically use precracked specimens tested under static loads [63], while other investigators advocating hydrogen-enhanced plasticity frequently use
cathodic charging tests [13] and/or tests in electron beams, which
ionize hydrogen atoms and thereby generate high hydrogen fugacities
at the specimen [64]. These high surface fugacities cause very sharp
hydrogen concentration gradients and can plastically deform the sample—even if hydrogen had no effect on dislocation motion. Such hydrogen-induced plastic deformation has been discussed by many investigators and is receiving increased attention. Other advocates of
embrittlement models that include dislocation transport of hydrogen
have conducted tests on smooth samples [50b]. This type of sample
maximizes the extent to which plastic deformation precedes failure,
thus increasing the probability of observing an effect of dislocation
motion. To minimize the importance of hydrogen-dislocation interactions, tests of precracked specimens under static load (or displacement) conditions are frequently used. Results of the fixed displacement tests indicate that although the presence of a gaseous hydrogen
environment may increase crack growth rates, hydrogen induced initiation of crack movement is difficult [62]. This difficulty may be
overcome by testing samples under static load rather than in fixed
crack opening displacement. The differences in results are probably associated with the fact that static load tests are generally conducted by loading the specimen while it is in a hydrogen atmosphere,
and the reverse is true in fixed displacement tests. The combined results of these two types of tests suggest that hydrogen may influence
the macroscopic (or microscopic) flow processes in metals and alloys.
To test this hypothesis, investigators have used torsional test samples because of the increased ease of dislocation motion. In a torsion
test the ratio of the maximum tensile to maximum shear stress is 1, as
opposed to simple tension, where the ratio is 2; thus the probability
for slip processes is enhanced by torsion testing.

The premise that hydrogen embrittlement is caused by localized
formation of a phase whose mechanical properties differ from those
of the matrix because of hydrogen enrichment is primarily supported
by experiments with hydride-forming alloys [17]. To generalize the
premise of high local hydrogen concentration to include embrittlement
of non-hydride-forming metals, results were obtained with cathodically charged samples, where forced precipitation of gaseous hydrogen, which is indeed a hydrogen-rich second phase, could occur.

These observations suggest that the mechanism under consideration frequently played an important role in the selection of test techniques for studying hydrogen embrittlement. The literature, however,

contains a wide variety of experimental observations that have been
reported by many investigators. A review of this literature indicates
that a universal hydrogen embrittlement mechanism is unlikely and
that any valid mechanism must explain, or at least be consistent with,
several observations obtained by a variety of tests.

Reported results from three types of tests, each used by a large
number of investigators, are generalized below to obtain criteria for
evaluating possible embrittlement mechanisms. The tests considered
were:

Tensile and delayed failure tests of precharged samples
Low strain rate tensile tests in gaseous hydrogen
Crack growth studies in precracked specimens exposed to hydro-
    gen

Results from tests that clearly involved hydride embrittlement, reaction, or hydrogen-induced blistering are not included in this generalization.

Delayed failure studies of precharged samples have shown that the
cracking process can be separated into nucleation and propagation
stages and that a critical applied stress is generally required to cause
failure. This was shown schematically in Fig. 11. The nucleation (or
incubation) process is reversible. Brittle fracture modes are typical
for samples broken by delayed failure, and there is little evidence of
macroscopic plastic strain preceding fracture. Many studies have
shown that temperature, pressure, and hydrogen segregation (e.g.,
[65]) are key factors in delayed failure processes.

Tensile tests of precharged samples show that the degree of embrittlement at any given temperature and strain rate increases with
increasing hydrogen content until, in some alloys, a maximum in embrittlement is observed. The test data also show that embrittlement
is temperature-dependent, often showing a maximum effect at the
temperature where the elongation is normally the greatest. The degree of embrittlement decreases with increasing strain rate. These
effects were illustrated in Fig. 14 and have been reproduced frequently.

Low strain rate studies of uncharged samples tested in gaseous
hydrogen have shown that the degree of embrittlement depends on
the pressure; however, several different functional relationships
have been observed (e.g., see [66]). Fractographic studies of a
variety of embrittled alloys have shown that both ductile and brittle-
type fracture modes are common. Stainless steels, for example, typically show ductile fracture even when significant hydrogen-induced
losses in ductility are observed, while nickel may fail in an apparently brittle-type mode with only minor losses in macroscopic ductility. Furthermore, the embrittlement observed during tensile tests

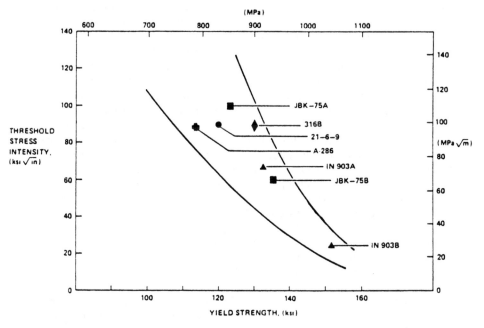

FIGURE 19 Effect of yield strength on threshold stress intensity for hydrogen cracking in austenitic alloys.

of hydrogen-charged nickel is minimized by the addition of a thoria dispersion, although this beneficial effect is not observed in studies of fatigue crack growth in hydrogen.

The potential for hydrogen embrittlement has been correlated with the dislocation substructure [72]. These studies indicate that co-planar dislocation motion enhances the susceptibility to hydrogen embrittlement. Other investigators, using a variety of alloys, have shown that for a given alloy or family of alloys, susceptibility to hydrogen cracking generally increases with increasing strength level [62]. This observation is even true when the threshold stress intensity for hydrogen-induced cracking was determined by delayed failure tests using Modified Wedge Opening Load specimens. Experimental correlations between K_{TH} and yield strength are shown in Fig. 19 and discussed in [42b].

Studies of the equilibrium aspects of hydrogen-induced cracking of steels demonstrated that a self-arrested crack in a MWOL specimen exposed to hydrogen could be made to propagate by simply increasing the hydrogen pressure. Conversely, it was also shown that for a propagating crack, a change in gas pressure was reflected by a change in crack rate [68]. Crack growth during fatigue in hydrogen is also pressure-dependent, and the dependence is not in general a simple function of pressure or applied stress intensity [69].

Crack growth under static loads at constant hydrogen pressures is discontinuous. Early studies involving monitoring of cathodically charged, delayed failure specimens showed sporadic changes in specimen electrical resistance [26]. Permeation studies with disk rupture specimens exposed to high-pressure hydrogen indicated discontinuous crack propagation [70]. Acoustic emission measurements also indicate an alternating cracking-incubation-cracking process [71].

Any hydrogen embrittlement mechanism applicable to a wide group of metals and alloys must be compatible with a range of observations and a variety of results. The following observations have been made by many investigators and therefore should be incorporated in or at least be compatible with any valid mechanism. These observations were first summarized in [62].

   Delayed failure in smooth, notched, and/or precracked specimens under conditions where little or no macroscopic strain precedes fracture
   Temperature and strain rate dependence during dynamic testing including the existence of embrittlement at temperatures as low as 77 K under impact conditions
   Hydrogen-induced ductility losses and slow crack growth with or without accompanying changes in fracture modes
   No single effect of pressure on embrittlement susceptibility
   Reinitiation of hydrogen-induced crack growth by small changes in gas pressure and no accompanying change in stress intensity
   Extensive embrittlement in uncharged, smooth bar samples tested in high-pressure hydrogen and increased embrittlement in notched bar samples
   Increases in the yield strength because of the absorption of hydrogen
   Inhibition of gas-phase hydrogen embrittlement by small quantities of impurities such as oxygen and water vapor

None of the currently proposed mechanisms as now formulated appears to be compatible with all the observations suggested. For example: (a) the gas pressure theory fails to explain reinitiation of hydrogen-induced cracking by small changes in pressure unless hydrogen absorption also increases the ease of plastic deformation; (b) the effect of hydrogen pressure on embrittlement suceptibility and the observation of ductile fracture are not consistent with the reduction of cohesive strength model; (c) the surface energy model does not explain how oxygen, which is more readily adsorbed than hydrogen and should therefore lower the surface energy more than hydrogen, inhibits embrittlement; (d) it is difficult to explain either the cleavage-like fracture or the observation of delayed failure by the hydrogen-induced ductility model; and (e) the formation of a hydrogen-rich phase is not likely to be inhibited by the presence of a dispersed phase. These considerations lead to the conclusion that two or more of the proposed mechanisms must be operative and that none of them is universally valid. Thus, either there is no universal embrittlement mechanism or a new theory must be developed.

In spite of this lack of a universally accepted theory for embrittlement, a phenomenological model for embrittlement may be developed. A potential model concludes that hydrogen embrittlement processes in metals and alloys can be understood by assuming that hydrogen absorption decreases the strength of various interfaces in the lattice. This decrease in strength will depend on the hydrogen content in the metal and the type of metallurgical interface. In general, the interfacial strength will decrease as the hydrogen content increases. This effect is shown in Fig. 20 for both grain boundaries and slip bands. For illustrative purposes the grain boundary strength is shown as higher than the slip band strength when the hydrogen content is low, but hydrogen is assumed to have more effect on the strength of the grain boundaries. Thus at a high hydrogen content grain boundaries are weaker than slip bands. Figure 20 also illustrates a potential hydrogen-induced strengthening of the material. This effect is not necessary but is consistent with studies which show that hydrogen absorption raises the flow stress of many metals and alloys.

For any metal or alloy represented by the illustration in Fig. 20, application of a load would cause yielding whenever the hydrogen content was less than that represented by the dashed line separating ductile fracture processes from mixed-mode fracture. This would be the case because the normal deformation processes of yielding, work hardening, and ductile fracture would occur at stresses lower than those necessary to cause hydrogen-induced fracture of slip bands or grain boundaries. (Clearly, hydrogen could affect particle-matrix interface strengths and thereby affect ductile rupture processes, but such effects are not discussed in this simplified model.) As the hydrogen content is increased beyond that represented by the first dashed line in Fig. 20, the failure mode is affected and hydrogen-induced fracture mode changes are observed. Whenever the hydrogen content is between the two vertical dashed lines, ductile rupture, slip band fracture, and grain boundary fracture may all occur. The increased hydrogen-induced reductions in interfacial strength make

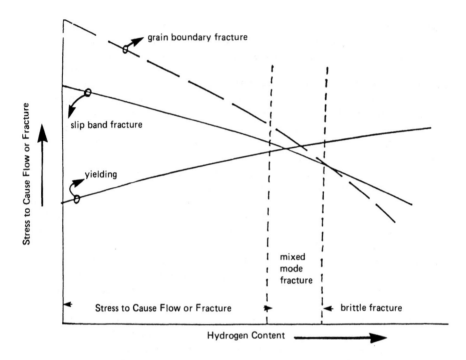

FIGURE 20 Sketch depicting effects of hydrogen on yield strength and interfacial strength of austenitic steels.

either slip band or grain boundary fracture probable at high hydrogen contents and ductile rupture processes are unlikely. This model also predicts several of the typical metallurgical effects on hydrogen embrittlement. For example, if the yield strength of the metal is increased by metallurgical processing, the yield curve would be shifted upward and would therefore intersect the slip band fracture at a lower hydrogen content. An increase in strength would therefore decrease the tolerance of a metal or alloy for hydrogen, a phenomenon often observed in hydrogen embrittlement testing.

The existence of delayed failure hydrogen embrittlement can also be explained with the proposed model. Assume that a sample is stressed so that the stress level, either the net section stress or the concentrated stress at a crack tip, is only slightly lower than the yield strength. The sample (or component in an industrial system) would support the stress. With time in service hydrogen would diffuse to the region of lattice dilation at the crack or flaw tip and the hydrogen content in that area would increase. (Service or test conditions could also increase the bulk hydrogen content.) Whenever the local hydrogen content exceeded the content represented by the interaction or the isostress line with one of the interfacial strength lines, hydrogen-induced fracture would result in that local area. This effect would cause discontinuous crack growth, as has been observed in many hydrogen embrittlement tests. Furthermore, thermomechanical treatments which lowered the interfacial strengths because of phenomena such as sulfur segregation or sensitization should also decrease the tolerance of the component for hydrogen and promote susceptibility to hydrogen embrittlement.

The relative values of yield strength and interfacial strength and the effects of hydrogen on both of these parameters will be important in determining the susceptibility of a metal or alloy to hydrogen embrittlement. These factors will be affected by grain size, temperature, dislocation substructure, particle size and distribution, and numerous other metallurgical factors. Because of these large-scale effects of strength and microstructure on hydrogen compatibility, selection of metals and alloys for hydrogen service is a difficult task.

7.5 SUMMARY

This chapter attempted to separate hydrogen embrittlement of metals into three categories:

Embrittlement resulting from blister formation
Embrittlement resulting from hydride formation
Generalized embrittlement

Clearly, hydrogen embrittlement is not a simple phenomenon, and many aspects of hydrogen-metal interactions remain unknown. However, within the above classifications, a basis for selection of materials for hydrogen service may be found. Only a small number of metals are hydride formers, and these should not be used in hydrogen atmospheres. Blister formation can be avoided by proper control of process variables, heat treating, and finishing operations. Selection of materials which are resistant to generalized hydrogen embrittlement is more difficult. In general, if an alloy is primarily a transition metal, the alloy should be assumed to be susceptible to embrittlement until proved otherwise. Specific material properties to minimize hydrogen embrittlement are not well established; however, low hydrogen-dislocation binding energies, high stacking fault energies, low strengths, and low hydrogen solubilities appear to be desirable.

REFERENCES

1. G. R. Caskey, G. R. Cole, and W. G. Holmes, Joint U.S.-Euratom Rept. CEND-152, Vol. II, pp. IV-E-1 to IV-E-23, 1962.
2. W. Beck, E. J. Jankowsky, and P. Fischer, *Hydrogen Stress Cracking of High Strength Steels*, NADC-MA-7140, Warminster, Pa., 1971.
3. R. P. Jewett, R. J. Walter, W. T. Chandler, and R. P. Frohmberg, *Hydrogen-Environment Embrittlement of Metals*, NASA CR-2163, Rocketdyne Division, North American Rockwell, Canoga Park, Calif., 1973.
4. C. G. Interrante, in *Hydrogen Problems in Steels*, American Society for Metals, Metals Park, Ohio, 1982, p. 3.
5. P. Cotterill, *Prog. Mater. Sci.*, *9*: 20 (1961).
6. J. D. Fast, *Interaction of Metals and Gases*, Academic, New York, 1965, p. 54.
7. B. A. Kolachev, *Hydrogen Embrittlement of Nonferrous Metals*, Israel Program for Scientific Translations, Jerusalem, 1968, p. 113.
8. J. D. Fast, *Interaction of Metals and Gases*, Academic, New York, 1965, p. 129.
9. R. I. Jaffee and N. E. Promisel, *The Science, Technology and Application of Titanium*, Pergamon, Elmsford, N.Y., 1970.
10. L. Christodoulou and H. M. Flower, in *Hydrogen Effects in Metals*, AIME, New York, 1980, p. 493.
11. R. L. Shamakian, A. R. Troiano, and R. F. Hehemann, in *Environment-Sensitive Fracture of Engineering Materials*, AIME, New York, 1979, p. 116.
12. A. J. Bursle and E. N. Pugh, in *Environment-Sensitive Fracture of Engineering Materials*, AIME, New York, 1979, p. 18.
13. R. A. Oriani, *Ber. Bunsenges. Phys. Chem.*, *76*: 848 (1972).
14. C. D. Beachen, *Metall. Trans.*, *3*: 437 (1972).
15. J. J. Gilman, paper presented at the Fifth Annual Spring Meeting, Institute of Metal Division, AIME, Philadelphia, May 29–June 1, 1973.
16. W. Pesch, T. Schober, and H. Wenzl, in *Hydrogen Effects in Metals*, AIME, New York, 1980, p. 923.
17. D. G. Westlake, *Trans. Am. Soc. Met.*, *62*: 1000 (1969).
18. R. Lagneborg, *J. Iron Steel Inst. London*, *207*: 363 (1969).
19. H. C. Rogers, *Science*, *159*: 1057 (1968).
20. I. M. Bernstein, *Mater. Sci. Eng.*, *6*: 1 (1970).
21. R. M. Vennet and G. S. Ansell, *Trans. Am. Soc. Met.*, *62*: 1007 (1969).
22. D. P. Williams and H. G. Nelson, *Metall. Trans.*, *2*: 1987 (1971).

23.   A. S. Tetelman, in *Proceedings of a Conference on the Fundamental Aspects of Stress Corrosion Cracking*, Ohio State Univ., Sept. 11—15, 1967, National Association of Corrosion Engineers, Houston, 1969, p. 446.
24.   N. J. Petch, *Philos. Mag.*, *1*: 331 (1956).
25.   D. P. Williams and H. G. Nelson, *Metall. Trans.*, *2*: 1987 (1971).
26.   A. R. Troiano, *Trans. Am. Soc. Met.*, *52*: 54 (1960).
27a. C. A. Zapffe and O. G. Specht, Jr., *Trans. Am. Soc. Met.*, *39*: 193 (1947).
27b. C. A. Edwards, *J. Iron Steel Inst.*, *110*: 9 (1924).
28.   *Current Solutions to Hydrogen Problems in Steels*, American Society for Metals, Metals Park, Ohio, 1982.
29a. F. N. Rhines and W. A. Anderson, *Trans. AIME*, *143*: 312 (1941).
29b. R. L. Klueh and W. W. Mullins, *Trans. AIME*, *242*: 237 (1968).
29c. L. C. Weiner, *Corrosion*, *17*: 137t (1961).
30a. *Hydrogen Damage*, American Society for Metals, Metals Park, Ohio, 1977, p. 391.
30b. P. J. Grobner and V. A. Biss, *Environmental Degradation of Materials in Hydrogen*, Virginia Polytechnic Inst., Blacksburg, 1981, p. 273.
31.   T. W. Wood and R. D. Daniels, *Trans. AIME*, *233*: 898 (1965).
32.   T. Enjo and T. Kuroda, in *Environmental Degradation of Materials in Hydrogen*, Virginia Polytechnic Inst., Blacksburg, 1981, p. 393.
33.   D. Hardie, *J. Nucl. Mater.*, *42*: 317 (1972).
34.   R. P. Marshall and M. R. Louthan, Jr., *Trans. Am. Soc. Met.*, *56*: 693 (1963).
35.   W. Evans and G. W. Parry, *Electrochem. Technol.*, *4*: 225 (1966).
36.   J. S. Bradbrook, G. W. Lorimer, and N. Ridley, *J. Nucl. Mater.*, *42*: 142 (1972).
37.   B. A. Kolachev, *Hydrogen Embrittlement of Nonferrous Metals*, Israel Program for Scientific Translations, Jerusalem, 1968, p. 42.
38.   J. C. Williams, in *Effects of Hydrogen on Behavior of Materials*, AIME, New York, 1976, p. 367.
39.   H. A. Wriedt and R. A. Oriani, *Acta Metall.*, *18*: 753 (1970).
40.   G. R. Caskey, Jr., *Hydrogen in Metals*, American Society for Metals, Metals Park, 1973, p. 465.
41a. O. Z. Rylski, *Prog. Mater. Sci.*, *9*: 20 (1961).
41b. D. O. Northwood, *Environment-Sensitive Fracture of Engineering Materials*, AIME, New York, 1979, p. 451.
41c. H. K. Birnbaum, *Environment-Sensitive Fracture of Engineering Materials*, AIME, New York, 1979, p. 327.
42a. A. R. Troiano, *Trans. Am. Soc. Met.*, *52*: 54 (1960).

42b. M. W. Perra, *Environmental Degradation of Engineering Ma-*
      terials in Hydrogen, Virginia Polytechnic Inst., Blacksburg,
      1979, p. 321.
43.   B. G. Johnson, *ASTM STP No. 345*, American Society for Test-
      ing and Materials (ASTM), Philadelphia, 1962.
44.   C. F. Barth and E. A. Steigerwald, *Metall. Trans.*, *1*: 3451
      (1970).
45.   R. A. Oriani, in *Proceedings of a Conference on Fundamental*
      Aspects of Stress Corrosion Cracking, Ohio State Univ., Sept.
      11-15, 1967, National Association of Corrosion Engineers,
      Houston, 1969, p. 32.
46.   T. Toh and W. M. Baldwin, Jr., in *Stress Corrosion Cracking*
      and Embrittlement (W. D. Robertson, ed.), Wiley, New York,
      1956, p. 176.
47.   M. Cornet, W. Raczynski, and S. Talbot-Besnard, *Mem. Sci.*
      Rev. Metall., *69*: 27 (1972).
48.   G. C. Smith, *Hydrogen in Metals*, American Society for Metals,
      Metals Park, Ohio, 1973, p. 485.
49.   J. A. Donovan, *Metall. Trans. A.*, *7B*: 1677 (1976).
50a. J. K. Tien, *Effect of Hydrogen on Behavior of Materials*, AIME,
      New York, 1976, p. 309.
50b. M. R. Louthan, Jr., G. R. Caskey, Jr., J. A. Donovan, and
      D. E. Rawl, Jr., *Mater. Sci. Eng.*, *10*: 357 (1972).
51.   H. E. McCoy, Jr., Effects of Hydrogen on the High Tempera-
      ture Flow and Fracture Characteristics of Metals, USAEC Rept.
      ORNL-3600, Oak Ridge National Laboratory, Oak Ridge, Tenn.,
      1964.
52.   A. W. Thompson and B. A. Wilcox, *Scr. Metall.*, *6*: 689 (1972).
53.   D. D. Perlmutter and B. F. Dodge, *Ind. Eng. Chem.*, *48*: 885
      (1956).
54.   H. C. Van Ness and B. F. Dodge, *Chem. Eng. Prog.*, *51*: 266
      (1955).
55.   R. H. Cavett and H. C. Van Ness, *Weld. J.*, *42*: 316 (1963).
56.   H. H. Johnson, in *Proceedings of Conference Fundamental As-*
      pects of Stress Corrosion Cracking, Ohio State Univ., Sept.
      11-15, 1967, National Association of Corrosion Engineers,
      Houston, 1969, p. 439.
57.   M. Hansen, *Constitution of Binary Alloys*, Second Ed., McGraw-
      Hill, New York, 1958.
58.   M. R. Louthan, Jr., *Trans. Am. Soc. Met.*, *57*: 1004 (1964).
59.   D. Eliezer et al., *Metall. Trans. A, 10A*: 935 (1979).
60.   W. Chu, J. Yao, and C. Hsiao, *Metall. Trans. A. 15A*: 729
      (1984).
61.   T. V. Venkatasubramanian and T. J. Baker, *Met. Sci.*, *18*: 241
      (1984).

62. M. R. Louthan, Jr., and R. P. McNitt, in *Effect of Hydrogen on Behavior of Materials* (A. W. Thompson and I. M. Bernstein, eds.), TMS-AIME, New York, 1976, p. 496.

63. R. A. Oriani and P. H. Josephic, *Acta Metall.*, *22*: 1065 (1974).

64. O. A. Onyewuenyi and J. P. Hirth, *Scr. Met.*, *15*: 113 (1981).

65. M. Gao, M. Lu, and R. P. Wei, *Metall. Trans. A*, *15A*: 735 (1984).

66. D. P. Williams and H. G. Nelson, *Metall. Trans.*, *1*: 63 (1970).

67. A. J. West, Jr., and M. R. Louthan, Jr., *Metall. Trans. A*, *13A*: 2049 (1982).

68. R. A. Oriani, in *Environmental Degradation of Engineering Materials in Hydrogen* (M. R. Louthan, Jr., R. P. McNitt, and R. D. Sisson, eds.), Virginia Polytechnic Inst., Blacksburg, 1981, p. 3.

69. N. R. Moody, An Evaluation of the Current Status of Hydrogen Embrittlement and Stress Corrosion Cracking in Steels, SAND 81-8259, Sandia National Laboratories, Livermore, Calif., 1981.

70. J. P. Fidelle, in *Hydrogen Problems in Steels* (C. G. Interrante and G. M. Pressouyre, eds.), American Society for Metals, Metals Park, Ohio, 1982, p. 449; H. Bartheleny, J. Bryselbout, and C. Bambe, *ibid.*, p. 383.

71. R. B. Clough and H. N. Wadley, *Metall. Trans. A*, *13A*: 1965 (1982).

72. A. J. West and M. R. Louthan, Jr., *Metall. Trans. A. 10A*: 1675 (1979).

8

CORROSION FATIGUE

DAVID J. DUQUETTE

Materials Engineering Department, Rensselaer Polytechnic Institute, Troy, New York

8.1 INTRODUCTION

Simultaneous exposure of metals and alloys to aggressive environments and cyclic stresses or strains generally results in a degradation of fatigue resistance. Even ambient, relatively unaggressive environments such as dry air at room temperature have been shown to increase crack propagation rates (when compared to experiments performed in vacuum) in a large number of metals and alloys. Addition of water vapor to the environment has been shown to further increase crack propagation rates. A large body of published literature is available on the specific effects of environment/fatigue interactions and much of the data has been summarized in reviews [1,2]. The purpose of this chapter is to present a few of the highlights of some of the past work, to discuss previously proposed mechanisms for crack initiation and propagation, and to propose several possible explanations for corrosion fatigue failures.

8.2 GENERAL FATIGUE MECHANISMS

Fatigue processes in metals and alloys can be categorized in several different ways, depending on the point of view of the scientist or engineer. For example, phenomenologically, the subject can be divided into precrack deformation, crack initiation, and crack propagation. There is certainly a continuum in behavior between these processes, although they are often treated independently. Another classification is related to the amount of plastic vs. elastic strain which a component experiences in service. If the stress level is such that the yield strength is exceeded, or if plastic strains are encountered (or controlled), the phenomenon is often referred to as low-cycle fatigue (LCF). On the other hand, if only elastic strains are experienced, the phenomenon is generally labeled high-cycle fatigue (HCF). These classifications are not precise, since for ductile pure metals plastic strains must be imposed to cause fatigue failures, although failure may require many millions of stress or strain cycles. Accordingly, the division between low-cycle fatigue and high-cycle fatigue is some-times arbitrarily set as 10^5 cycles to failure.

Still another classification of fatigue behavior is a comparison be-tween the number of cycles to failure for smooth surface geometries (or machined notches) and for crack propagation from well-charac-terized defect geometries such as precracks. The latter classifica-tion is related to the application of fracture mechanics to cyclically loaded materials and is characterized by the stress or strain intensi-ties of defects.

Addressing the first classification mode, it has been well estab-lished that, for pure single-phase metals and alloys, precrack defor-mation requires a considerable amount of plastic deformation, which results in a metastable dislocation substructure. For annealed ma-terials, cyclic hardening occurs, reaching a plateau or "saturation" in stress or strain (depending on deformation mode control) early in fatigue life (typically 1—10%). For work-hardened materials, cyclic softening is observed and a plateau in stress or strain is also ob-served after only a fraction of fatigue life. At the onset of this plateau region, the metastable dislocation substructure is locally al-tered, and fatigue-related slip bands form. These are called *persis-tent* slip bands (or PSBs) to distinguish them from the slip bands which are formed when the initial stress or strain is applied. It was noted in early experiments that, if a metal was electropolished after the plateau region was realized, only the fatigue-related slip bands reappeared. Thus they were termed *persistent*. Continued cycling in the plateau region often leads to the development of *intrusions* and *extrusions* and a notch-peak topography in the surface of the metal. Crack nucleation is generally associated with this geometry and grows into the PSBs in a crystallographic direction. This mode of cracking is generally called stage I to distinguish it from later crack propaga-tion, which nominally occurs normal to the stress or strain direction, is generally not macroscopically crystallographically oriented, and is called stage II. In single crystals stage I cracking can, in some cases, dominate the entire fatigue crack propagation process. However, in polycrystals it is generally limited to one to three grains at the metal surface and is sometimes virtually not observable.

For many engineering alloys and for polyphase materials, macro-scopic plastic deformation is not a requirement for fatigue crack ini-tiation. Precrack deformation and crack initiation in these instances generally occur at strain localizations such as grain boundaries, inclusions, and second-phase particles. In these cases PSBs and intrusions/extrusions are generally not observed, and crack initiation appears to occur on undeformed surfaces when examined at low magnifications. Transmission electron microscopy experiments have shown, however, that PSBs are, in fact, found locally and that the regions of crack initiation are quite similar to those observed for pure single-phase metals.

The driving force for the transition from stage I to stage II cracking is not well understood, but is probably related to the availability of secondary slip systems as stress and strain are concentrated at the tip of a growing stage I crack.

The distinction between LCF and HCF, as indicated, is not well defined, and some investigations have suggested that LCF experiments, where cyclic strain rather than cyclic stress is controlled, can be used to model the deformed material at the leading edge of a growing stage II crack. In most engineering applications, however, metals and alloys are not deformed beyond their elastic limits and LCF is accordingly not often observed. An exception to this statement is the problem of elevated-temperature fatigue, where thermal cycling may lead to significant plastic strain ranges. Low-cycle fatigue (as defined in the engineering sense of less than 10^5 cycles to failure) is not generally considered to be significantly affected by environment and will not be treated in this chapter.

The concept of applying concepts of fracture mechanics to fatigue problems has received a considerable amount of attention during the past 20–25 years, and they have been used widely to examine environmental effects on crack growth. It has been well established that, for materials which have appropriate geometries and loading conditions, a crack leading edge is dominated by plane strain conditions, and a stress intensity factor K can be related to the applied stress and crack (or defect) length and geometry:

$$K = C \sigma \sqrt{a}$$

where C is a constant which is a function of crack and/or specimen geometry, σ is the nominal applied stress level, and a is the crack or defect length. This relationship assumes that the crack tip region is essentially only elastically loaded, which appears to be a good approximation for many engineering alloys of reasonable toughness. Under fatigue loading conditions, the driving force for crack growth is the cyclic stress range and the equation can be rewritten as

$$\Delta K = C \, \Delta \sigma \sqrt{a}$$

where only tensile components of the applied stress are considered since only those components open the crack. In the absence of environmental considerations, a very large number of alloys obey the relationship

$$\frac{da}{dN} = C' \, \Delta K^n$$

where da/dN is crack growth per cycle, n is a constant, which is usually between 2 and 4, and C' is a material-sensitive constant. In this form the equation is generally referred to as the Paris law. There have been many attempts to derive the exponent from first principles, but it is still primarily an empirical constant, determined by experimentation. Mildly corrosive environments may affect either C' or n, but in aggressive environments the relationships between the stress intensity range and crack propagation rates are not well characterized.

The following sections address the effects of environment or the various processes which define fatigue crack initiation and early growth.

8.3 ENVIRONMENTAL CONSIDERATIONS

Environments which affect fatigue behavior span the range from apparently benign to severely corrosive. For example, it has been shown that a few parts of H_2 per million in a vacuum of 10^{-8} torr can result in significant increases in grain boundary crack propagation rates in bicrystals of nickel. At the other end of the spectrum, cyclic stressing of alloys in environments which lead to stress corrosion cracking under static loading conditions can result in crack propagation rates several orders of magnitude larger than simple stress corrosion.

In general, research in corrosion fatigue has concentrated on two principal environmental regimes: gaseous and aqueous. An attempt has been made to bring these two regimes together in cases where hydrogen-assisted cracking has been identified as the cause of increased crack growth.

With few exceptions, the more aggressive a given environment, the lower is the corrosion fatigue resistance. While stress corrosion environments certainly lead to enhanced corrosion fatigue crack growth, virtually all corrosive environments also result in enhanced crack growth. Since the corrosion fatigue phenomenon is generally (but not universally) thought to be different for gaseous and aqueous environments, the two problems are addressed separately in this chapter.

8.4 MECHANISMS OF CORROSION FATIGUE IN GASEOUS ENVIRONMENTS

Metals and alloys exposed to gaseous environments such as oxygen or water vapor generally show accelerated fatigue crack growth when compared to experiments conducted in vacuo. Inert gases tend to show behavior similar to that observed in vacuo. Figure 1 shows the effects of air, oxygen, and water vapor on fatigue crack propagation in a high-strength Al alloy. The partial pressure of the gaseous environment is also an important variable for corrosion fatigue behavior. Figure 2 shows typical S-shaped curves for lead in air and in various levels of vacuum. It shows that there appears to be a critical partial pressure of O_2 below which fatigue resistance is unaffected by environment, as well as an upper partial pressure above which increasing partial pressures of the gaseous species do not further reduce fatigue resistance. This behavior may be explained by considering the rate of surface coverage of a growing crack compared to the generation of new surface at the crack tip. A mathematical model has been developed based on kinetic gas theory [5], where the critical pressure is expressed as

$$P_c = \frac{3}{8}\left(\frac{1}{r}\right)^2 \frac{(MT)^{1/2}}{(3.5 \times 10^{22})\, At} \tag{1}$$

where

P_c = critical pressure
$1/r$ = ratio of crack length to width
A = cross-sectional area of the impinging molecule
M = molecular weight of the adsorbate
T = absolute temperature
t = time required to form a monolayer of adsorbate on the freshly formed surface

Results with this type of formulation have not been entirely successful since the model is essentially a steady-state one; discrepancies of three orders of magnitude have been observed.

If, however, corrections are made for the fact that a fatigue crack is open for only part of the loading cycle, the model appears to be more reasonable [5]. For example, the time of exposure for each increment of crack growth is:

$$t = \frac{X}{2(da/dN)f}\ \sec \tag{2}$$

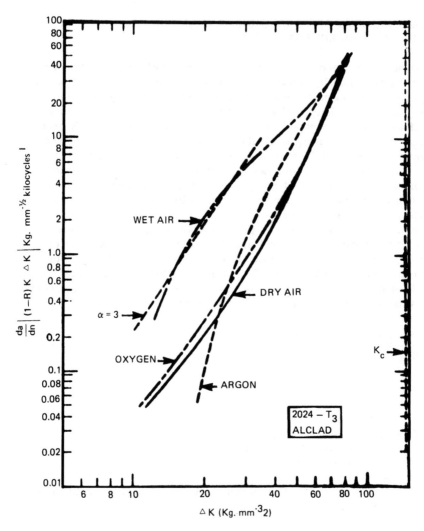

FIGURE 1 Fatigue crack propagation of 2024-T3 alloy in wet air, dry air, argon, and oxygen [3].

where

X	= interatomic spacing $\simeq 3$ Å
da/dN	= crack propagation rate
f	= frequency of applied stress

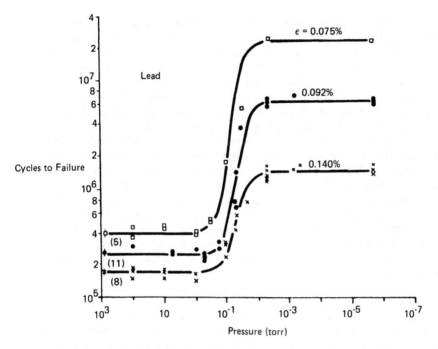

FIGURE 2 Typical S-shaped curve showing the effect of oxygen partial pressure on fatigue life [4].

For metals, the density of atoms in newly created surface $\simeq 1.5 \times 10^{15}$ cm^{-2}, and from kinetic gas theory the impingement rate is

$$\nu = \frac{3.5 \times 10^{22} p \text{ sec}^{-1}}{1.5 \times 10^{15} (MT)^{1/2}} \tag{3}$$

where p is gas pressure and M and T have been defined. Thus, saturation of the fresh surface occurs when

$$t = \frac{1}{\nu} \tag{4}$$

or

$$P_c = 2 \times 10^2 \left(\frac{da}{dN}\right) f \text{ torr} \tag{5}$$

an equation which agrees well with published data for the critical pressure P_c [6].

The specific mechanisms for how the gaseous species act to reduce fatigue resistance are generally divided into several categories: (a) interference with otherwise reversible slip, (b) prevention of slip band crack rewelding, (c) surface energy reduction at the crack tip due to gas-phase adsorption, and (d) bulk oxide interference with slip processes. There has been at least one suggestion that the transition from slip to stage I cracking is associated with oxygen which is dissolved in developing slip bands and thus prevents re-welding of crack surfaces [7]. However, noble metals such as gold show no effect of oxygen, suggesting that chemical interaction with the metal or alloy is required for accelerated crack growth [8]. Ex-periments conducted on Al, Cu, and Ni showed that the atmospheric effect decreases with increasing crack length, suggesting that a re-welding mechanism could not explain delayed cracking in vacuum, since higher local stresses and increased cold work should intensify rewelding effects. It was also noted that, for an Al alloy, N_2 was as effective as vacuum in preventing accelerated crack growth, implying that neither interference with a rewelding process (which would be accomplished by physisorption) nor gas phase adsorption could be an acceptable model [9].

Bulk oxide interference with slip processes has also been sug-gested to explain the effect of gaseous environmental fatigue. Fujita [10], for example, suggested that a thin adherent oxide, rapidly formed on newly emerging slip steps and at the crack tip, reduces the degree of reversibility, thus accelerating crack growth. Pelloux [11] also noted that aluminum alloys tested in vacuum do not show the striations associated with stage II crack growth normally observed when specimens are cyclically stressed in air (Fig. 3a) and proposed that completely reversible slip must occur in vacuum. In air, on the other hand, the fracture surface oxidizes, striations are developed due to irreversible slip, and crack growth is accelerated (Fig. 3b). Meyn [12] also observed the absence of striations and slip markings in high-strength aluminum alloys tested in vacuum.

Grosskreutz [13,14] showed that the slip character of aluminum is markedly altered in vacuum and associated this observation with changes in properties of the oxide film formed on aluminum. At 10^{-9} torr formation of slip bands was markedly suppressed compared to slip band formation in air. Numerous dislocation dipoles were observed in the near-surface region, and the effect of atmosphere was associated with oxidation of emerging surface slip steps leading to a local "work-hardening" of surface dislocation sources, thus reducing or prevent-ing slip band reversibility and accordingly leading to delayed crack initiation in vacuum. An extension of this model was also used to ex-plain accelerated crack growth. Later experiments to further under-stand this effect, however, showed that the mechanical properties of the oxide film are markedly increased in vacuum, the Young's modulus

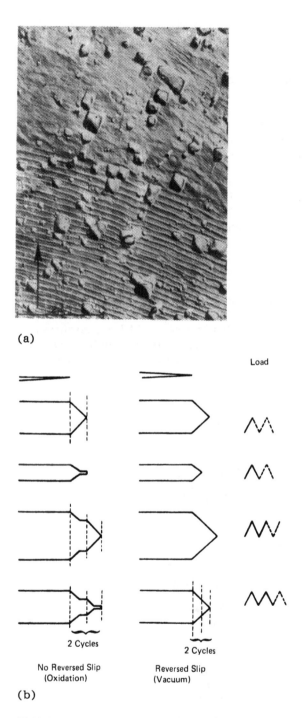

(a)

Load

2 Cycles

2 Cycles

No Reversed Slip
(Oxidation)

Reversed Slip
(Vacuum)

(b)

FIGURE 3 a) Fracture surface of 2024-T3 Al alloy tested in air (lower)
and vacuum (upper). Etch pits indicate crack growth on a (100) plane
in a <110> direction. b) Model to explain effects observed in a) [11].

of the thin oxide film being increased by a factor of 4 [15]. Thus the stronger preexisting film observed in vacuum is responsible for the reduction of surface slip and, conversely, the weaker film in air is easily ruptured by emerging slip steps and crack initiation and propagation are accordingly accelerated. The change in modulus and strength of the oxide film was associated with absorbed water vapor from the ambient environment, aluminum oxide being highly hygroscopic.

A similar mechanism based on a strong oxide film to explain accelerated crack initiation and growth in an aluminum alloy was postulated by Shen et al. [16]. Noting that the number of cycles to failure in an 1100 aluminum alloy was independent of partial pressure of oxygen above 3×10^{-2} torr and below 10^{-2} torr, these investigators suggested that surface regions are strengthened by oxide films. Specifically, surface films prevent dislocation escape through the alloy free surface and accordingly lead to large accumulations of dislocation debris in near-surface regions. The formation of cavities and voids (as suggested by Wood et al. [17] to explain fatigue crack growth) is thus enhanced and crack propagation rates increase. At low pressures, oxidation of newly created surfaces is slower, dislocation escape from the surface is more common, and cavity formation and void linkage are delayed (Fig. 4).

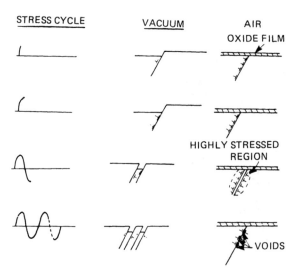

FIGURE 4 Model of void nucleation under oxide films to accelerate crack initiation in gaseous environments [16].

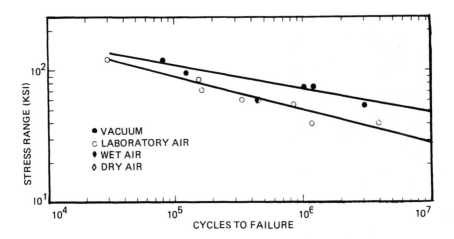

FIGURE 5 Effect of vacuum on the fatigue life of nickel-base super-alloy single crystal [18].

The majority of metals and alloys considered in environmental studies fail predominantly by stage II fatigue cracking (cracking normal to the stress axis) with a large degree of associated plasticity. These materials generally show converging fatigue resistance with lower applied stresses. Some authors, however, have suggested that the stage I mode of cracking (along slip bands) is more sensitive to environmental effects [7,9]. Since stage I cracks extend only a few micrometers or a grain diameter from the surface in most materials, this conclusion has been difficult to support. It has been shown that extensive stage I fractures occur in high-strength nickel alloy single crystals because of the highly planar nature of plastic deformation in that class of alloys, and, significantly, single crystals of these alloys show marked differences in fatigue life and diverging rather than converging fatigue resistance at lower applied stresses in air than in vacuum (Fig. 5).

Apparent localized cleavage occurs at the crack tip in low stress range tests in air, while ductile (dimpled) rupture occurs in vacuum and at high strains (Fig. 6). One approach to understanding this behavior is that of the Kelly et al. [19] criterion for whether cracking in crystalline materials will occur in a ductile or a brittle manner. These authors proposed that cleavage will occur if the ratio R of the normal stress to the applied shear stress at the crack tip, σ/τ, is greater than the ratio of the theoretical fracture stress to the theoretical shear stress, σ_{th}/τ_{th}, for the perfect crystal of the material. In general, $R > \sigma_{th}/\tau_{th}$ for body-centered cubic (bcc) metals, alkali

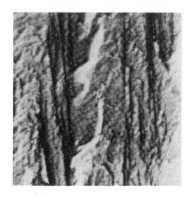

(a)

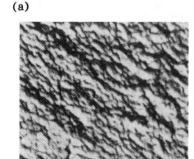

(b)

FIGURE 6 Effect of air (a) and vacuum (b) on stage I fatigue fracture surface appearance of a single-crystal nickel-base alloy [18].

halides, and diamond, while $R < \sigma_{th}/\tau_{th}$ for face-centered cubic (fcc)
metals, which is in agreement with the relative tendency of these materials to cleave. The buildup of edge dislocation dipoles observed in
the case of superalloy single crystals can be viewed as increasing σ,
and the lowering of the surface energy by oxygen adsorption as re-
ducing σ_{th}, both effects tending to make $R > \sigma_{th}/\tau_{th}$ and thus promoting local cleavage.

Since stage I fracture in nickel-base superalloys and in other alloy systems appears to involve a type of localized cyclic cleavage, it
is also possible to extend the Griffith-Orowan criterion to this situation. In addition to the applied stress σ_a, there is a local stress σ_p

near the crack tip resulting from an appropriate distribution of dislocations or other defects in the plastic zone. The magnitude of σ_p
will decrease with distance from the crack tip in accordance with a
reduced dislocation density in going from the crack tip to the edge
of the plastic zone. In order to consider the effect of this stress in
a modified Griffith-Orowan criterion, we define an equivalent stress
σ_e applied at infinity that gives the same strain energy release during crack growth as that provided by σ_p near the crack tip. The
relevant equation is then

$$\sigma_f = \sigma_a + \sigma_e = \frac{C\sqrt{E\left(\gamma_s - \gamma_{ads} + \gamma_p\right)}}{(1 - \nu^2)\pi L} \tag{6}$$

where $-\gamma_{ads}$ is the lowering of surface energy due to gas adsorption.
For the majority of high-cycle fatigue experiments, σ_a may be considered fixed by the maximum stress in the cycle and σ_p is a local
enhancement of stress at the crack tip. For nickel-base alloys this
stress enhancement has been attributed to an array of dislocation dipoles in the crack plane [18]. As a first approximation $\sigma_a + \sigma_e$ may
be considered constant in each cycle prior to crack extension. Once
a crack begins to propagate, it runs into a region of lower σ_p (and
σ_e) as a result of the gradient in defect structure with distance from
the crack tip and thus comes to a stop in each cycle.

When σ_a is low, $\gamma_p \to 0$ and γ_{ads} can be a significant fraction of
$\gamma_s + \gamma_p$. In support of this model, convergence of fatigue curves is
observed at high stress levels and fracture surfaces appear dimpled
independent of environment.

This model may be used to consider both stage I and stage II fracture in other materials where localized cleavage occurs at the crack
tip. Slip plane fractures that are highly reflective and exhibit fracture steps and river lines have also been observed in a number of Al
alloys that exhibit planar slip [20,21]. In support of this model, it
is interesting to note that nickel-base superalloy single crystals cycled
at ultrasonic frequencies in air also show increased fatigue lives and
dimpled fracture surfaces [22]. At these test frequencies, where failure occurs in a matter of minutes, oxygen presumably adsorbs at a
lower rate than the crack growth rate and the fatigue crack accordingly propagates in a virtual vacuum.

Although this model appears to be sound where limited ductility is
observed, such as is presumably the case for stage I microcracks in
the majority of materials and for both stage I and stage II cracks in
high-strength precipitation-hardened alloys, it seems unlikely that it
would be valid for stage II cracks in more ductile materials. The criticism of a surface energy reduction mechanism in these cases generally
hinges on the fact that the strain energy associated with large plastic

zones in the vicinity of growing cracks should be orders of magnitude greater than the amount of energy reduction attributed to surface adsorption. This criticism would appear to be well founded and, based on the conflicting observations of a number of investigators, it must be concluded that no general mechanism to explain the effect of environment on stage II crack propagation in ductile metals and alloys can yet be accepted unequivocally. Perhaps the most promising models are those which suggest interference with slip reversibility due to a thin corrosion product interacting with emerging slip bands.

8.5 MECHANISMS OF CORROSION FATIGUE IN AQUEOUS ENVIRONMENTS

Theories of aqueous corrosion fatigue have generally relied on one or more of the following mechanisms: (a) stress concentration at the base of hemispherical pits created by the corrosive medium, (b) electrochemical attack at plastically deformed areas of metal with non-deformed metal acting as cathode, (c) electrochemical attack at ruptures in an otherwise protective surface film, and (d) lowering of surface energy of the metal due to environmental adsorption and increased propagation of microcracks.

Early investigators of corrosion fatigue [23,24] favored the stress-concentration pit theory. Their conclusions were based on physical examination of failed specimens, which revealed a number of very large cracks originating at large hemispherical pits at the metal surface.

Pit formation in metals and alloys in aggressive environments undoubtedly does lead to a reduction in fatigue life. However, it is important to note that the corrosion fatigue phenomenon also occurs in environments where pitting does not occur. For example, low-carbon steels are highly susceptible to corrosion fatigue in acid solutions where pits are not observed [25,26]. In addition, reduced fatigue lives can be induced in steel specimens by the application of small anodic currents in deaerated solutions where pits do not form. Fatigue tests performed in 3% NaCl + NaOH solution of pH 12, where only a few randomly distributed pits were observed, showed fatigue limits identical with those observed in air [27]. Results of this kind are perhaps not unexpected since corrosion-induced pits tend to be hemispherical in nature and the stress intensity factor associated with surface-connected hemispherical defects is not large.

In order to examine the corrosion fatigue crack initiation process, low-carbon steels were sectioned and examined metallographically. Although some hemispherical pits were observed in the specimen surface, no cracking could be attributed to their presence. Rather, accelerated corrosion of initiated stage I cracks was noted, with a deep

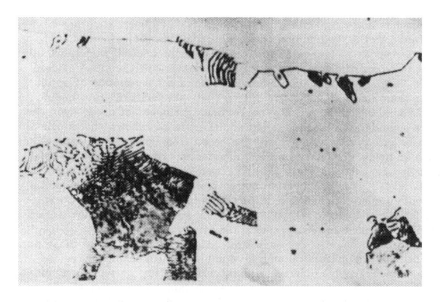

FIGURE 7 Crystallographic pitting (growing) of cyclically deformed steel in 3% NaCl at 4% of total life. Note parallel pits in individual grains [28].

"pitlike" configuration being oriented at approximately 45° to the specimen surface (Fig. 7). No fatigue cracks were observed emanating from these pits, and examination of specimens cycled for longer periods showed that the extent of growth of initiated fatigue cracks was always equivalent to "pit" depth, with no "normal" fatigue cracks associated with pits. It may be concluded, then, that in many cases the pits observed at failure by previous observers are not the cause of corrosion fatigue cracking but rather the result [28].

On the basis of a series of corrosion fatigue experiments performed on cold-worked and annealed steel wires, Whitwham and Evans [29] suggested that failure is due to distorted metal acting as anode with undistorted metal acting as cathode, with very fine cracks advancing by a combination of electrochemical and mechanical action.

Surface film rupture as the principal cause of the corrosion fatigue phenomenon of steels has also been proposed [30]. Simnad and Evans [31,32] suggested that film rupture might be important in neutral solutions but that structural changes in the metal predominate in acid solutions (distorted metal as anode). The electrode potential of a steel has been observed to become more active in fatigue tests, with a higher rate of change being noted at higher stresses. This potential drop continues throughout a particular alternating stress experiment, but reaches a steady state in static tests. This observation has been attributed to the destruction of a protective film [33].

Experiments on low-carbon steels in chloride showed that there appears to be a critical corrosion rate associated with the onset of corrosion fatigue and that this corrosion rate is not a function of applied stress level [27]. It has been shown that this "critical" rate has no fundamental meaning, since general corrosion does not occur over the entire specimen surface [34]. In addition, it was shown that steels could be effectively cathodically protected from corrosion fatigue either above or below the fatigue limit at a potential corresponding to that normally observed for cathodic protection independent of the applied stress level. This observation suggests that there is no fundamental thermodynamic shift in the equilibrium potential of the iron. Corrosion fatigue was also observed to occur in acid solutions where adherent films are unstable, suggesting that film rupture cannot be accepted as a general mechanism for corrosion fatigue. In addition to these data obtained for the effects of corrosive environments on fatigue resistance, metallographic observations of emerging slip patterns are shown in Fig. 8. In air, metallographic cross sections obtained at 4% of total life show small intrusion-extrusion pairs. Under conditions of active corrosion, the intrusion-extrusion pairs appear to be larger and broader. These observations suggest that there is a fundamental interaction between emerging slip bands and the corrosive environment and that the slip bands (which normally became stage I cracks) are preferentially corroded. In fact, extensive corrosion product was associated with areas of maximum stress under cantilever bend loading. Regions of low stress were essentially uncorroded, indicating a type of sacrificial anode cathodic protection.

To examine these phenomena in more detail, experiments were performed on polycrystalline pure copper, which does not exhibit a reduction in fatigue resistance under free corrosion conditions. Figure 9 shows that increasing corrosion rates, by applying anodic currents, results in a reduction of fatigue resistance in a manner similar to that observed for steels, with a "critical" corrosion rate also being observed. (At very high corrosion rates, there is a reversal in fatigue resistance due to rapid blunting of nascent cracks.) Examination of the free surfaces showed that, while slip band crack initiation was observed in air, free corrosion resulted in mixed transgranular-intergranular cracking and applied anodic currents resulted in totally intergranular failures [35].

In addition, slip bands were shown to be preferentially corroded, and there appeared to be an increase in both the magnitude and density of emerging slip bands. These data indicate that it is the local rather than the general corrosion rate which is critical. Figure 10 shows metallographic cross sections of the copper and clearly shows

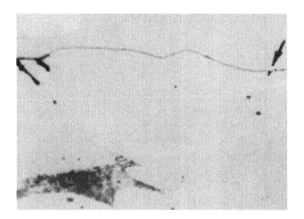

(a)

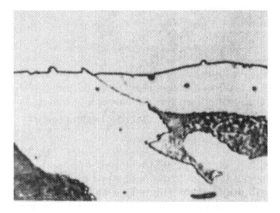

(b)

FIGURE 8 Surface of low-carbon steel cyclically stressed to 4% of total life in: a) air and b) aerated NaCl solution, showing comparative
development of intrusions and extrusions. Dark lines at left of air
specimen are stage I fatigue cracks [28].

the intergranular nature of the corrosion fatigue cracking and the increase in slip offset height under corrosion fatigue conditions [38].

Corrosion fatigue experiments conducted on single crystals of identical orientation showed an increase in corrosion fatigue resistance under conditions of applied current (Table 1) and a distinct increase in
the magnitude of slip step offsets (Table 2).

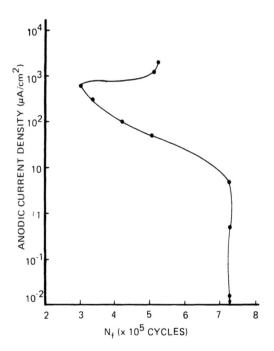

FIGURE 9 Effect of applied anodic currents on fatigue resistance of copper in chloride solutions [36].

The character of the slip offsets is also affected, with clusters of persistent slip bands occurring in air and a more uniform distribution of active dissolution (Fig. 11) [34].

Analysis of the results obtained for steels and copper indicates that corrosion fatigue crack initiation occurs by two steps: (a) persistent slip bands are preferentially attacked, leading to stress intensification and subsequent early crack propagation, and (b) corrosion results in the formation of numerous persistent slip bands, which produce numerous crack initiation sites. The latter presumably occurs because metal atoms associated with mobile dislocations in the persistent slip bands are more active than surrounding metal atoms, leading to strain relief in the surface and significant increases in surface deformation. A simple model of this process is shown schematically in Fig. 12. Rollins and Pyle [37,38] have also observed active current spikes associated with slip band emergence in passive stainless steels, with subsequent repassivation occurring if strains are held constant. Current spikes are observed in both tension and compression modes

TABLE 1 Copper Single Crystal Fatigue Behavior Versus Environment

Environment	Applied σ	Resolved shear stress	N_f (cycles)	Crack initiation (% N_f)
Air	65.5 MN/m^2 (9.5 ksi)	30.5 MN/m^2 (4.4 ksi)	1.4×10^6	2–4
0.5 N NaCl 100 μA/cm^2	65.6 MN/m^2 (9.5 ksi)	30.5 MN/m^2 (4.4 ksi)	4.8×10^6	25–30

mode of deformation, although the effect is greater in the tensile half of the cycle.

The observation that polycrystalline copper fails in an intergranular manner under corrosion fatigue conditions suggests that the enhanced deformation associated with grain boundaries causes enhanced preferential dissolution of the grain boundaries relative to emerging slip bands. Single crystals of copper show improved corrosion fatigue resistance due to crack blunting. In steel, on the other hand, preferential grain boundary attack is not observed and corrosion fatigue failures are accordingly transgranular.

The fatigue resistance of high-strength aluminum alloys is also severely affected by corrosive solutions, particularly chloride solutions, and this behavior has generally been attributed to either preferential

TABLE 2 Copper Single Crystal Slip Offset Heights and Slip Band Density

Environment	Number of cycles	Slip offset height (Å, ±150)	Slip band density (cm^{-1})
Air	10^4	800	—
1000 μA/cm^2	10^4	1300	—
Air	10^5	1500	160
1000 μA/cm^2	10^5	2200	420

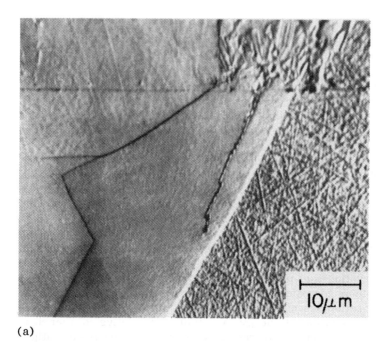

(a)

FIGURE 10 Comparative cross sections of polycrystalline copper showing stage I cracking in air (a) and intergranular cracking (b) with ac-
centuated surface deformation under applied anodic current conditions
[36].

dissolution at the tips of growing cracks or preferential adsorption of
a damaging ionic species [39–41]. Experiments on a 7075-T6 commercial alloy and on a high-purity analog of the alloy, Al5.0Zn2.5Mg1.6Cu,
indicate that a third mechanism, localized hydrogen embrittlement, may
be responsible for the poor corrosion fatigue resistance of these alloys. For example, Fig. 13 shows the results of fatigue tests performed on the 7075 alloy under simultaneous exposure to cycle stresses
and a corrosive environment (curve C) compared to tests performed in
laboratory air (curve A). If specimens are precorroded and tested in
laboratory air, there is also a significant reduction in fatigue resistance (curve C) [42]. The reduction in life at low N_f is associated
with pits which form at nonmetallic inclusions. If the alloy is resolu-
tionized and aged, equivalent to a low-temperature bake, a significant amount of fatigue resistance is regained, indicating at least partial reversibility of the damaging phenomenon and strongly suggesting

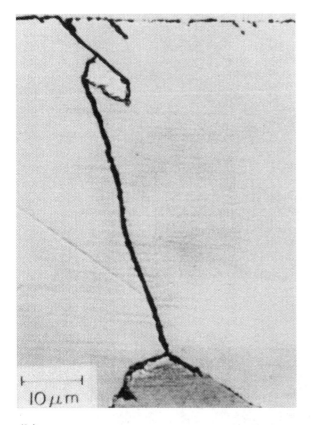

(b)

a solid-solution effect arising from the environmental interaction. In addition, the reversibility in fatigue resistance is a function of baking time (Table 3). Hydrogen is the only stable gaseous species at the corrosion potential of the alloy, and gas bubbles were observed to emanate from growing cracks.

It was previously observed that halide ions are particularly damaging to the fatigue behavior of Al alloys. However, if the alloy is cathodically charged during stressing, sulfate ion proves to be equally damaging, particularly at long N_f. At lower N_f the slight decrease observed in Cl^- solutions appears to be associated with damage to the passive film (Fig. 14). In SO_4^{2-} solutions a crack must initiate to break the protective film to allow access to the bulk alloy [43,44]. Cathodic charging of the high-purity analog of the 7075 alloy also

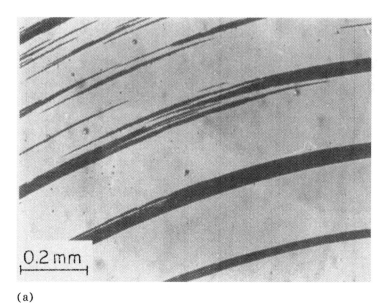

(a)

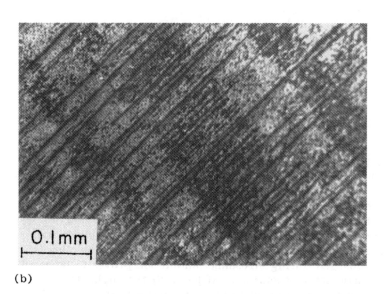

(b)

FIGURE 11 Comparative surfaces of single-crystal copper tested in air (a) and under applied anodic current conditions (b) [34].

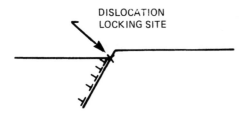

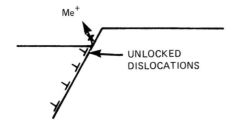

FIGURE 12 Schematic diagram of model to explain increased surface deformation and early crack nucleation in corrosion fatigue.

resulted in a reduction in fatigue resistance. In many cases, fatigue crack initiation in the equiaxed grain high-purity alloy is intergranular, and Table 4 shows that at more active cathodic potentials there is a tendency toward a higher percentage of transgranular cracking.

The relative amount of integranular cracking is also a function of applied stress levels, and for the most active condition, lower stresses lead to an increased amount of intergranular cracking (Table 4). These data indicate that there is a relationship between stress corrosion cracking of the alloy and corrosion fatigue and that there is a competition between transport of hydrogen to grain boundaries and interaction of hydrogen with growing transgranular cracks. At long lives or for rapidly propagating cracks hydrogen cannot diffuse to grain boundaries, and thus it interacts with the tips of growing transgranular cracks. Stress corrosion cracking of a similar alloy was associated with hydrogen cracking by Montgrain and Swann [45].

It has been argued that bulk aluminum alloys should not be hydrogen-embrittled since the solubility and diffusion rates of hydrogen in aluminum are relatively low. However, the subcritical crack growth in aluminum alloys need only propagate into the region immediately

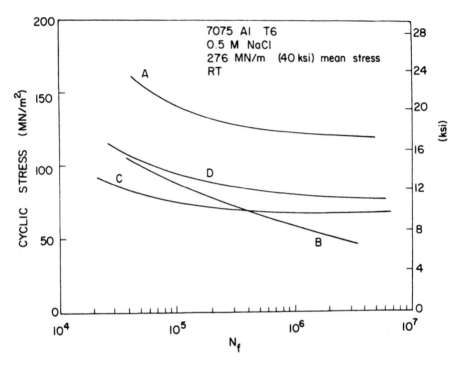

FIGURE 13 Effect of precorrosion and subsequent heat treatment on fatigue behavior of 7075-T6 Al alloy [42]: A, air; B, corrosion; C, precorroded/air fatigue; D, precorroded/heat-treated/air fatigue.

ahead of a growing crack to affect propagation rates. It has also been suggested that accelerated transport of hydrogen might occur by dislocation motion [46,47], and tensile tests of the high-purity alloy under cathodic charging conditions in 1 N H_2SO_4 show serrated yielding, in contrast to tests performed in air. Figure 15 shows these results, where the alloy is initially yielded to rupture and protective oxide film, allowed to stress relax while being charged, and subsequently restrained [48].

To summarize the Al alloy results, it appears that corrosion reactions liberate hydrogen, which effectively embrittles the region in the vicinity of a crack tip. The specific details of the embrittlement are not known, but it appears that dislocation transport of the hydrogen is involved. It has been speculated that hydrogen may collect at the semicoherent precipitate-matrix interface, thus explaining the reported {100} fracture plane; however, a great deal more research will have to be performed before a more definitive answer will be available.

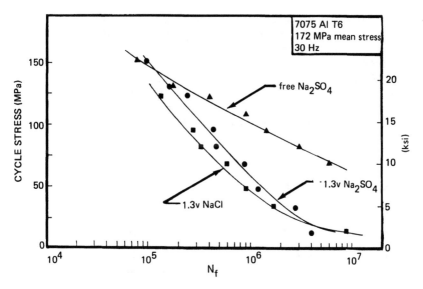

FIGURE 14 Effect of cathodic charging on the fatigue behavior of 7075-T 6 Al alloy in Cl⁻ and SO_4^{2-} solutions [44].

8.6 SUMMARY

Corrosion fatigue in gaseous or aqueous environments is, at best, a highly complex interaction between environment and metal or alloy mechanical properties. It appears that there is no central, unifying mechanism which governs these phenomena. Rather, the aggressiveness of the environment and the mechanical response of the metal or alloy interact so that classes of metal/environment interactions tend to occur. For example, it appears that in gaseous environments

TABLE 3 Effects of 24-hr Precorrosion in 0.5 N NaCl and Subsequent Reheat Treatment on the Fatigue Resistance of 7075 Alloy in Air[a]

	Resolutionizing time (hr)			
	0	3	6	24
N_f	3×10^4	8.5×10^4	10.1×10^4	$>13 \times 10^6$

[a] $\sigma_m = 276$ MN/m² (40 ksi); $\Delta\sigma = \pm 96$ MN/m² (14 ksi).

TABLE 4 Effect of Potential and Cyclic Stress Level on
Fracture Mode of Al5.5Zn2.5Mg1.5Cu Alloy in 0.5 N NaCl

Cyclic stress (±) at σ_m = 207 MN/m² (MN/m²)	Potential (V vs. SCE)	Intergranular failure (%)
68	-0.780 (free corrosion)	23
	-1.300	5
	-1.750	0
55	-1.750	0
41	-1.750	10
28	-1.750	26
17	-1.750	35

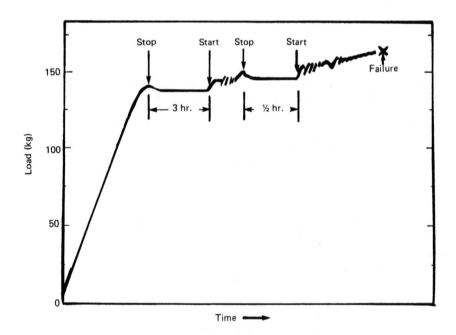

FIGURE 15 Interrupted tensile load-time curve for Al5.0Zn2.5Mg1.5-
Cu alloy under cathodic charging conditions (as processed, M_2
charged) [44].

stage I cracking is often primarily affected by molecular adsorption, while stage II crack acceleration is due to interference with slip reversability. In aqueous environments metal atoms associated with mobile dislocations are preferentially dissolved, leading to attack of emerging slip bands and early crack initiation. This preferential corrosion also causes enhanced surface deformation and numerous crack initiation sites. Alloys which may be subject to hydrogen embrittlement may obey still another mechanism for premature failure, with a corrosion reaction supplying hydrogen to the plastic zones of propagating cracks, leading to a form of localized embrittlement.

ACKNOWLEDGMENT

The author acknowledges the support of the Office of Naval Research under contract N00014-75-C-0466.

REFERENCES

1. O. F. Devereux, A. J. McEvily, and R. W. Staehle, eds., *Corrosion Fatigue: Chemistry, Mechanics, and Microstructure*, National Association of Corrosion Engineers, Houston, 1972.
2. D. J. Duquette, *Rev. Coat. Corros.*, *1*: 187 (1973).
3. A. Hartman and J. Schijve, Natl. Bur. Stand. NLR-MP-68001-u, 1968.
4. K. U. Snowden, *Acta Metall.*, *12*: 295 (1964).
5. M. J. Hordon, *Acta Metall.*, *14*: 1173 (1966).
6. A. H. Smith, P. Shahinian, and M. Achter, *Trans. TMS-AIME*, *245*: 947 (1969).
7. N. Thompson, N. J. Wadsworth, and N. Louat, *Philos. Mag.*, *1*: 113 (1956).
8. N. J. Wadsworth, *Philos. Mag.*, *6*: 387 (1961).
9. C. Laird and G. C. Smith, *Philos. Mag.*, *8*: 1945 (1963).
10. F. E. Fujita, *Acta Metall.*, *6*: 543 (1958).
11. R. M. N. Pelloux, *Trans. ASM*, *62*: 281 (1969).
12. D. A. Meyn, *Trans. ASM*, *61*: 53 (1968).
13. J. C. Grosskreutz and C. Q. Bowles, *Environment-Sensitive Mechanical Behavior*, Gordon & Breach, London, 1963, p. 67.
14. J. C. Grosskreutz, in *Fracture 1969, Proceedings International Conference on Fracture*, Chapman & Hall, Brighton, 1969, p. 731.
15. J. C. Grosskreutz, *Surf. Sci.*, *8*: 173 (1967).
16. H. Shen, S. E. Podlaseck, and I. R. Kramer, *Acta Metall.*, *14*: 341 (1966).

17. W. A. Wood, S. M. Cousland, and K. R. Sargant, *Acta Metall.*, *11*: 643 (1963).
18. D. J. Duquette and M. Gell, *Metall. Trans.*, *2*: 1325 (1971).
19. A. Kelly, W. R. Tyson, and A. H. Cottrell, *Philos. Mag.*, *15*: 567 (1971).
20. K. R. L. Thompson and J. V. Craig, *Metall. Trans.*, *1*: 1047 (1970).
21. C. A. Stubbington and P. J. E. Forsyth, *Metallurgia*, *74*: 15 (1966).
22. J. K. Tien and R. P. Gamble, *Metall. Trans.*, *2*: 1933 (1971).
23. D. J. McAdam, Jr., and G. W. Geil, *Proc. Am. Soc. Test. Mater.*, *41*: 696 (1941).
24. B. B. Wescott, *Mech. Eng.*, *60*: 813–829 (1938).
25. H. Spahn, *Metalloberflache*, *16*: 299 (1962).
26. M. T. Simnad and U. R. Evans, *Proc. R. Soc. London A188*: 372 (1947).
27. D. J. Duquette and H. H. Uhlig, *Trans. ASM*, *62*: 839 (1969).
28. D. J. Duquette and H. H. Uhlig, *Trans. ASM*, *61*: 449 (1968).
29. D. Whitwham and U. R. Evans, *J. Iron Steel Inst.*, *165*: 72 (1950).
30. K. Laute, *Oberflachentech*, *10*: 281 (1933).
31. M. T. Simnad and U. R. Evans, *J. Iron Steel Inst.*, *156*: 531 (1947).
32. M. T. Simnad and U. R. Evans, *Proc. R. Soc. London A188*: 372 (1947).
33. A. V. Ryabchenkov, *Zh. Fiz. Khim.*, *26*: 542 (1952).
34. H. N. Hahn and D. J. Duquette, *Acta Metall.*, *26*: 279 (1978).
35. H. Masuda and D. J. Duquette, *Metall. Trans.*, *6A*: 87 (1975).
36. H. N. Hahn, Ph.D. thesis, Rensselaer Polytechnic Inst., Jan. 1977.
37. V. Rollins and T. Pyle, *Nature*, *254*: 322 (1975).
38. T. Pyle, V. Rollins, and D. Howard, *J. Electrochem. Soc.*, *122*: 1445 (1975).
39. R. E. Stoltz and R. M. Pelloux, *Metall. Trans.*, *3*: 2433 (1972).
40. R. M. Pelloux, in *Fracture 1969, Proceedings, International Conference on Fracture*, Chapman & Hall, Brighton, London, 1969.
41. R. J. Selines and R. M. Pelloux, *Metall. Trans.*, *3*: 2525 (1972).
42. E. F. Smith, III, R. Jacko, and D. J. Duquette, in *Effect of Hydrogen on Behavior of Materials* (A. W. Thompson and I. M. Bernstein, eds.), AIME, New York, 1976, p. 218.
43. R. J. Jacko and D. J. Duquette, *Metall. Trans.*, *17A*: 339 (1986).
44. E. F. Smith, III, R. J. Jacko, and D. J. Duquette, in *Proceedings 2nd International Congress on Hydrogen in Metals*, paper 3C1, Paris, 1977.

45. L. Montgrain and P. R. Swann, in *Hydrogen in Metals* (I. M. Bernstein and A. W. Thompson, eds.), American Society for Metals, Metals Park, Ohio, 1974, p. 575.
46. J. K. Tien, in *Effects of Hydrogen on Behavior of Materials* (A. W. Thompson and I. M. Bernstein, eds.), ASM, Metals Park, Ohio, 1974, p. 207.
47. H. H. Johnson and J. P. Hirth, *Metall. Trans.*, 7A: 1543 (1976).
48. E. F. Smith, III, Ph.D thesis, Rensselaer Polytechnic Inst., March 1977.

9

MECHANISMS OF DIFFUSION-CONTROLLED HIGH-TEMPERATURE OXIDATION OF METALS

GREGORY J. YUREK

Department of Materials Science and Engineering, Massachusetts Institute of Technology, Cambridge, Massachusetts

9.1 INTRODUCTION

There are many forms of oxidation and corrosion of materials at high temperatures, each of which depends on the material itself, the environment, the temperature, etc. It is not possible to survey the phenomenology and mechanisms of all forms of high-temperature corrosion of materials in a single chapter. The present chapter focuses on the mechanisms of diffusion-controlled growth of scales formed on pure metals in oxidizing gases at elevated temperature (T $\gtrsim$ 300°C). Although this subject is only one small part of the field of high-temperature oxidation and corrosion, it is, in fact, a very important area for investigation. The ultimate goal of most research dealing with high-temperature oxidation and corrosion is the development of materials that form protective scales on their surfaces, thereby ensuring that they are degraded very slowly. Basically, protective surface scales are diffusion barriers between the material and the environment. An understanding of the mechanisms of diffusion-controlled growth of scales is, therefore, fundamental to virtually all investigations of high-temperature oxidation and corrosion. The key features of diffusion-controlled scale growth can be discussed by considering scale growth on pure metals, which narrows the subject considerably and allows for a fairly complete discussion in one chapter.

This chapter is organized into three main sections. The first deals with forms of high-temperature corrosion and is intended to acquaint the reader with the variety of forms that exist and to point out some of the similarities with the forms of aqueous corrosion. This is followed by a brief section on the general features of alloy oxidation as it is related to the contents of this chapter. The bulk of the chapter is devoted to a review of Wagner's model for the diffusion-controlled oxidation of pure metals and a discussion of the assumptions underlying that model. The influence of short-circuit diffusion paths on the kinetics of diffusion-controlled oxidation of metals is discussed in the final section.

The subject of the final portion of this chapter is of considerable importance. Oxidation kinetics should always be related to actual scale microstructures, and modern investigations of oxidation mechanisms do, in fact, rely heavily on the use of analytical tools (e.g., scanning and transmission electron microscopy, Auger electron spectroscopy) to characterize scale composition and microstructure (i.e., scale constitution). Studies that combine kinetic data with a thorough characterization of the corresponding scales typically lead to a better understanding of oxidation mechanisms than studies that rely on only one of these experimental approaches. Still better insight into oxidation mechanisms can usually be attained when characterization and kinetic data are correlated with analytical models of the oxidation process.

Oxidation kinetics and scale constitution should also be related to the constitution of the material on which the scale forms. Indeed, since processing affects the structure of a material, elucidation of the relationships between oxidation kinetics, scale constitution, and the constitution of the material being corroded may eventually lead to an understanding of how to better exploit materials processing to improve the corrosion resistance of materials at elevated temperatures.

9.1.1 Forms of High-Temperature Corrosion

Metals, alloys, and ceramics may experience severe environmental degradation when they come into contact with gaseous, liquid, or solid phases at elevated temperatures. There are many forms of environmental degradation, or corrosion, of materials at high temperatures. Aside from direct volatilization, high-temperature corrosion of metals and alloys always involves an oxidation process; i.e., metal atoms lose electrons and enter a solid, liquid, or gaseous phase. The corrosion of ceramic materials may involve further oxidation (FeO $\rightarrow$ Fe_3O_4), a conversion from one type of compound to another (SiC $\rightarrow$ SiO_2), direct volatilization ($SiO_2 \rightarrow SiO$ vapor), or "simple" dissolution, as occurs when refractories dissolve into molten slags.

There is usually some confusion of the terms oxidation and hot corrosion when referring to environmental degradation of materials at high temperatures. The term *oxidation* is typically reserved for the conversion of metallic materials to solid oxidation products by interaction with hot oxidizing gases or gas mixtures. The term *hot corrosion* is reserved for the attack of metals and alloys by molten salts and slags, and the direct dissolution of refractories or glasses into molten salts, slags, or metals is usually referred to simply as corrosion.

Oxidation and hot corrosion of metals and alloys may give rise to uniform, pitting, or intergranular attack, the morphology of attack being generally the same as that for uniform, pitting, or intergranular attack in aqueous systems. Intergranular attack at high temperatures is a localized type of *internal oxidation*, a process that involves the dissolution and diffusion of metalloids into an alloy leading to the formation of oxidation products (oxides, sulfides, carbides, etc.) within the alloy matrix. Hydrogen attack is similar in nature except that it involves the dissolution of hydrogen in an alloy at elevated temperatures and its subsequent reaction with carbon to form methane gas at very high pressures, which can lead to the initiation and propagation of cracks in the alloy.

The rate of crack propagation in superalloys under conditions of tensile and cyclic (fatigue) loading is enhanced in the presence of oxygen- or sulfur-bearing gases. These occurrences of environmentally assisted cracking are analogous to stress-corrosion cracking and corrosion fatigue in aqueous media because of the synergistic effects of stress and the environment on the enhanced rate of crack propagation.

Finally, severe degradation of metallic and ceramic components at high temperatures can occur by erosion. The rate of degradation in some cases is enhanced considerably when erosion and corrosion occur simultaneously; however, in many instances, it appears that either erosion or corrosion predominates, depending on variables such as the composition of the environment and the particulate matter, the velocity of the particles, and the angle of impingement.

9.1.2 General Features of High-Temperature Oxidation of Alloys

High-temperature corrosion problems are usually combated by changes in process conditions, changes in design, implementation of a replacement schedule for components that corrode at known rates, use of coatings, or selection of more corrosion-resistant materials. Corrosion-resistant alloys employed for service in high-temperature oxidizing environments are designed to form protective scales on their surfaces. Protective scales are those that form rapidly on the alloy surface, that are very good solid-state diffusion barriers between the alloy and the environment, and that are very resistant to spallation. The protective scales typically found on oxidation-resistant high-temperature alloys are Cr_2O_3, Al_2O_3, and SiO_2. These oxides are very stable relative to other oxides and to other compounds (carbides, sulfides, etc.), and solid-state diffusion in them is very slow relative to

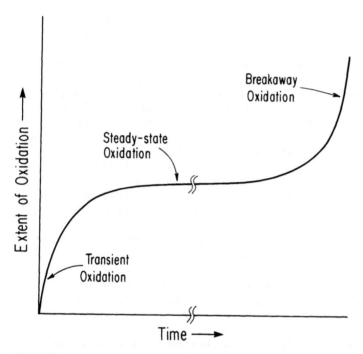

FIGURE 1 Schematic representation of the extent of oxidation (e.g., weight change per unit area or scale thickness plus depth of internal oxidation) as a function of time for alloys that develop protective oxide scales during the steady-state period of oxidation.

that in the oxides of typical base metals (Fe, Co, and Ni) used for high-temperature alloys.

Three stages of oxidation are exhibited typically by alloys that form protective oxide scales [1]; see Fig. 1. During the initial or transient stage of oxidation, oxides of essentially all the reactive elements in the alloy may form on the alloy surface, the amount of each being roughly proportional to the concentration of the element in the alloy [2]. The extent of oxidation during the transient stage depends on the relative amounts and rates of growth of oxides formed. For example, if the constitution of a Ni-Cr-Al alloy is such that the amount of NiO nucleated is large relative to the amounts of Cr_2O_3 and Al_2O_3, the extent of oxidation before steady state is reached will be greater than if less NiO nucleated on the surface. This is because the growth of NiO is much faster than that of either Cr_2O_3 or Al_2O_3.

The extent of oxidation during the transient stage also depends on the rate at which slower-growing oxides can develop at the interface between the alloy and faster-growing oxide nuclei, a phenomenon that is usually referred to as *healing*. The rate of healing, or the time at which steady-state oxidation begins, is a function of alloy constitution, temperature, and the nature and composition of the environment to which the alloy is exposed. The extent of formation and the nature of the transient oxidation products can have long-term effects on the oxidation resistance of an alloy. For example, higher stresses or a more detrimental stress distribution in a non-uniform scale developed during the transient stage of oxidation may give rise to a shorter period of steady-state oxidation.

The steady-state period of oxidation occurs when the rate of oxidation is controlled by the growth of a protective oxide scale, which invariably is a process controlled by solid-state diffusion in the scale. The discussion in this chapter focuses on the mechanisms of diffusion-controlled scale formation on pure metals. The aim is to review the current understanding of diffusion-controlled scale growth and to discuss the effects of short-circuit diffusion paths on the kinetics of steady-state scaling. An understanding of the basic mechanisms of diffusion-controlled scale growth on pure metals provides a sound basis for interpreting oxidation mechanisms during the steady-state period of oxidation for more complex cases (e.g., oxidation of alloys in contact with multicomponent gaseous, liquid, or solid phases). Also, an understanding of the basic mechanisms of scale growth during the steady-state period of oxidation may be important in helping to elucidate the relationships between scale growth and the onset of breakaway behavior.

Growth stresses develop in scales as they thicken and may eventually cause failure of the scales and the onset of breakaway oxidation by one or more stress relief mechanisms (blistering, tensile or shear cracking of the scale, etc.). The magnitude of the growth stresses is related to scale thickness, the amount of alloy consumed, the rate of vacancy injection and annihilation in the alloy, the development of concentration gradients in the alloy and in the scale, the degree of formation of scale within the scale, etc. The rate of scale growth during the steady-state period of oxidation may, in principle, be related to each of the latter features, and hence quantitative models for scale growth may eventually be useful in predicting the onset of breakaway oxidation.

The onset of breakaway behavior is also related to the penetration of protective oxide scales by cations (Fe, Ni, Mn, etc.) from the base metal or anions (S, Cl, C, etc.) from the environment. An understanding of the mechanisms of diffusion-controlled growth of protective oxide scales in the absence of these species is of prime importance

in elucidating the mechanisms by which they influence the breakdown
of protective scales when they are present.

9.2 MECHANISMS OF DIFFUSION-CONTROLLED
   SCALE GROWTH

9.2.1 The Parabolic Rate Law

The rate of thickening of a protective scale during the steady-state
period of oxidation is usually described by the parabolic rate law, i.e.,

$$d(\Delta x)/dt = k_p/\Delta x \tag{1}$$

where Δx is the instantaneous scale thickness, t is time, and k_p is the
parabolic rate constant. Integration of Eq. (1) yields

$$\Delta x^2 = 2k_p t + C \tag{2}$$

where C is an integration constant. A plot of Δx^2 vs. time yields a
straight line if the parabolic rate law describes the experimental data.
The slope of such a plot yields k_p, which is used as a relative measure of the rate of oxidation of metals and alloys. Representative
values of the parabolic rate constant for the oxidation of several metals [3–7] are shown in Fig. 2. It is obvious from Fig. 2 that alloys
for high-temperature service should be designed to form scales of
Cr_2O_3, SiO_2, or $\alpha\text{-}Al_2O_3$ instead of the oxides of the base metals iron,
nickel, and cobalt.

The value of C in Eq. (2) is virtually zero if the extent of oxidation during the transient stage is negligible. Even when diffusion-
controlled scale growth occurs from virtually the beginning of the
oxidation process, a nonzero value of C results if short-circuit diffusion contributes to the growth process and if the volume fraction
of short-circuit diffusion paths changes with time.

The rate of oxidation may also be measured by using a thermo-
gravimetric apparatus (TGA) to determine continuously the change
in weight of a specimen as a function of exposure time. To normal-
ize the TGA data, the weight change Δm is divided by the original
specimen area A_0, or the instantaneous specimen area if the dimensions of the specimen change significantly during the experiment [8].
If the metal or alloy oxidizes according to a parabolic rate law, the
following equation may be used to describe the data:

$$(\Delta m/A_0)^2 = k_g t + C' \tag{3}$$

where k_g is the parabolic rate constant determined gravimetrically.
The constant C' is to be interpreted similarly to the constant C in
Eq. (2).

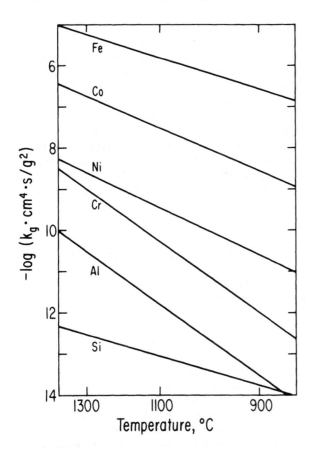

FIGURE 2 Representative values of the parabolic rate constants for the growth of several oxides: iron oxidized in pure oxygen at 1 atm pressure to yield a multilayered scale comprising FeO, Fe_3O_4, and Fe_2O_3 [3]; cobalt oxidized in pure oxygen at 0.21 atm pressure to form CoO [4]; nickel oxidized in pure oxygen at 1 atm pressure to form NiO [5]; chromium oxidized in oxygen at 0.1 to 1 atm to form Cr_2O_3, midrange of data compiled by Hindam and Whittle [6]; oxidation of Fe-, Ni-, or Co-base alloys that formed α-Al_2O_3 scales at oxygen partial pressures in the range 0.1 to 1 atm, midrange of data compiled by Hindam and Whittle [6]; oxidation of silicon in pure oxygen at 1 atm to form SiO_2 [7].

Note that the coefficient of $k_g t$ in Eq. (3) is unity while the coefficient of $k_p t$ in Eq. (2) is 2. The latter coefficient results from integration of Eq. (1), which is the usual starting point for a discussion of parabolic oxidation kinetics because it shows directly the inverse

dependence of oxidation rate on scale thickness. Equation (3), however, has its roots in the paper of Pilling and Bedworth [9], who
noted that when the square of experimentally determined values of
($\Delta m/A$) is plotted as a function of time a straight line often results.
This difference in the coefficients in Eqs. (2) and (3) is often a
source of confusion in the literature because not all investigators
use this convention; i.e., some delete the factor of 2 in Eq. (2)
while others add a factor of 2 in Eq. (3). Caution should be employed to determine the convention used by each investigator before
reported rate constants are interpreted or compared with other data.

The relationship between k_g and k_p depends on the convention
employed. In the present dase, the appropriate relationship is

$$k_p = (1/2)(V^m/A_x)^2 k_g \tag{3a}$$

where A_x is the atomic weight of oxidant X and V^m is the molar volume of the scale *per mole of oxidant X*.

Experimentally determined values of the parabolic rate constant
provide a measure of the rate of diffusion of reactants in a scale.
The rate constant reflects the effects of the constitution of the scale,
growth stresses in the scale, temperature, total pressure, and gradients in the chemical potentials of the reactants across the scale on
solid-state diffusion in the scale. The constitution of a metal or alloy generally has an influence on the constitution of the scale formed
on it, and thus the effects of the former are also reflected in the parabolic rate constant. Unfortunately, no model takes into account the
effects of all these variables on the magnitude of the parabolic rate
constant. Wagner's model [10—12] for the diffusion-controlled oxidation of metals leads, however, to an expression for the parabolic rate
constant that does reflect both the transport properties of the scale
and the thermodynamic driving force for scale formation. It can also
be modified to take into account the effects of short-circuit diffusion
paths on oxidation kinetics. Wagner's model provides, therefore, a
very useful basis for understanding the mechanisms of diffusion-controlled scale growth.

The remainder of this section is directed to reviewing the important features of Wagner's theory for the diffusion-controlled growth
of scales on pure metals. Particular attention is paid to the major
assumptions in Wagner's model. For the purpose of illustration, the
oxidation of a pure metal Me in a gas phase X_2 to form a uniform, compact, adherent layer of $Me_{\alpha+\delta}X$ is considered; see Fig. 3. The value
of α is the ratio of Me to X in the stoichiometric compound and is given
by the ratio $|z_X|/z_{Me}$, where z_X and z_{Me} are the valences of anions
and cations, respectively, on normal lattice sites in the stoichiometric
compound. The deviation from stoichiometry is represented by δ,
which is positive for metal-excess compounds and negative for metal-
deficient compounds.

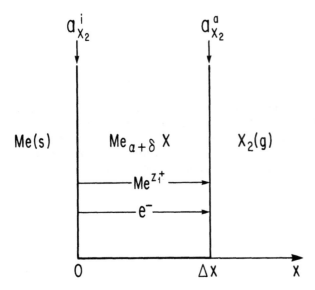

FIGURE 3 Schematic representation of growth of an oxidation product $Me_{\alpha+\delta}X$ on metal Me in gas X_2. It is assumed that the most mobile components in the scale are cations of valence z_1^+ and electrons. The activity of X_2, a_{X_2}, is defined as $P_{X_2}/P^{\circ}_{X_2}$, where $P^{\circ}_{X_2}$ is 1 atm. Superscripts i and a refer to the metal/scale and scale/gas interfaces, respectively.

9.2.2 Wagner's Model for Diffusion–Controlled Scale Growth

In his original derivation of an expression for the parabolic rate constant, Wagner [10,11] assumed that ions and electrons migrate independently in a growing scale; that is, the rate of transport of each component (cation, anion, or electron) is proportional to the gradient in the electrochemical potential for that component, but independent of the electrochemical potentials for the other components. This is best illustrated [12] by expressing the fluxes of the components in the form typically employed in irreversible thermodynamics. To elucidate several important points in the derivation of expressions for the parabolic rate constant, the growth of a scale in which cations and electrons are the most mobile species (Fig. 3) will be considered here. The resulting expressions can be altered readily to account for anion transport.

For an isothermal, isobaric, isotropic single-phase system in which only cations and electrons are mobile, the fluxes for one-dimensional transport are

$$j_1 = L_{11}(d\eta_1/dx) + L_{12}(d\eta_2/dx) \tag{4}$$

and

$$j_2 = L_{21}(d\eta_1/dx) + L_{22}(d\eta_2/dx) \tag{5}$$

where 1 and 2 refer to cations and electrons, respectively, j_i is the flux of i in moles per unit area time, η_i is the electrochemical potential of i, L_{11}, L_{22}, L_{12}, and L_{21} are the general transport coefficients, and x is distance measured from the metal/scale interface. According to the Onsager relationship, $L_{12} = L_{21}$. Independent migration of cations and electrons requires that $L_{12} = L_{21} = 0$, in which case

$$j_1 = L_{11}(d\eta_1/dx) \tag{6}$$

and

$$j_2 = L_{22}(d\eta_2/dx) \tag{7}$$

Independent migration of cations and electrons does not preclude, however, the occurrence of *ambipolar diffusion* in electronically conducting compounds. When cations and electrons migrate independently, the difference in the mobilities of electrons and ions yields a local electrical potential gradient along the direction of diffusion. Because electrons are generally much more mobile than cations and may typically precede cations by a few lattice parameters as both migrate through a growing scale, a local electrostatic potential gradient is set up between the cations and electrons that tends to decrease the rate of migration of the electrons and increase that of the cations. This effect is, in fact, accounted for in the transport equations by relating the fluxes of the migrating components to their electrochemical potential gradients, instead of just their chemical potential gradients.

The electrochemical potential of species i, η_i, is equal to $\mu_i + z_i F\phi$, where μ_i and z_i are the chemical potential and charge of species i, F is the Faraday constant, and ϕ is the average electrical potential in a plane at distance x from the metal/scale interface. The chemical and electrochemical potentials of ions and electrons in compound $Me_{\alpha+\delta}X$ can be related to the chemical potentials of elements Me and X, through the assumption of internal equilibrium; i.e., at any plane parallel to the metal/scale interface, the equilibria

$$Me^{z_1+} + z_1 e^- = Me \tag{8}$$

and

$$X^{z_2-} - z_3 e^- = X \tag{9}$$

are assumed to be established. Hence, at any plane in the scale parallel to the metal/scale interface,

$$\mu_1 + z_1\mu_2 = \eta_1 + z_1\eta_2 = \mu_{Me} \tag{10}$$

and

$$\mu_3 - z_3\mu_2 = \eta_3 - z_3\eta_2 = \mu_X \tag{11}$$

In Eqs. (8)−(11), species 1, 2, and 3 refer to cations, electrons, and anions, respectively, and $z_2 = -1.0$. In addition to internal equilibrium within the scale, local equilibrium is assumed to exist between the metal and the scale and between the scale and the oxidant.

The molar flux of metal Me, j_{Me}, in the oxidation product $Me_{\alpha+\delta}X$ is equal to the molar flux of component 1 (cations). In the absence of the development of a space charge layer in the oxidation product, the flux of cations must be equivalent to the flux of electrons; hence,

$$j_{Me} = j_1 = j_2/z_1 \tag{12}$$

where $z_1 = z_{Me}$, the latter being the valence of cations Me in $Me_{\alpha+\delta}X$ with $\delta = 0$; e.g., $z_1 = 3$ for Cr_2O_3, $z_1 = 2$ for CoO, etc. If, as appears to be the case, Cr_2O_3 is an n-type semiconducting compound containing an excess of metal at low oxygen pressures and at temperatures below about 1100°C, the most important defects are likely to be $Cr_i^{\cdot\cdot}$, $Cr_i^{\cdot\cdot\cdot}$, and free (conduction) electrons. Motion of $Cr_i^{\cdot}$ is equivalent to coupled motion ($L_{12} = L_{21} \neq 0$) of a trivalent chromium cation and one electron. In p-type CoO, the major defects appear to be $V_{Co}^{\prime\prime}$, $V_{Co}^{\prime}$, and electron holes, $h^{\cdot}$; migration of $V_{Co}^{\prime}$ is equivalent to coupled migration of a divalent cobalt cation and one electron. The effective charge of the major point defect in the compound has no effect on the value of z_1 used in Eq. (12).

The values of j_{Me} and j_2 ($= j_{Me}z_1$) can be calculated from Eqs. (4) and (5) for the general case in which coupled motion of ions and electrons occurs. To eliminate the gradient in the electrochemical potential of component 1 in the latter equations in favor of the gradient in the chemical potential of Me, which has well-defined values at a given temperature, pressure, and deviation from stoichiometry, Eq. (10) may be differentiated with respect to x at constant temperature and pressure to yield

$$d\eta_1/dx = d\mu_{Me}/dx - z_1 \, d\eta_2/dx \tag{13}$$

whereupon one obtains [12]

$$j_{Me} = L_{11} \, d\mu_{Me}/dx + (L_{12} - z_1L_{11}) \, d\eta_2/dx \tag{14}$$

and

$$j_2 = z_1 j_{Me} = L_{12} d\mu_{Me}/dx + (L_{22} - z_1 L_{12}) d\eta_2/dx \tag{15}$$

Combining Eqs. (14) and (15) to derive an expression for $d\eta_2/dx$ and using the latter expression to solve for j_{Me}, one obtains

$$j_{Me} = [(L_{11}L_{22} - L_{12}{}^2)/(z_1{}^2 L_{11} - 2z_1 L_{12} + L_{22})](d\mu_{Me}/dx) \tag{16}$$

Equation (16) is a general transport equation that applies whether $Me_{\alpha+\delta} X$ is an electronic, ionic, or mixed conductor. For the special case in which $Me_{\alpha+\delta} X$ is an electronic conductor ($\sigma_1 \ll \sigma$, where σ is the electrical conductivity and σ_1 is the partial ionic conductivity for component 1 in $Me_{\alpha+\delta} X$), one has

$$|d\mu_{Me}/dx| \gg |d\eta_2/dx| \qquad \sigma_1 \ll \sigma \tag{17}$$

Thus, Eq. (16) becomes

$$j_{Me} \approx L_{11} d\mu_{Me}/dx \qquad \sigma_1 \ll \sigma \tag{18}$$

Equation (18) is the same as Eq. (6), which was derived on the basis of independent migration of ions and electrons ($L_{12} = L_{21} = 0$), because $d\eta_1/dx \approx d\mu_{Me}/dx$ when Eq. (17) is valid. Thus, the expressions for the fluxes of ions and electrons in an electronic conducting compound are the same whether the migration of these components is coupled or not. As will be shown below, the *calculated* value of the cation flux is the same whether $L_{12} = L_{21} = 0$ or not, so long as L_{11} is evaluated in terms of a chemical or self-diffusion coefficient. If, however, L_{11} is evaluated in terms of the partial ionic conductivity for cations, then the calculated value of the flux does depend on whether or not coupled migration of ions and electrons occurs. For coupled migration, care must be taken to employ the proper partial ionic conductivity for cations [12]. This point will be considered again later.

Expressions for the parabolic rate constant may be derived starting with Eq. (16) for an ionic or mixed conductor or Eq. (18) for an electronic conductor. It is, however, useful to first derive an expression for k_p, starting with Fick's first law [12]. In so doing, one obtains relationships between the chemical diffusion coefficient and the transport coefficients. Also, it becomes evident in such a derivation that no special provisions are required to account for coupled migration of ions and electrons when chemical diffusivity data are employed to calculate oxidation rates.

According to Fick's first law, the unidirectional flux of Me across a compact, pore-free scale of $Me_{\alpha+\delta} X$ on Me is given by

$$j_{Me} = -\widetilde{D}(dc_{Me}/dx) \tag{19}$$

where $\widetilde{D}$ is the chemical diffusion coefficient for component Me and c_{Me} is the concentration of Me in moles per unit volume of $M_{\alpha+\delta}X$. For small deviations from stoichiometry ($\delta \cong 0$), $c_{Me} = (\alpha+\delta)/V^m$, where V^m is the molar volume of stoichiometric $Me_{\alpha+\delta}X$. Noting that $dc_{Me}/dx = (1/V^m)(d\delta/dx)$, Eq. (19) becomes

$$j_{Me} = -(\widetilde{D}/V^m)(d\delta/dx) \tag{20}$$

or, in terms of the chemical potential gradient of Me,

$$j_{Me} = -(\widetilde{D}/V^m)(\partial\delta/\partial\mu_{Me})(d\mu_{Me}/dx) \tag{21}$$

where the partial derivative of δ with respect to μ_{Me} is to be taken at constant temperature and pressure.

Comparison of Eq. (21) with Eqs. (16) and (18) yields

$$\widetilde{D} = -\frac{L_{11}L_{22} - L_{12}^2}{z_1^2 L_{11} - 2z_1 L_{12} + L_{22}} \frac{V^m}{\partial\delta/\partial\mu_{Me}} \tag{22}$$

for the general case and, for an electronic conductor,

$$\widetilde{D} \cong -L_{11}V^m/(\partial\delta/\partial\mu_{Me}) \qquad \sigma_1 \ll \sigma \tag{23}$$

Equations (22) and (23) are expressions for the chemical diffusivity of component Me in the compound $Me_{\alpha+\delta}X$ at a particular temperature and total pressure. Values of $\widetilde{D}$ may be measured in relaxation experiments [13,14].

The parabolic rate constant for growth of $Me_{\alpha+\delta}X$ on metal Me in contact with oxidant X may be calculated using chemical diffusivity data. Assuming that $\delta \ll \alpha$, changes in the Me/X ratio in a growing layer of $Me_{\alpha+\delta}X$ can be disregarded and j_{Me} is virtually independent of x (this assumption of quasi-steady-state scale growth will be examined again later in this chapter). Hence, integrating Eq. (21) over the scale thickness, Δx,

$$j_{Me} \int_0^{\Delta x} dx = j_{Me} \Delta x = -(1/V^m) \int_{\mu_{Me}^i}^{\mu_{Me}^a} \widetilde{D}(\partial\delta/\partial\mu_{Me}) \, d\mu_{Me} \tag{24}$$

where μ_{Me}^i and μ_{Me}^a are the chemical potentials of Me at the metal/ scale and scale/gas interfaces, respectively.

The rate of thickening of the scale is given by

$$d(\Delta x)/dt = j_{Me}(V^m/\alpha) \tag{25}$$

which, upon comparison with Eq. (1), yields the following expression for the parabolic rate constant k_p [12]:

$$k_p = -(1/\alpha) \int_{\mu_{Me}^i}^{\mu_{Me}^a} \widetilde{D}(\partial\delta/\partial\mu_{Me}) \, d\mu_{Me} \tag{26}$$

The values of δ and $\widetilde{D}$ in ionic compounds are usually measured as functions of the partial pressure of X_2 molecules, P_{X_2}, rather than μ_{Me}. For the equilibrium

$$\alpha Me(s) + \frac{1}{2}X_2(g) = Me_\alpha X(s) \tag{27}$$

one has

$$\alpha\mu_{Me} + \frac{1}{2}\mu_{X_2} = \mu_{Me_\alpha X} \tag{28}$$

Noting that $\mu_{Me_\alpha X}$ is essentially constant for $\delta \ll \alpha$, and $\mu_{X_2} = \mu_{X_2}^\circ + RT \ln a_{X_2}$, differentiation of Eq. (28) yields

$$\alpha d\mu_{Me} + (RT/2) \, d \ln P_{X_2} = 0 \tag{29}$$

where $a_{X_2} = P_{X_2}^\circ/P_{X_2}$, $P_{X_2}^\circ$ being taken as 1 atmosphere. Employing Eq. (20) in Eq. (26), one obtains [12]

$$k_p = -(1/\alpha) \int_{P_{X_2}^i}^{P_{X_2}^a} \widetilde{D}(\partial\delta/\partial \ln P_{X_2}) \, d \ln P_{X_2} \tag{30}$$

where $P_{X_2}^i$ is the partial pressure of X_2 for equilibrium between Me and $Me_{\alpha+\delta}X$ at the temperature of interest and $P_{X_2}^a$ is the partial pressure of X_2 in the gas phase. Equation (30) is completely general and applies whether $Me_{\alpha+\delta}X$ is an ionic or electronic conductor and whether or not coupled migration of ions and electrons occurs in $Me_{\alpha+\delta}X$.

Calculation of k_p according to Eq. (30) may be accomplished if $\widetilde{D}$ and $(\partial\delta/\partial \ln P_{X_2})$ have been measured as a function of P_{X_2}. The value of $(\partial\delta/\partial \ln P_{X_2})$ is expected to be virtually independent of the structure (grain size, dislocation density, etc.) of the compound, but $\widetilde{D}$ is expected to be structure-sensitive. Hence, calculated and experimental values of k_p may differ if $\widetilde{D}$ is measured in relaxation experiments with single crystals and fine-grained polycrystalline $Me_{\alpha+\delta}X$ grows on Me during an oxidation experiment.

The parabolic rate constant may also be calculated using data from tracer diffusion experiments. The tracer diffusion coefficient for component Me, D_{Me}^T, is related to the transport coefficient L_{11} according to [12,15]

$$D_{Me}^T / f_{Me} = -V^m R T L_{11}/\alpha = D_{Me} \tag{31}$$

where f_{Me} is the correlation factor, which accounts for correlated elementary steps of diffusing tracer atoms, and D_{Me} is the self-diffusion coefficient. For an electronic conductor, Eq. (31) may be inserted into Eq. (18), which yields

$$j_{Me} = -\frac{D_{Me}^T}{f_{Me}} \frac{\alpha}{V^m R T} \frac{d\mu_{Me}}{dx} = -\frac{D_{Me}\alpha}{V^m R T} \frac{d\mu_{Me}}{dx} \qquad \sigma_1 << \sigma \tag{32}$$

Equation (32) may be used to calculate the parabolic rate constant for the oxidation of a metal, using the same procedure that was employed to derive Eq. (30). The result is [12]

$$k_p = (1/2\alpha) \int_{P_{X_2}^i}^{P_{X_2}^a} (D_{Me}^T / f_{Me}) \, d \ln P_{X_2} \qquad \sigma_1 << \sigma \tag{33}$$

which is in accord with Wagner's previous derivation of the relationship between the parabolic rate constant and the self-diffusion coefficient for cations [11]. Note that nowhere in the derivations of Eq. (30) or (33) was it assumed that ions and electrons migrate independently. If coupled migration of ions and electrons occurs in the compound of interest, that will be reflected in the values of the diffusion coefficients and their dependences on P_{X_2}. The effective charge of the point defect by which cations diffuse in $Me_{\alpha+\delta}X$ also does not appear explicitly in Eq. (30) or (33). If, however, one assumes a simple point defect model for the compound (one predominant type of point defect with effective charge z), one can show that

$$k_p = [(1 + z)/\alpha] (^a D_{Me}^T / f_{Me}) \qquad P_{X_2}^a \gg P_{X_2}^i \qquad (30a)$$

for a p-type semiconducting compound, and

$$k_p = [(1 + z)/\alpha] (^i D_{Me}^T / f_{Me}) \qquad P_{X_2}^a \gg P_{X_2}^i \qquad (30b)$$

for an n-type semiconducting compound. The quantities $^a D_{Me}^T$ and $^i D_{Me}^T$ are the tracer diffusion coefficients for Me at $P_{X_2}^a$ and $P_{X_2}^i$, respectively. Similar relationships have been derived by Kofstad [16]. Equations (30a) and (30b) are not applicable when more than one type of defect, or one type of defect with different effective charges, contribute significantly to the diffusion process. This point will be reconsidered in Section 9.3.

If one also considers the transport of anions, component 3, in an electronically conducting compound, then

$$j_X = -(D_X/V^m RT)(d\mu_X/dx) \qquad \sigma_1, \sigma_3 \ll \sigma \qquad (34)$$

The net diffusion-controlled rate of thickening of a scale $Me_{\alpha+\delta}X$ is given by

$$d(\Delta x)/dt = j_{Me}(V^m/\alpha) + j_X V^m \qquad (35)$$

whereupon

$$k_p = 1/2 \int_{P_{X_2}^i}^{P_{X_2}^a} [(D_{Me}/\alpha) + D_X] \, d \ln P_{X_2} \qquad (36)$$

in agreement with a previous derivation of Wagner [11].

Finally, the parabolic rate constant can be calculated from conductivity data. In this case, however, care must be taken to use the proper types of partial conductivities. Wagner [12] pointed out that conductivity measurements in which metallic electrodes are employed (suppressed transfer of ions at the electrodes) may yield partial ionic conductivities, σ_1, that differ from those, σ_1', obtained using ionic conductors for the electrodes (suppressed transfer of electrons at the electrodes). In particular, Wagner demonstrated that, for samples of uniform composition,

$$\sigma_1 = -z_1^2 F^2 L_{11}(1 - L_{12}/z_1 L_{11}) \qquad (37)$$

for ion-blocking (metallic) electrodes, and

$$\sigma_1' = -z_1{}^2 F^2 L_{11}(1 - L_{12}^2/L_{11}L_{22}) \tag{38}$$

for electron-blocking (ionic) electrodes.

For electronic conductors ($\sigma_1 \ll \sigma_2, \sigma$), Eq. (38) becomes [12]

$$\sigma_1' = -z_1{}^2 F^2 L_{11} \qquad \sigma_1' \ll \sigma \text{ and } z_1 L_{11} \ll L_{22} \tag{39}$$

while Eq. (37) remains unchanged in this limiting case. Measurement of σ_1' for an electronic conductor gives a direct measure of the transport coefficient L_{11}; however, measurement of σ_1 does not yield directly a value for L_{11} unless the cross-coefficient, L_{12}, is equal to zero. In the latter case, $\sigma_1 = \sigma_1'$.

Equation (39) may be used to derive an expression for the parabolic rate constant for growth of an electronically conducting scale in terms of a partial ionic conductivity. Insertion of Eq. (39) into Eq. (18) yields

$$j_{Me} = -(\sigma_1'/z_1{}^2 F^2)(d\mu_{Me}/dx) \qquad \sigma_1 \ll \sigma \tag{40}$$

Under the assumptions of steady-state scale growth and local equilibrium, one can readily show that [12]

$$k_p = (V^m RT/2\alpha^2 z_1{}^2 F^2) \int_{P_{X_2}^i}^{P_{X_2}^a} \sigma_1' \, d \ln P_{X_2} \qquad \sigma_1' \ll \sigma \tag{41}$$

Equation (41) is the same as that derived in 1933 by Wagner [10] only when $\sigma_1 = \sigma_1'$, i.e., when independent migration of ions and electrons occurs as Wagner originally assumed. The use of Eq. (41) does not require the assumption of independent motion of ions and electrons because σ_1' yields a direct measure of L_{11}.

Morin [17] pointed out correctly that discrepancies occur when Wagner's original equation for the parabolic rate constant in terms of partial ionic conductivities is applied to scales in which associated defects (e.g., V_{Co}' in CoO or $Cr_i^{\cdot\cdot\cdot}$ in Cr_2O_3) are important. In this case, independent migration of ions and electrons does not occur. These discrepancies are avoided, however, through use of Eq. (41). Use of Eq. (30), (33), or (36) to calculate k_p from diffusion coefficients measured as a function of P_{X_2} is also independent of whether or not associated defects are important.

9.2.3 Review of the Assumptions in the Wagner Model

A number of assumptions were made in deriving Eqs. (30), (33), (36) and (41) for the parabolic rate constant for the oxidation of a metal. The assumptions are:

1. The scale is a homogeneous diffusion barrier.
2. Only a single oxidation product forms.
3. The fluxes of ions and electrons are unidirectional and independent of distance for a given scale thickness (quasi-steady-state scale growth).
4. Local equilibrium is maintained at the metal/scale and scale/gas interfaces and throughout the thickness of the scale.

It seems worthwhile to review these assumptions and to comment on their validity. First, however, it should be noted that it was *not* necessary to assume in deriving Eqs. (30), (33), (36), and (41) that ions and electrons migrate independently. Independent migration of ions and electrons ($L_{12} = L_{21} = 0$) was one of the original assumptions of the Wagner model for diffusion-controlled scale growth [10,11]. Indeed, in his 1951 paper [11], Wagner stated that one of the main uses of his expression for the parabolic rate constant would be to test this assumption. As demonstrated in Wagner's 1975 paper [12], it is generally not necessary to make this assumption for electronically conducting compounds. The only exception to the latter statement occurs when the rate constant is calculated using partial ionic conductivities that were measured using metallic (ion-blocking) electrodes.

Assumption No. 1

The first assumption listed above is that the scale is a homogeneous diffusion barrier, which means that it is free of voids, pores, and fissures, that the metal/scale and scale/gas interfaces are essentially flat relative to the thickness of the scale, and that the scale is adherent to the metal. Figure 3 is a schematic representation of such a scale. The formation of scales that are truly uniform diffusion barriers usually does not occur in practice. It is necessary, therefore, to decide in which cases deviations from the ideal scale morphology preclude use of Wagner's model. For example, loss of scale adherence does not necessarily mean that Wagner's model cannot be employed to describe the continued thickening of the scale as long as the scale remains intact and vapor-phase transport across the gap between the metal and the scale is rapid relative to solid-state diffusion in the scale.

When scales do become detached from the metal substrate, however, they often undergo considerable wrinkling and buckling. An example of a wrinkled scale of alumina, Al_2O_3, formed on an Fe-Al alloy [18] is

FIGURE 4 SEM secondary electron image of the surface of an Al_2O_3 scale formed on an Fe-6 wt% Al alloy in pure oxygen at 1 atm total pressure after 20 hr at 1000°C [18].

shown in Fig. 4. The occurrence of wrinkled chromia and alumina scales on alloys is quite common and is usually assumed to be the result of lateral growth of the oxide, which, in turn, is caused by formation of oxide within the scale. In cases where the scale adheres strongly to the alloy, growth stresses often cause plastic deformation of the metallic substrate as well as of the scale itself [19]. The amount of deformation or lateral growth that these scales undergo is striking, and it is questionable whether Wagner's model for diffusion-controlled scale growth can be applied in these cases.

The scales formed on metals and alloys at high temperatures are polycrystalline unless special precautions are taken to ensure epitaxial growth of monocrystalline scales. In many cases, the scales may be very fine grained, as shown in Fig. 5, and the grains may grow during the oxidation process. If the temperature of oxidation is less than about one-half to two-thirds of the absolute melting temperature of the scale, then a significant portion of the ion fluxes through the scale is likely to occur along grain boundaries. As demonstrated in Section 9.4, Wagner's model for diffusion-controlled scale growth can be altered to take into account the effects of grain boundary diffusion, including the effects of grain growth, on the magnitude of the parabolic rate constant.

0.1 μm

FIGURE 5 TEM bright-field image of a Cr_2O_3 scale formed in pure chromium after 40 min at 900°C in an H_2–H_2O gas mixture having an equilibrium oxygen partial pressure of 10^{-19} atm. The section of the Cr_2O_3 scale shown was located near and oriented parallel to the Cr/Cr_2O_3 interface [20].

FIGURE 6 SEM secondary electron image of the surface of a Cr_2O_3 scale formed on a pure chromium single crystal having a (110) orientation. The chromium was electropolished and then annealed in hydrogen at 1025°C before oxidation. The specimen was exposed for 1 hr to an H_2-H_2O gas mixture ($P_{O_2} = 10^{-17}$ atm), and then for 20 hr to an Ar-5% O_2 gas mixture. Growth ledges of the type shown form on (110) Cr under both reducing and oxidizing conditions [20].

Line defects may also influence transport of ions through scales. Figure 6 is a photomicrograph of a portion of the scale/gas interface for a chromia scale formed on a (110) orientation of a high-purity single crystal of chromium [20]. Special care was taken in this case during specimen preparation and during the oxidation experiment to promote formation of monocrystalline chromia. The ledges shown in Fig. 6 are believed to be associated with the emergence of screw dislocations from the surface of the scale. It is assumed that transport of chromium cations along the cores of the dislocations is more rapid than in the lattice of chromia. Wagner's model can be altered according to the methods outlined in Section 9.4 to take into account contributions from dislocation pipe diffusion if the dislocation density in the scale is known.

Diffusion of cations through scales predominantly along surface or line defects may give rise to a very uneven scale/gas interface. As described in more detail in Section 9.4.2, grain boundary diffusion plays a major role in the growth of the CoO and Co_3O_4 scales that form on pure cobalt at 700°C at an oxygen partial pressure of 1 atm. The ridges of oxide at the scale/gas interface shown in Fig. 7a formed over oxide grain boundaries. As shown in Fig. 7b, the height of the ridges is small relative to the total thickness of the scales. Indeed, Wagner's model, which was altered to take into account contributions from grain boundary diffusion, was used successfully to describe quantitatively the diffusion-controlled rate of growth of the Co_3O_4 layer [21,22].

A case of more extensive "roughening" of the scale/gas interface is shown in Fig. 8, which is a scanning electron microscope (SEM) image of the surface of a chromia scale formed on pure chromium at 1025°C in an H_2-H_2O gas mixture with an oxygen partial pressure of 10^{-17} atm [20]. The surface is covered by "whiskers" of oxide, which are observed when the chromium is mechanically polished, but not when it is electropolished or electropolished and lightly etched. It appears, therefore, that the whiskers are associated with structural features in the scale that are inherited from the metal substrate. It has been proposed [23,24] that whisker growth, which is observed in many other cases of metal oxidation, occurs along screw dislocations that emerge from the surface of a scale. Transport of cations along the cores of the dislocations is assumed to be very rapid relative to lattice diffusion, which would allow for extensive growth of the whiskers relative to growth of the scale itself. Voss et al. [25] have shown micrographs of hollow hematite blades formed during the oxidation of iron. They suggested that surface diffusion of iron cations through the hollow cores of the blades accounted for their rapid growth.

Rapp [24] has suggested that, depending on temperature and thus on the relative rates of cation diffusion along dislocation pipes, grain boundaries, and in the lattice of a scale, dislocations in scales may give rise to whiskers, pyramidal growth, or pits at the scale/gas interface. Whisker growth occurs when dislocation pipe diffusion of cations is rapid relative to lattice or grain boundary diffusion.

In addition to point, line, and surface defects, growing scales on metals may contain voids, pores, and fissures. Voids are isolated cavities in scales, whereas pores penetrate the entire scale. The term fissure or fault may be used to describe cracks (tensile or shear) or blisters. The latter are always induced by growth or applied stresses. Voids and pores may be induced by stresses, or may form under stress-free conditions. For example, nucleation of voids at the metal/scale interface is probably aided by growth stresses.

The development of pores in sulfide scales during their growth has been demonstrated convincingly by Mrowec and co-workers [26].

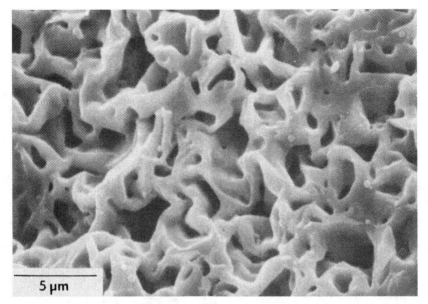

(a)

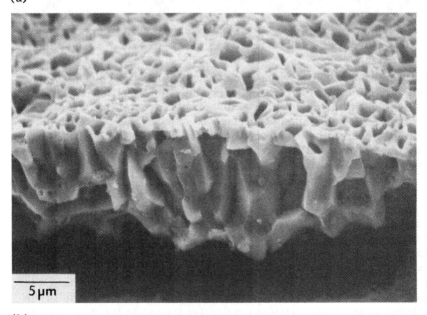

(b)

FIGURE 7 SEM secondary electron images of the Co_3O_4 scales formed on pure cobalt at 800°C: a) topography of the scale-gas interface after 21 hr in Ar-5% O_2 showing ridges of oxide formed over oxide grain boundaries; b) fracture section through Co_3O_4 (the CoO phase was dissolved in a bromine-methanol solution) after 21 hr in pure oxygen.

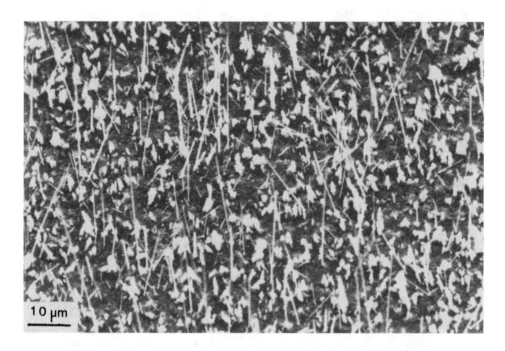

FIGURE 8 SEM secondary electron image of the surface of a Cr_2O_3
scale formed on a (110) Cr single crystal at 1025°C after 45 hr in an
H_2-H_2O gas mixture having an equilibrium oxygen partial pressure of
10^{-17} atm. The chromium specimen was polished through 1 μm diamond paste before being inserted directly into the hot gas mixture.
Shown are blades and whiskers (white features) growing from a polycrystalline Cr_2O_3 scale into the gas phase [20].

Although direct observation of the pores themselves was not made,
their presence was deduced from the results of autoradiographic experiments. The pores always formed along grain boundaries in the
scales, and their development was explained in terms of the "dissociative mechanism" [26,27]. According to the latter mechanism, the
development of a separation between a metal and a scale leads to dissociation of the scale. Oxidant molecules formed in the dissociation
process migrate across the gap to the metal surface, where they react with the metal to form new oxide. Dissociation of the scale is
proposed to occur preferentially along grain boundaries in the scale,
and this may lead eventually to formation of pores through the thickness of the scales. The process also yields a layer of porous scale
adjacent to the metal, by a mechanism that is not clear, thereby giving

rise to scales with a "duplex" microstructure (outer compact scale
with fine pores along grain boundaries and a porous inner scale).
Metallographic evidence for the development of pores along grain
boundaries in Cu_2O scales growing on copper has been presented
by Raynaud et al. [28].

Hobbs et al. [29] used transmission electron microscopy (TEM)
to study the inner layer of duplex scales formed on high-purity
nickel and reported that the inner layer is really a compact, fine-
grained oxide containing very few, if any, voids; see Fig. 9. They
contended, in fact, that the voids observed in NiO scales by others
using optical or scanning electron microscopy were artifacts of specimen preparation. They also reported that pores do not exist along
the grain boundaries of the outer portion of the scale. One must be
careful, however, in comparing the work of various investigators in
deciding whether duplex scales form at temperature or are simply
artifacts. In particular, it appears that impurities tend to promote
formation of duplex scales (e.g., C or Cr in Ni forming NiO scales),
and various investigators do not always use metals of the same purity.

It is questionable whether the dissociative mechanism can be applied to the growth of protective chromia, alumina, or silica scales.
The dissociative mechanism requires that scale growth occur predominantly by cation diffusion, and it appears that this criterion is
not met for alumina and silica and perhaps for chromia under certain
conditions. Furthermore, if a void formed at the interface between a
chromia scale and its metallic substrate, the rate of transport of chromium vapor across the void could be greater than that of oxygen-
bearing gases, thereby precluding the onset of pore formation by a
dissociative mechanism.

It appears that voids can originate either at the metal/scale or at
the scale/gas interface. In the former case, voids may develop by a
mechanism of vacancy condensation when scale growth involves cation
diffusion. The latter may occur by a vacancy, interstitial, or grain
boundary diffusion mechanism. If anion diffusion predominates, the
associated stresses may give rise to creep cavitation in the substrate.

Voids may develop at the scale/gas interface when cations diffuse
to that interface mainly along grain boundaries in the oxide. Cavities develop over grain interiors as ridges grow over the grain boundaries. Cavities may be shut off from the gas phase by sideways growth
of ridges, which creates a void in the oxide [21].

Void formation within a scale may require growth or applied stresses
(creep cavitation). It has been suggested in the literature, however,
that voids can nucleate within scales that become supersaturated with
respect to vacancies. It is not at all clear, however, that vacancy supersaturation can occur in growing scales. Wagner [30] demonstrated
that only small deviations from internal equilibrium typically occur in

rise to scales with a "duplex" microstructure (outer compact scale with fine pores along grain boundaries and a porous inner scale). Metallographic evidence for the development of pores along grain boundaries in Cu_2O scales growing on copper has been presented by Raynaud et al. [28].

Hobbs et al. [29] used transmission electron microscopy (TEM) to study the inner layer of duplex scales formed on high-purity nickel and reported that the inner layer is really a compact, fine-grained oxide containing very few, if any, voids; see Fig. 9. They contended, in fact, that the voids observed in NiO scales by others using optical or scanning electron microscopy were artifacts of specimen preparation. They also reported that pores do not exist along the grain boundaries of the outer portion of the scale. One must be careful, however, in comparing the work of various investigators in deciding whether duplex scales form at temperature or are simply artifacts. In particular, it appears that impurities tend to promote formation of duplex scales (e.g., C or Cr in Ni forming NiO scales), and various investigators do not always use metals of the same purity.

It is questionable whether the dissociative mechanism can be applied to the growth of protective chromia, alumina, or silica scales. The dissociative mechanism requires that scale growth occur predominantly by cation diffusion, and it appears that this criterion is not met for alumina and silica and perhaps for chromia under certain conditions. Furthermore, if a void formed at the interface between a chromia scale and its metallic substrate, the rate of transport of chromium vapor across the void could be greater than that of oxygen-bearing gases, thereby precluding the onset of pore formation by a dissociative mechanism.

It appears that voids can originate either at the metal/scale or at the scale/gas interface. In the former case, voids may develop by a mechanism of vacancy condensation when scale growth involves cation diffusion. The latter may occur by a vacancy, interstitial, or grain boundary diffusion mechanism. If anion diffusion predominates, the associated stresses may give rise to creep cavitation in the substrate.

Voids may develop at the scale/gas interface when cations diffuse to that interface mainly along grain boundaries in the oxide. Cavities develop over grain interiors as ridges grow over the grain boundaries. Cavities may be shut off from the gas phase by sideways growth of ridges, which creates a void in the oxide [21].

Void formation within a scale may require growth or applied stresses (creep cavitation). It has been suggested in the literature, however, that voids can nucleate within scales that become supersaturated with respect to vacancies. It is not at all clear, however, that vacancy supersaturation can occur in growing scales. Wagner [30] demonstrated that only small deviations from internal equilibrium typically occur in

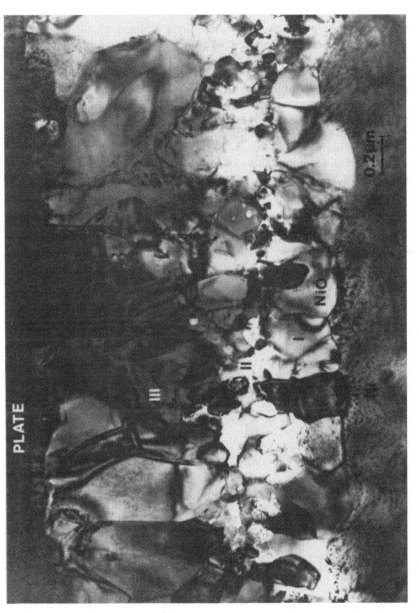

FIGURE 9 TEM bright-field image of a cross section through an NiO scale formed on pure nickel after 15 min in pure oxygen at a total pressure of 0.03 atm at 1000°C. The oxidized specimen was plated with nickel before being thinned to electron transparency. Note the very fine-grained oxide and the submicrometer cavities near the Ni/NiO interface [29].

to extend the scale/gas interface—very little is used to adjust the overall composition of the scale. Also, the values of $\alpha+\delta$ and the molar volume of $Me_{\alpha+\delta}X$ are essentially constant across the thickness of the scale.

Hirth and Rapp [33] examined the assumption of quasi-steady-state scale growth and suggested modifications of the expressions for the parabolic rate constant for scales that have large composition ranges and large variations in molar volume (e.g., FeO, FeS, MnO, CrS). Their treatment allows for a flux that depends on diffusion distance, $j_{Me} = j_{Me}(x)$, and employs values of the Me/X ratio, $\overline{(\alpha+\delta)}$, and the molar volume of $Me_{\alpha+\delta}X$, $\overline{V^m}$, that are averages over the scale thickness.

The average value of the cation flux is given by

$$\overline{j}_{Me} = 1/\Delta x \int_0^{\Delta x} j_{Me}(x) \; dx = f j_{Me}(Me/Me_{\alpha+\delta}X) \tag{42}$$

with

$$f = 1 + [1/\overline{c}_{Me}(\Delta x)^2] \int_0^{\Delta x} [\overline{c}_{Me}(x) - c_{Me}(x)]x \; dx \tag{43}$$

where $\overline{c}_{Me}$ is the average value of c_{Me} over the scale thickness, $\overline{c}_{Me}(x)$ is the average value of c_{Me} in the interval between $x = 0$ and $x = x$, and $c_{Me}(x)$ is the value of c_{Me} at a distance x from the metal/scale interface. The rate thickening of the scale would then be given by

$$d \; \Delta x/dt = \overline{j}_{Me} \overline{V^m}/(\overline{\alpha+\delta})f = \overline{j}_{Me}/\overline{c}_{Me} f \tag{44}$$

Inserting Eq. (44) into Eq. (1) and employing Fick's first law, Eq. (19), for $j_{Me}(x)$ in Eq. (42), one obtains

$$k_p = [\overline{V^m}/(\overline{\alpha+\delta})f] \int_{c_{Me}^i}^{c_{Me}^a} D_{Me} \; dc_{Me} \tag{45}$$

where c_{Me} and c_{Me} are the concentrations of Me in $Me_{\alpha+\delta}X$ at the metal/scale and scale/gas interfaces, respectively. The latter equation has the same form as that derived earlier for the parabolic rate constant, Eq. (26). It differs in that the values of $\overline{V^m}$ and $\overline{\alpha+\delta}$ are averages over the scale thickness and the weighting factor f is included to account for the fact that as Me is added locally to the scale, the volume increase per mole of Me is more heavily weighted in regions where $V^m(x)$ is large [39].

Evaluation of the weighting factor f and the average values of V^m and $(\alpha+\delta)$ requires knowledge of the concentration gradient in the growing scale. Hirth and Rapp [39] evaluated f for some hypothetical cases and concluded that it is near unity unless the composition gradient has an extreme deviation from linearity. The modified expression for the parabolic rate constant has not been tested experimentally. Indeed, for a single-phase scale comprising a compound that exhibits a wide deviation from stoichiometry and large variations in the molar volume, and therefore rapid diffusion, surface reactions are likely to be rate-limiting, thereby yielding nonparabolic oxidation behavior—e.g., oxidation of iron to FeO in CO/CO_2 gas mixtures [40]. The modified expression for k_p may, however, be applied to the growth of compounds such as FeO under conditions where additional compounds (Fe_3O_4, Fe_2O_3) may form at the scale/gas interface, thereby precluding a reaction at the scale/gas interface as the rate-limiting step. The treatment by Hirth and Rapp is also useful in illustrating that the assumption of quasi-steady-state scale growth is valid as long as $\bar{c}_{Me}(x) \approx c_{Me}(x)$ at any distance x from the metal/scale interface. In this case, which occurs for small deviations from stoichiometry, the factor f is equal to unity and the quantities $\overline{V^m}$ and $\overline{(\alpha+\delta)}$ may be replaced, respectively, by the molar volume and the Me/X ratio for stoichiometric MeX.

Assumption No. 4

The fourth assumption of the Wagner model is that equilibrium is maintained at the metal/scale and scale/gas interfaces and throughout the thickness of the scale. If local equilibrium is not achieved at the metal/scale or scale/gas interface, nonparabolic oxidation kinetics are observed. If the rate of thickening of a compact, pore-free, adherent scale were limited, for example, by the rate of a chemical reaction at the scale/gas interface or by gas-phase diffusion, then the scale would not be in equilibrium with the gas phase [40,41]. Furthermore, if the oxidation rate were limited entirely by a surface reaction or by gas-phase diffusion, then the gradient in the chemical potential of metal or metalloid across the scale would be negligible and the oxidation process would obey a linear rate law. As scale thickening proceeded, however, solid-state diffusion in the scale would eventually play a role in the rate of scale thickening because the diffusion flux is inversely proportional to scale thickness. As the rate-limiting step changed from a surface reaction or gas-phase diffusion to solid-state diffusion, the oxidation kinetics would change from linear to parabolic behavior. The duration of the transition stage depends decisively on the properties of the scale [41]. If the duration of the transition stage is long relative to the total oxidation time, a plot of $(\Delta m/A)^2$ or

Δx^2 vs. time may appear to be linear, indicating parabolic oxidation
behavior. Many authors have, in fact, succumbed to the temptation
to fit a straight line to data obtained in the transition stage. The resulting parabolic rate constant is, of course, not a true parabolic rate
constant, and comparisons of such rate constants should not be made
with values calculated using the Wagner model.

The assumption that local equilibrium exists throughout the growing scale is introduced through Eqs. (8)−(11); i.e., it is assumed
that the electrochemical potentials of cations, anions, and electrons
and the chemical potentials of the metal and metalloid have definite,
well-defined values. Fromhold et al. [42] and Morin [17] questioned
whether it is possible to apply equations based on thermodynamic concepts to the growth of scales on metals. Wagner [30] demonstrated
that only small deviations from internal equilibrium typically occur in
growing scales because the relaxation times for defect equilibration
are normally much shorter than the reaction times for oxide layer formation on metals [15]. The effect is more significant in compounds
with lower defect concentrations.

9.3 APPLICATION OF WAGNER'S MODEL

Wagner's model for diffusion-controlled oxidation of metals has been
compared in the literature with experimentally determined parabolic
rate constants for the oxidation of a large number of metals. In this
chapter, in order to demonstrate that the essential features of Wagner's model are correct, comparisons of calculated and experimental
rate constants will be made for only two cases, i.e., oxidation of cobalt to form CoO and oxidation of FeO to form Fe_3O_4. Discussions of
additional cases may be found elsewhere [8,16,32,43].

9.3.1 Oxidation of Cobalt

The oxidation of cobalt at temperatures above approximately 925°C
yields a single-phase scale of CoO. Although the partial ionic conductivity of cobalt ions in CoO has not been measured, the chemical
and tracer diffusion coefficients for cobalt and oxygen in CoO have
been measured by many investigators. Only tracer diffusion data
will be used here to calculate the parabolic rate constant. Since
$D_{Co} \gg D_O$ in CoO [44], Eq. (33) may be used to calculate the parabolic rate constant.

The tracer diffusion for Co in CoO is plotted in Fig. 10 as a function of oxygen partial pressure at 1000°C. The plot of log D_{Co}^T vs.
log P_{O_2} was made using Dieckmann's [45] expression for the dependence of D_{Co}^T on temperature and the deviation from stoichiometry in
CoO. Values of D_{Co}^T at 1000°C measured by other investigators [46–48]

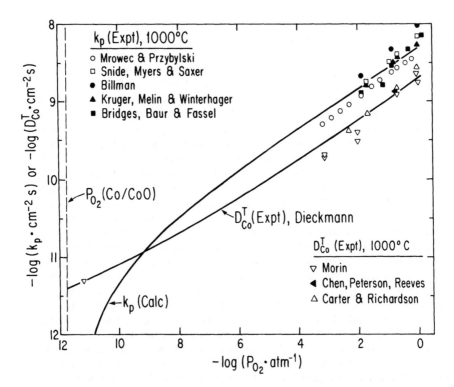

FIGURE 10 Comparison of calculated and experimental values of the parabolic rate constant for the oxidation of cobalt as a function of oxygen partial pressure at 1000°C. The curve labeled D_{Co}^T(Expt) was determined by Dieckmann [45] to fit experimental data for the tracer diffusion coefficient.

agree quite well with those obtained by Dieckmann. To use Eq. (33) to calculate k_p, one must know the correlation factor, f_{Co}, for correlated jumps of tracer cations in CoO. This requires knowledge of the diffusion mechanism, and thus represents a deviation from a truly phenomenological approach to calculating the parabolic rate constant. For CoO, the majority point defects appear quite clearly to be cation vacancies [45]. Since diffusion occurs on a face-centered-cubic cation sublattice in CoO (rock salt crystal structure), the appropriate value of f_{Co} is 0.78.

Upon inserting into Eq. (33) the appropriate values of D_{Co}^T, f_{Co}, and $P_{O_2}^{\frac{1}{2}}$ (1.86 × 10^{-12} atm) for 1000°C, one can calculate the parabolic rate constant as a function of oxygen partial pressure. The results of such a calculation are presented in Fig. 10. Also included

in Fig. 10 are experimentally determined values of k_p [49–53]. Gravimetric parabolic rate constants, k_g, were determined experimentally.
Values of k_g, with units of g^2/cm^4 sec, were converted to k_p, with
units of cm^2/sec, by multiplying the former rate constant by $(1/2)$
$(V_{CoO}/16)^2$, Eq. (3a), where $V_{CoO} = 11.62$ $cm^3/mole$ CoO.

The agreement between the experimental and calculated values of
k_p is quite good and appears to be limited mainly by the precision of
the measurements of D_{Co}^T and k_p. Both the experimental and calculated values of k_p have about the same dependence on P_{O_2}, as demonstrated in Fig. 10. It is concluded, therefore, that the essential
features of Wagner's model are verified. Deviations between experimental and calculated values of k_p for other systems are due, apparently, to deviations from the basic assumptions of Wagner's model.

Because Wagner's oxidation model is phenomenological in nature,
it is not required to know the details of the defect structure of the
oxidation product to employ the model. All that is needed is the experimentally determined dependence of D_{Me}^T, $\tilde{D}$, or σ_1' on P_{X_2}. When
Eq. (30) is used, the dependence of δ on P_{X_2} is also needed. Use of
D_{Me} requires a value for f_{Me}, which requires knowledge of the predominant type of defects in the oxidation product (i.e., vacancies or
interstitials) and, if an interstitialcy diffusion mechanism is involved,
which of the interstitialcy diffusion mechanisms predominates.

In the case of CoO, Dieckmann [45] has shown that the majority
point defects are either singly or doubly charged cation vacancies,
depending on the oxygen partial pressure, and that ideal mixing of
defects can be assumed. Neutral cation vacancies and interstitial cobalt cations may, however, play a minor role in determining the properties of CoO; e.g., Gesmundo and Viani [54] have argued that data
for $D_{Co}^T(CoO)$ are not inconsistent with the presence of a significant
concentration of doubly charged interstitial cobalt cations at oxygen
pressures approaching the decomposition pressure of CoO. The latter authors showed that when a model that includes cobalt interstitials is fit to the diffusion data, a slightly higher value is calculated
for the parabolic rate constant at low oxygen pressures (e.g., $P_{O_2} <$
10^{-6} atm at 1000°C). No data for k_p are available to compare with
the calculated values at these low oxygen partial pressures.

It may be noted in Fig. 10 that the slope of log D_{Co}^T (CoO) vs.
log P_{O_2} varies with P_O reflecting the change with P_O in the relative concentrations of the various point defects in CoO. Also, both
experimental values of D_{Co}^T and calculated values of k_p exhibit about
the same dependence on P_{O_2} at high values of P_{O_2}, but not at low
oxygen pressures. The variation of k_p with P_{O_2} does not reflect the
subtle changes in D_{Co}^T with P_{O_2} under reducing conditions because
k_p decreases rapidly as the driving force for oxidation approaches
zero $(P_{O_2}^a \rightarrow P_{O_2}^i)$.

9.3.2 Oxidation of Wustite to Magnetite

The major point defects in magnetite, Fe_3O_4, at low oxygen pressures are interstitial iron cations, while at higher oxygen partial pressures cation vacancies are the predominant point defects [55–57]. Unlike the case of CoO, the diffusion coefficient for cations in Fe_3O_4 exhibits a marked change in its dependence on P_{O_2} with a change in the predominant type of defect; i.e., at a constant temperature, the dependence of the tracer diffusion coefficient on P_{O_2} can be described approximately by the following relationship [55–57]:

$$D_{Fe}^T(Fe_3O_4) = D_I^T P_{O_2}^{-2/3} + D_V^T P_{O_2}^{-2/3} \tag{46}$$

where

$$D_I^T = 7.94 \times 10^7 \exp[-613,200 \ (J/mole)/RT] \tag{47}$$

and

$$D_V^T = 3.98 \times 10^{-11} \exp[139,600 \ (J/mole)/RT] \tag{48}$$

Subscripts V and I refer to diffusion by a vacancy or interstitial-type diffusion mechanism, respectively. Because of the marked variation of the tracer diffusion coefficient with P_{O_2}, it is useful to examine the variation of k_p with P_{O_2} for the oxidation of wustite, $Fe_\nu O$, to magnetite.

The oxidation of wustite to magnetite has been studied by a number of investigators [58–61]. A model is shown in Fig. 11 for the growth of magnetite with an Fe/O ratio of 0.75 on wustite that had previously been annealed at a P_{O_2} close to that for equilibrium between wustite and magnetite (Fe/O ratio = ν'). Magnetite grows at both the wustite/magnetite and magnetite/gas interfaces [58]. Assuming that $D_{Fe} \gg D_O$, one can derive [59,60] the following expression for the gravimetrically determined parabolic rate constant, using Wagner's model for diffusion-controlled scale growth:

$$k_g = \frac{\nu' - 0.75}{0.75\nu'} \left(\frac{16}{V_{Fe_{0.75}O}} \right)^2 \int_{P_{O_2}^i}^{P_{O_2}^a} (D_{Fe}^T/f_{Fe}) \ d \ln P_{O_2} \tag{49}$$

The values of D_I^T and D_V^T in Eq. (46) must be divided by the appropriate correlation factor, f_I or f_V, respectively. The exact

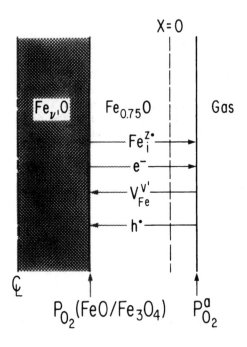

FIGURE 11 Model for the oxidation of wustite, $Fe_\nu O$, to magnetite, $Fe_{0.75}O$; ν' is the Fe/O ratio in wustite saturated with respect to magnetite; $x = 0$ is the original surface of the wustite specimen.

mechanisms of diffusion in magnetite are not known. If vacancy diffusion occurs only over octahedral lattice sites, $f_V = 0.56$; however, if vacancy diffusion occurs only over tetrahedral lattice sites, $f_V = 0.5$ [62]. Using data for cation tracer diffusion and point defect concentrations in magnetite, Dieckmann [63] demonstrated that f_V is 0.56 in the temperature range of about 900–1100°C. He also showed that f_V decreases slightly as the temperature is increased above 1100°C, apparently because there is a greater contribution from vacancy diffusion over the tetrahedral sublattice at higher temperatures.

For a simple interstitial diffusion mechanism, $f_I = 1.0$. If, however, an interstitialcy mechanism prevails, f_I is less than unity and the exact value of f_I depends on the details of the diffusion steps. Using the geometric correlation factors calculated by Peterson et al. [62], Dieckmann [63] concluded that the value of f_I varied from 0.345 for noncollinear jumps of the type $Fe_I(tet) \rightarrow Fe(oct) \rightarrow Fe_I(tet)$ to 0.72 for jumps of the type $Fe_I(oct) \rightarrow Fe(tet) \rightarrow Fe_I(oct)$. The symbols Fe(oct) and Fe(tet) represent iron cations on normally occupied octahedral and

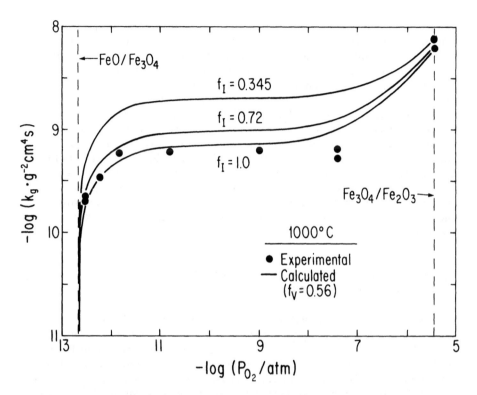

FIGURE 12 Comparison of calculated and experimental [59,60] parabolic rate constants for the oxidation of wustite to magnetite at 1000°C in CO/CO_2 gas mixtures. The two points at P_{O_2} (Fe_3O_4/Fe_2O_3) were obtained by oxidizing wustite in an Ar-5% O_2 gas mixture, which yielded a thin layer of hematite at the oxide/gas interface.

tetrahedral sites, while $Fe_I(tet)$ and $Fe_I(oct)$ represent iron cations on normally unoccupied interstitial sites.

Peterson et al. [62] concluded that interstitialcy jumps of the type $Fe_I(oct) \rightarrow Fe(tet) \rightarrow Fe_I(oct)$ were the most probable based on geometric arguments. Dieckmann [63] demonstrated, however, that noncollinear interstitialcy jumps of the type $Fe_I(tet) \rightarrow Fe(oct) \rightarrow Fe_I(tet)$ with $f_I = 0.345$ predominate at lower temperatures (850–1100°C), while an interstitial mechanism with $f_I = 1.0$ seems to prevail at much higher temperatures ($\sim$1300–1400°C).

Using $f_V = 0.56$ and $f_I = 1.0$, 0.72, and 0.345, one calculates the dependence of k_g on P_{O_2} shown in Fig. 12. The interesting result is that the parabolic rate constant is independent of the driving force for oxidation for about four orders of magnitude in the oxygen partial pressure. This occurs at low values of P_{O_2} because diffusion by interstitial cations predominates. At higher values of P_{O_2}, the parabolic rate constant increases as P_{O_2} increases because diffusion by cation vacancies predominates. The parabolic rate constant drops sharply to zero as P_{O_2} approaches P_{O_2} (FeO/Fe_3O_4).

The essential features of this model, which, aside from mass balances and the inclusion of appropriate correlation factors, is based on Wagner's model for diffusion-controlled scale growth, have been confirmed experimentally [59,60] as indicated in Fig. 12. It should be noted, however, that Dieckmann et al. [61] performed the same kind of experiments and observed nonparabolic oxidation kinetics, which they attributed to a slow reaction at the magnetite/gas interface. The reasons for the discrepancies between the experimental results are not apparent.

9.4 EFFECTS OF SCALE MICROSTRUCTURE ON OXIDATION KINETICS

There are often wide variations in oxidation kinetics of metals and alloys that are oxidized, apparently, under exactly the same conditions. Aside from the effects of impurities, which may vary from one batch of "pure" metal to the next, the variations in oxidation kinetics can be ascribed to differences in the microstructures of the scales formed. The differences in microstructures arise because of slight differences in surface preparation and the procedure used to carry out an oxidation experiment, and because of effects of the crystallographic orientation of the metallic phase on the microstructure of the oxidation product. The microstructural features of scales that affect oxidation kinetics include voids, pores, second-phase, particles, low- and high-angle grain boundaries, dislocations, etc. Each of these microstructural features can influence transport processes in the scale and, hence, oxidation kinetics.

The microstructure of a scale may also influence the magnitude of the growth stresses and/or stress distribution in a growing scale, which may have an effect on the transport properties of a scale during the steady-state period of oxidation. The mechanisms of generation and relief of stresses in a growing scale, which one would expect to be affected by scale microstructure, can also determine the time at which the transition from steady-state to breakaway oxidation occurs.

This section is divided into two parts. Evidence for the influence of scale microstructure on oxidation kinetics is discussed in the first part, and models for the effects of "short-circuit diffusion" on the diffusion-controlled oxidation of metals are described in the second.

9.4.1 Effects of Crystallographic Orientation, Surface Preparation, and Oxidation Procedure

The influence of crystallographic orientation of the metal substrate on oxidation kinetics has been studied by a number of investigators [64–69]. The occurrence of "oxidation-rate anisotropy" has been demonstrated dramatically by the development of interference color patterns on single-crystal spheres of various metals [64,65]. The variations in color of the oxide films that develop on different crystallographic orientations of the metal are caused by variations in oxide film thickness.

Cathcart [65] reported that the rate of growth of thin films of NiO on various crystallographic planes of nickel decreased in the order (110), (100), (111), and (311). The NiO formed on (110) nickel was polycrystalline; however, the NiO formed on the other three orientations was related epitaxially to the substrate. The oxide on (311) nickel was a single crystal although it contained low-angle boundaries. Twin boundaries in the NiO formed on (111) and (100) nickel were incoherent. More incoherent twin boundaries were found in the NiO on (100) nickel, thereby accounting for the faster rate of growth on the (100) nickel; i.e., diffusion of ions along incoherent twin boundaries in NiO was faster than diffusion in the lattice of NiO.

Anistropy of oxidation of nickel has been observed to occur for NiO scales as thick as 15 µm at 800°C [68]. Oxidation-rate anistropy has also been demonstrated for polycrystalline metals; i.e., thick polycrystalline oxide forms on some metal grains, while monocrystalline oxide forms on other metal grains of the same specimen [70,71].

The crystallographic orientation of a metal generally has an effect on oxidation kinetics only if the metal is well annealed and if the oxidation procedure does not give rise to a high rate of nucleation of oxide on the metal surface. Oxidation-rate anisotropy of *annealed* polycrystalline nickel [71] occurs when a nickel specimen, which is coated with an oxide film formed at room temperature, is heated in the presence of oxygen. If an annealed nickel specimen is heated first in hydrogen to avoid film formation, the oxide that develops is fine-grained and uniform in thickness [71]. Under the former conditions, epitaxial growth of NiO on some grains of nickel apparently occurs during heatup and subsequent oxidation. When nickel is heated first in hydrogen, however, a large number of randomly oriented NiO nuclei form on the bare nickel surface regardless of crystallographic orientation, and hence the oxide scale that develops is uniform in thickness. The rate of oxidation is higher when the entire scale comprises fine grains of NiO because the grain boundaries act as short-circuit paths for diffusion.

Cold-worked (abraded) iron oxidizes faster than electropolished or annealed iron in the temperature range 400–600°C because the grain size of the oxide formed on the cold-worked metal is finer [72]. The oxidation rate increases as the degree of cold work increases, but at 650°C the difference in the oxidation rate of cold-worked and annealed iron is negligible [72]. Effects of cold work on the oxidation rate of nickel, however, have been observed for temperatures up to 1270°C [73].

9.4.2 Models for the Effects of Short-Circuit Diffusion on Oxidation Kinetics

A number of investigators have examined the effects of short-circuit diffusion within scales on the kinetics of metal oxidation [21,22,69, 74–78]. At temperatures less than approximately one-half the melting point of an oxide, diffusion along low- or high-angle boundaries or dislocations may make a significant contribution to the net transport of ions through a growing scale. The general effect of enhanced transport by short-circuit diffusion paths is an increase in the rate of oxidation relative to the case in which transport of ions through the lattice of the oxidation product is the rate-controlling step.

If the number of short-circuit diffusion paths decreases (e.g., by oxide grain growth) simultaneously with scale thickening, the relative contributions of short-circuit and lattice diffusion are expected to change with time. As a result, a plot of the square of the scale thickness or the weight change per unit area of a specimen versus time will not be a straight line even though solid-state diffusion in the scale controls the rate of scale growth. The slope of such a plot would decrease as time increased, indicating a decreasing number of short-circuit diffusion paths with increasing time.

An example of the effect of short-circuit diffusion on oxidation kinetics is shown in Fig. 13, which depicts kinetic data for the oxidation of pure cobalt at 700°C in pure oxygen at 1 atm total pressure [22]. Before the specimens were exposed to oxygen at 700°C, they were given either a long-time anneal (36 hr at 1000°C, 3–5 hr at 700°C) or a short-time anneal (5 hr at 1000°C, 3–5 hr at 700°C). The preannealing affected the oxidation kinetics in a reproducible fashion.

The plots in Fig. 13 are not straight lines. Assuming that the rate of scale growth is controlled by solid-state diffusion in the scale and that the contribution of short-circuit diffusion is temporal, the kinetic data may be characterized by an instantaneous parabolic rate constant, k_g^i, which is defined by

$$k_g^i = d(\Delta m/A)^2/dt \tag{50}$$

The value of k_g^i is the instantaneous slope of a plot of $(\Delta m/A)^2$ versus time. It may be determined graphically or analytically. If the kinetic data can be fit to an equation of the type

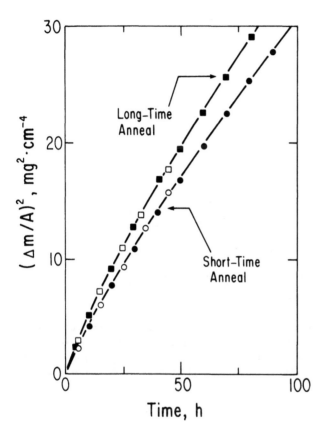

FIGURE 13 Square of the weight change per unit area vs. reaction time for cobalt specimens that had received a long-time (squares) or a short-time (circles) preoxidation anneal; oxidation in pure oxygen at 1 atm, 700°C, specimens polished through 1-μm diamond paste [22].

$$(\Delta m/A)^n = at + b \qquad (51)$$

where n, a, and b are constants, then

$$(\Delta m/A)^2 = (at + b)^{2/n} \qquad (52)$$

Upon differentiating Eq. (52) with respect to t and inserting the result into Eq. (50), one obtains

$$k_g^i = (2a/n)(at + b)^{(2 - n)/n} \qquad n \neq n(t) \qquad (53)$$

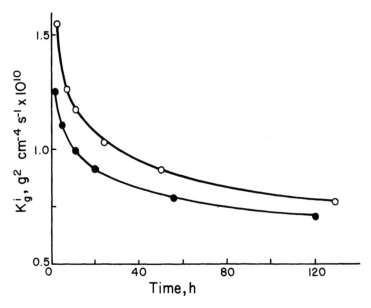

FIGURE 14 Dependence of the instantaneous parabolic rate constant on reaction time for the oxidation of cobalt at 700°C in pure oxygen at 1 atm; k_g^i is the slope of the plot of $(\Delta m/A)^2$ vs. time (Fig. 13) at a particular time of reaction. $\circ$ = Long-time anneal, $\bullet$ = short-time anneal [22].

The restriction $n \neq n(t)$ means that this analytical technique can be employed for periods of oxidation during which Eq. (51) describes all the kinetic data. Eventually, n may approach a value of 2.0, indicating true parabolic kinetics, if the density of short-circuit diffusion paths becomes constant or too low to contribute significantly to the net transport of ions through the scale.

The data in Fig. 13 were fit to a straight line using linear regression analysis, which yielded values of n that were constant up to 129 hr of oxidation, the longest time an experiment was run (n = 2.42 or 2.32 for specimens that had received the long- or short-term annealing pretreatment, respectively). Values of k_g^i determined by this technique are presented in Fig. 14.

The effect of short-circuit diffusion on oxidation kinetics may be modeled quantitatively by using the approach pioneered by Smeltzer and co-workers [68,69,75]. That approach is expanded in the following to include the effects of short-circuit diffusion by both anions and cations. For simplicity, it is assumed that grain boundaries are the only effective short-circuit diffusion paths.

The expression for the parabolic rate constant for the growth of $Me_\alpha X$ on metal Me, Eq. (36), may be modified by replacing the self-diffusion coefficients for cations and anions by effective self-diffusion coefficients, D_{Me}^{eff} and D_X^{eff}, which represent contributions from both lattice and grain boundary diffusion, i.e.,

$$k_p = (1/2) \int_{P_{X_2}^i}^{P_{X_2}^a} [(D_{Me}^{eff}/\alpha) + D_X^{eff}] \; d \ln P_{X_2} \tag{54}$$

The effective self-diffusion coefficients may be defined by [79]

$$D_i^{eff} = f_t D_i^b + (1 - f_t) D_i^l \tag{55}$$

where D_i^b and D_i^l are the grain boundary and lattice diffusion coefficients for species i, respectively, and f_t is the fraction of the cross-sectional area of the oxide along which grain boundary diffusion can occur. The subscript t is used to emphasize that f may be a function of time. The latter quantity is defined by

$$f_t = gd/\lambda_t \tag{56}$$

where g is a factor that depends on the shape of the grains (e.g., $g = 2$ for square grains), d is the grain boundary width, and λ_t is the grain size of the scale. If λ_t is a function of time, then the effective diffusion coefficients are time-dependent and k_p in Eq. (54) should be replaced by k_p^i.

The results of grain growth studies in materials can generally be fit to an equation of the form [80]

$$\lambda_t^m - \lambda_0^m = G(t - t_0) \tag{57}$$

where λ_t and λ_0 are average grain sizes at time t and at time t_0, which is the time at which grain growth starts, m is a real positive number, and G is the grain growth constant. Usually, m is equal to unity when recrystallization occurs; when grain growth occurs, m can be a number between 2 and 10 [81--83]. General cases have been treated by Matsunga and Homma [76], who have taken into account both recrystallization and grain growth and a distribution of grain sizes in a growing scale.

Upon insertion of Eq. (55) into Eq. (54), in which k_p is replaced by k_p^i, and noting that $(1 - f_t) \approx 1.0$, one obtains

$$k_p^i = \frac{1}{2} \int_{P_{X_2}^i}^{P_{X_2}^a} \left(\frac{D_{Me}^l}{\alpha} + D_X^l \right) d \ln P_{X_2}$$

$$+ \frac{1}{2} \int_{P_{X_2}^i}^{P_{X_2}^a} f_t \left(\frac{D_{Me}^b}{\alpha} + D_X^b \right) d \ln P_{X_2} \qquad (58)$$

or, assuming that f_t is not a function of P_{X_2},

$$k_p^i = k_p^l + f_t k_p^b \qquad (59)$$

where k_p^l is the parabolic rate constant for the case in which lattice diffusion controls the rate, Eq. (36), and k_p^b is the parabolic rate constant for the case in which grain boundary diffusion controls the rate of oxidation. Note that k_p is the sum of k_p^l and $f_t k_p^b$ because lattice and grain boundary diffusion occur in parallel in the scale.

Equation (58) may be rewritten as

$$k_p = k_p^l + \frac{1}{2} \int_{P_{X_2}^i}^{P_{X_2}^a} f_t D_{Me}^l \left(\frac{D_{Me}^b}{D_{Me}^l \alpha} + \frac{D_X^b}{D_{Me}^l} \right) d \ln P_{X_2} \qquad (60)$$

The latter equation is quite general and should be useful in examining the relative effects of grain boundary and lattice diffusion of both cations and anions in growing scales. To demonstrate the salient features of the model under consideration, it is useful to examine some limiting cases. If, for example, $D_{Me}^b \gg D_X^b$, $D_{Me}^l \gg D_X^l$, D_{Me}^l and D_{Me}^b have the same dependences on P_{X_2}, and f_t is not a function of P_{X_2}, one obtains

$$k_p^i = k_p^l [1 + f_t (D_{Me}^b / D_{Me}^l)] \qquad (61)$$

where k_p is now given by Eq. (33).

Upon inserting Eqs. (56) and (57) into Eq. (61), one obtains

$$k_p^i = k_p^l \left\{ 1 + (D_{Me}^b / D_{Me}^l)(gd) / [G(t - t_0) + \lambda_0^m]^{1/m} \right\} \qquad (62)$$

Note that k_p^i and k_p^l may be replaced by k_g^i and k_g^l with no other changes.

An assumption used in deriving Eq. (61) was that D_{Me}^l and D_{Me}^b have the same dependence on P_{X_2}, although it is not necessary to make this assumption to solve Eq. (58) or (60). If this assumption is verified by experimental results, however, it implies that the concentrations of point defects in grain boundaries (or other short-circuit diffusion paths) and in the lattice have the same dependence on P_{X_2}. In this case, the typical observation that short-circuit diffusion is faster than lattice diffusion may be interpreted in terms of a higher mobility of ions in short-circuit paths. However, this conclusion ignores the possibility of segregation of impurities to short-circuit diffusion paths and possible effects of such impurities on defect concentrations.

The oxidation of some metals has been observed to obey a cubic rate las; that is, in Eq. (5) n = 3.0. If n = 3.0, then, according to Eq. (53), $k_g^i \propto ^{-1/3}$ if at >> b. The latter dependence of k_g^i on oxidation time is obtained from Eq. (62) if m = 3, $G(t - t_0) \gg \lambda_0$, and $t \gg t_0$. Thus, the cubic rate law may be associated with grain boundary diffusion and a specific grain growth exponent. In most cases, the actual kinetic data can be fit just as well, if not better, by other values of n. Most investigators have, however, chosen to round off to the value n = 3.0.

The oxidation of cobalt in pure oxygen at 1 atm pressure and 700°C yields a two-phase layered scale comprising CoO and Co_3O_4 [21]. The grain size of the Co_3O_4 layer increases with increasing time of oxidation as shown in Fig. 15; however, the grain size of the Co_3O_4 is not a function of the preannealing treatment of the cobalt. The straight line in Fig. 15 can be given by [22]

$$\lambda_t = 0.82t^{1/4} \tag{63}$$

i.e., t_0 and λ_0 in Eq. (57) are zero. Inserting Eq. (63) into Eq. (62) to calculate $k_p^i(Co_{0.75}O)$, the instantaneous parabolic rate constant for the growth of the spinel phase, one obtains

$$k_p^i(Co_{0.75}O) = k_p^l(Co_{0.75}O)\left[1 + 1.22gd\left(\frac{D_{Co}^b}{D_{Co}^l}\right)t^{-1/4}\right] \tag{64}$$

Hence, a plot of experimental values of $k_p^i(Co_{0.75}O)$ vs. $t^{-1/4}$ should be a straight line if, in fact, $k_p^i(Co_{0.75}O)$ decreases as time increases because of oxide grain growth. This was found to be the case [22] as illustrated in Fig. 16.

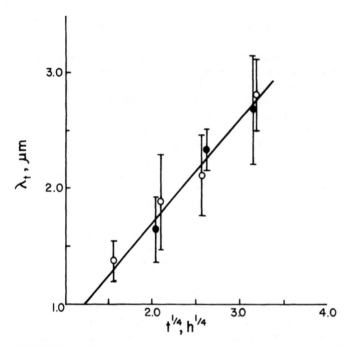

FIGURE 15 Dependence of the average grain size of the spinel layer (Co_3O_4) on reaction time for oxidation of cobalt at 700°C in pure oxygen at 1 atm. ○ = Long-time anneal, ● = short-time anneal [22].

As a second limiting case, it may be assumed that f_t is not a function of P_{X_2} and that $D_X^b \gg D_{Me}^b$ and $D_{Me}^l \gg D_X^l$, in which case Eq. (60) becomes

$$k_p^i = k_p^l + (1/2)f_t \int_{P_{X_2}^i}^{P_{X_2}^a} D_{Me}^l (D_X^b/D_{Me}^l) \, d \ln P_{X_2} \tag{65}$$

where k_p is given by Eq. (33).

In this instance, it does not seem reasonable to assume that D_X^b/D_{Me}^l is independent of P_{X_2}. If D_X^b has not been measured as a function of P_{X_2}, which would allow a direct solution of Eq. (65) if D_{Me}^l were known as a function of P_{X_2}, two other approaches might prove useful; i.e., D_X^b has the same dependence on P_{X_2} as the majority anionic point defects in $Me_{\alpha+\delta}X$ or D_X^b is independent of P_{X_2}. In the latter case, Eq. (65) becomes

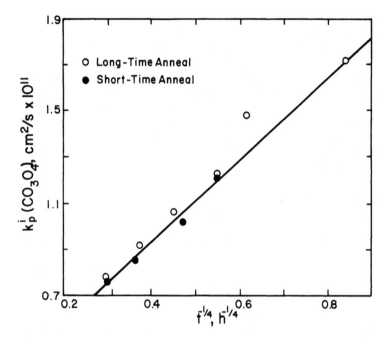

FIGURE 16 Dependence of the instantaneous parabolic rate constant on time for the thickening of the spinel layer during the oxidation of cobalt at 700°C in pure oxygen at 1 atm. ○ = Long-time anneal, ● = short-time anneal [22].

$$k_p^i = k_p^l + (f_t D_X^b/2) \ln(P_{X_2}^a/P_{X_2}^i) \qquad (66)$$

Equation (66) is analogous to Eq. (50) with $k_p^b = (D_X^b/2) \ln(P_{X_2}^a/P_{X_2}^i)$. To examine the relative contributions of lattice diffusion of cations and grain boundary diffusion of anions to the value of k_p^i, it is useful to replace k_p^l in Eq. (66) with Eq. (30a) or (30b). Inserting Eq. (30a) into Eq. (66), that is, assuming that $Me_{\alpha+\delta}X$ is a p-type semiconducting compound and that $P_{X_2}^a \gg P_{X_2}^i$, one obtains

$$k_p^i = [(1 + z)/\alpha](^a D_{Me}^l) + (f_t D_X^b/2) \ln(P_{X_2}^a/P_{X_2}^i) \qquad (67)$$

where $(^a D_{Me}^T/f_{Me})$ in Eq. (30a) has been replaced by the lattice self-diffusion coefficient for cations. Typically, $z \approx 2.0$, $\alpha \approx 1.0$, $f_t \approx 10^{-3}$, and $(D_X^b/D_{Me}^l) \approx 10^2$ to 10^5, consistent with one of the assumptions used to derive Eq. (65). If, for example, $D_{Me}^l \approx 10^{-12}$ cm^2/sec, $D_X^b/D_{Me}^l \approx 10^2$, $P_{X_2} \approx 1$ atm, and $P_{X_2}^i = 10^{-20}$ atm (values of 10^{-10} to 10^{-30} atm would be typical), then $k_p^l \approx f_t k_p^b$. In this case, transport of both anions and cations contributes significantly to scale growth, and the possibility of the formation of $Me_{\alpha+\delta}X$ within the existing layer of scale must be considered. In the latter case, large growth stresses may develop in the scale, giving rise to an early onset of breakaway oxidation behavior. Such a mechanism has been suggested for the formation of Al_2O_3 and Cr_2O_3 scales on alloys [84,85].

If, on the other hand, $k_p^l \ll f_t k_p^b$ (e.g., when $D_X^b/D_{Me}^l \approx 10^5$), then Eq. (66) becomes

$$k_p^i = (f_t D_X^b/2) \ln(P_{X_2}^a/P_{X_2}^i) \tag{68}$$

The latter equation may be used to evaluate D_X^b if f_t and k_p^i have been measured.

In the limiting cases of Eq. (60) considered here, it has been assumed that f_t is independent of P_{X_2}. It has been observed, however, that the grain size of Cr_2O_3 formed on pure chromium is a function of P_{O_2} [86,87]. Thus, it may be necessary in some cases to determine experimentally the dependence of f_t on both time and P_{X_2} to allow kinetic data to be interpreted correctly.

9.5 SUMMARY

The ultimate goal of most research in the area of high-temperature oxidation and corrosion is the development of materials that exhibit excellent resistance to oxidation for long periods of time at elevated temperatures. Materials are sought that exhibit very long times of steady-state, diffusion-controlled oxidation before the onset of breakaway oxidation kinetics (see Fig. 1).

Materials that exhibit good resistance to oxidation are those that develop protective scales on their surfaces. Such scales, which are usually oxides, provide a good barrier between the alloy and the environment, and they are very adherent to the metallic substrate under isothermal and athermal conditions of exposure. Scales that are good barrier layers grow by a process of solid-state diffusion, and thus the topic of diffusion-controlled scale growth is one of major importance to research in the area of high-temperature oxidation and corrosion.

This chapter has focused on the mechanisms of diffusion-controlled scale growth on pure metals. Wagner's model for diffusion-controlled scale growth was reviewed with emphasis placed on clearing up a number of misconceptions commonly encountered in the literature. The general applicability of Wagner's model was reviewed and the importance of scale microstructure in determining oxidation kinetics was emphasized.

The objective of this chapter was to provide readers with information to help develop a good understanding of the mechanisms of diffusion-controlled scale growth. Other topics of importance in the area of high-temperature oxidation and corrosion, which include alloy oxidation, stress effects, scale adhesion, multioxidant gaseous environments, hot corrosion, etc., were not reviewed in this chapter. Indeed, a single chapter would be needed for each of these topics, and others of importance in this field, if the discussions were to be of any consequence. It is hoped, however, that the information presented in this chapter will also provide a good basis for developing an understanding of oxidation mechanisms in the more complex cases.

ACKNOWLEDGMENT

The author thanks the Division of Materials Sciences, Office of Basic Energy Sciences, Department of Energy, for partial financial support during the preparation of this chapter (DOE contract DE-AC02-79ER10507).

REFERENCES

1. G. Wood, in *Oxidation of Metals and Alloys*, American Society for Metals, Metals Park, Ohio, 1971, p. 201.
2. D. P. Whittle and G. C. Wood, *J. Inst. Met.*, *96*: 115 (1968).
3. M. H. Davies, M. T. Simnad, and C. E. Birchenall, *Trans. A.I.M.E.*, *191*, 889 (1951).
4. S. Mrowec and K. Przybylski, *Oxid. Met.*, *11*: 365 (1977).
5. K. Fueki and J. B. Wagner, Jr., *J. Electrochem. Soc.*, *112*: 384 (1965).
6. H. Hindam and D. P. Whittle, *Oxid. Met.*, *18*: 245 (1982).
7. K. Motzfeldt, *Acta Chem. Scand.*, *18*: 1596 (1964).
8. S. Mrowec and T. Werber, *Gas Corrosion of Metals*, 1975, translated from the Polish, NTIS PB283054T, National Technical Information Service, Springfield, Va., p. 132.
9. N. B. Pilling and R. E. Bedworth, *J. Inst. Met.*, *29*: 529 (1923).
10. C. Wagner, *Z. Phys. Chem.*, *B21*: 25 (1933).
11. C. Wagner, in *Atom Movements*, American Society for Metals, Metals Park, Ohio, 1951, p. 153.
12. C. Wagner, *Prog. Solid-State Chem.*, *10*: 3 (1975).
13. J. B. Price and J. B. Wagner, Jr., *Z. Phys. Chem.*, *49*: 257 (1966).

14. P. E. Childs and J. B. Wagner, Jr., in *Heterogeneous Kinetics* (G. R. Belton and W. R. Worrell, eds.), Plenum, New York, 1970, p. 269.
15. H. Schmalzried, *Solid State Reactions*, 2nd ed., Verlag Chemie, Deerfield Beach, Fla., 1981.
16. P. Kofstad, *High-Temperature Oxidation of Metals*, Wiley, New York, 1966.
17. F. Morin, *Oxid. Met.*, *6*: 65 (1973).
18. D. B. Noble and G. J. Yurek, Massachusetts Institute of Technology, Cambridge, unpublished results, Aug. 1983.
19. D. Caplan and G. I. Sproule, *Oxid. Met.*, *9*: 459 (1975).
20. M. LaBranche and G. J. Yurek, Massachusetts Institute of Technology, Cambridge, unpublished results, Aug. 1983.
21. H. S. Hsu and G. J. Yurek, *Oxid. Met.*, *17*: 55 (1982).
22. G. J. Yurek and H.-S. Hsu, in *Proceedings, Third JIM International Symposium on High Temperature Corrosion of Metals and Alloys*, Mt. Fuji, Japan, Nov. 17–20, 1982, p. 141.
23. E. A. Gulbransen, *Mem. Sci. Rev. Met.*, *62*: 253 (1965).
24. R. A. Rapp, *Metall. Trans. A.*, *15A*: 765 (1984).
25. D. A. Voss, E. P. Butler, and T. E. Mitchell, *Metall. Trans. A*, *13A*: 929 (1982).
26. S. Mrowec, in *Proceedings, Third JIM International Symposium on High Temperature Corrosion of Metals and Alloys*, Mt. Fuji, Japan, Nov. 17–20, 1982, p. 69.
27. A. Dravinieks and H. McDonald, *J. Electrochem. Soc.*, *94*: 139 (1948).
28. G. M. Raynaud, W. A. T. Clark, and R. A. Rapp, *Metall. Trans. A*, *15A*: 573 (1984).
29. L. W. Hobbs, H. T. Sawhill, and M. T. Tinker, in *Proceedings, Third JIM International Symposium on High Temperature Corrosion of Metals and Alloys*, Mt. Fuji, Japan, Nov. 17–20, 1982, p. 115.
30. C. Wagner, *Ber. Bunsenges. Phys. Chem.*, *78*: 611 (1974).
31. G. J. Yurek and H. Schmalzred, *Ber. Bunsenges. Phys. Chem.*, *79*: 255 (1975).
32. P. Kofstad, in *High Temperature Corrosion* (R. A. Rapp, ed.), Houston, 1983.
33. G. J. Yurek, J. P. Hirth, and R. A. Rapp, *Oxid. Met.*, *8*: 265 (1974).
34. F. Gesmundo and F. Viani, *Corros. Sci.*, *18*: 217, 231 (1978).
35. S. R. Shatynski, R. A. Rapp, and J. P. Hirth, *Acta Met.*, *24*: 1071 (1976).
36. R. D. Shaw and R. Rolls, *Corros. Sci.*, *14*: 443 (1974).
37. G. Garnaud and R. A. Rapp, *Oxid. Met.*, *11*: 193 (1977).
38. G. Garnaud, *Oxid. Met.*, *11*: 127 (1977).
39. J. P. Hirth and R. A. Rapp, *Oxid. Met.*, *11*: 57 (1977).

40. F. Pettit, R. Yinger, and J. B. Wagner, Jr., *Acta Met.*, 8: 617 (1960).
41. C. Wagner, *Ber. Bunsenges. Phys. Chem.*, 70: 775 (1966).
42. A. T. Fromhold, S. R. Coriell, and J. Kruger, *J. Phys. Soc. Jpn.*, 34: 1452 (1973).
43. K. Hauffe, *Oxidation of Metals*, Plenum, New York, 1965.
44. W. K. Chen and R. A. Jackson, *J. Phys. Chem. Sol.*, 30: 1309 (1969).
45. R. Dieckmann, *Z. Phys. Chem.*, 107: 189 (1977).
46. R. E. Carter and F. D. Richardson, *Trans AIME*, 200: 1244 (1954).
47. F. Morin, *Can. Met. Q.*, 14: 105 (1975).
48. W. K. Chen, N. L. Peterson, and W. T. Reeves, *Phys. Rev.*, 186: 887 (1969).
49. D. W. Bridges, J. P. Baur, and W. M. Fassell, Jr., *J. Electrochem. Soc.*, 103: 614 (1956).
50. J. Kruger, A. Melin, and H. Winterhager, *Cobalt*, 33: 176 (1966).
51. J. A. Snide, J. R. Myers, and R. K. Saxer, *Cobalt*, 36: 157 (1967).
52. S. Mrowec and K. Przybylski, *Oxid. Met.*, 11: 365 (1977).
53. F. R. Billman, *J. Electrochem. Soc.*, 119: 1198 (1972).
54. F. Gesmundo and F. Viani, *J. Electrochem. Soc.*, 128: 460 (1961).
55. R. Dieckmann and H. Schmalzried, *Ber. Bunsenges. Phys. Chem.*, 81: 344 (1977).
56. R. Dieckmann and H. Schmalzried, *Ber. Bunsenges Phys. Chem.*, 81: 414 (1977).
57. R. Dieckmann, *Ber. Bunsenges. Phys. Chem.*, 86: 112 (1982).
58. M. A. Davies, M. T. Simnad, and C. E. Birchenall, *Trans. AIME*, 197: 1250 (1953).
59. G. J. Yurek and C. E. Meyers, 109th AIME Annual Meeting, Las Vegas, Nev., Nov. 24–28, 1980.
60. C. E. Meyers, S.M. thesis, Massachusetts Institute of Technology, Cambridge, June 1980.
61. R. Dieckmann, H. Schmalried, and T. O. Mason, *Arch. Eisenhuettenwes.*, 52: 211 (1981).
62. N. L. Peterson, W. K. Chen, and D. Wolf, *J. Phys. Chem. Solids*, 41: 709 (1980).
63. R. Dieckmann, Habilitationschrift, Univ. of Hannover, West Germany, 1983, pp. 203–220.
64. F. W. Young, Jr., J. V. Cathcart, and A. T. Gwathmey, *Acta Met.*, 4: 145 (1956).
65. J. V. Cathcart, in *Oxidation of Metals and Alloys*, American Society for Metals, Metals Park, Ohio, 1971, p. 30.
66. R. Herchl, N. N. Khoi, R. Homma, and W. W. Smeltzer, *Oxid-Met.*, 4: 35 (1972).

67. M. J. Graham, R. J. Hussey, and M. Cohen, *J. Electrochem. Soc.*, *120*: 1523 (1975).
68. N. N. Khoi, W. W. Smeltzer, and J. D. Embury, *J. Electrochem. Soc.*, *122*: 1495 (1975).
69. R. Herchl, N. N. Khoi, T. Homma, and W. W. Smeltzer, *Oxid. Met.*, *4*: 35 (1972).
70. D. Caplan and G. I. Sproule, *Oxid. Met.*, *9*: 459 (1975).
71. M. J. Graham, G. I. Sproule, D. Caplan, and M. Cohen, *J. Electrochem. Soc.*, *119*: 883 (1972).
72. D. Caplan and M. Cohen, *Corros. Sci.*, *6*: 321 (1966).
73. D. Caplan, M. J. Graham, and M. Cohen, *J. Electrochem. Soc.*, *119*: 1205 (1972).
74. W. W. Smeltzer, R. R. Haering, and J. S. Kirkaldy, *Acta Met.*, *9*: 880 (1961).
75. J. M. Perrow, W. W. Smeltzer, and J. D. Embury, *Acta Met.*, *16*: 1209 (1968).
76. S. Matsunga and T. Homma, *Oxid. Met.*, *10*: 361 (1976).
77. A. Atkinson and R. I. Taylor, *Philos. Mag.*, *A45*: 566 (1982).
78. A. Atkinson, R. I. Taylor, and A. E. Hughes, *Philos. Mag.*, *A45*: 823 (1982).
79. E. W. Hart, *Acta Met.*, *5*: 597 (1975).
80. C. J. Simpson, W. C. Winegard, and K. T. Aust, in *Grain Boundary Structure and Properties* (G. A. Chadwick and D. A. Smith, eds.), Academic, New York, 1976, p. 200.
81. B. B. Rath and H. Hu, *Trans. AIME*, *245*: 1577 (1969).
82. J. P. Drolet and A. G. Balibois, *Met. Trans.*, *2*: 53 (1971).
83. G. T. Higgins, *J. Mater. Sci.*, *8*: 143 (1974).
84. J. Stringer, B. A. Wilcox, and R. Jaffee, *Oxid. Met.*, *5*: 11 (1972).
85. F. A. Golightly, F. H. Stott, and G. C. Wood, *Oxid. Met.*, *10*: 163 (1976).
86. K. P. Lillerud and P. Kofstad, *J. Electrochem. Soc.*, *127*: 2397, 2410 (1980).
87. M. LaBranche and G. J. Yurek, Massachusetts Institute of Technology, Cambridge, unpublished research, 1983.

INDEX

Milton Keynes UK
Ingram Content Group UK Ltd.
UKHW021901071024
449327UK00021B/1599